AF780749

Novel Anti-Cancer Agents and Cellular Targets and Their Mechanism(s) of Action

Novel Anti-Cancer Agents and Cellular Targets and Their Mechanism(s) of Action

Editor

Simon J. Allison

MDPI • Basel • Beijing • Wuhan • Barcelona • Belgrade • Manchester • Tokyo • Cluj • Tianjin

Editor
Simon J. Allison
University of Huddersfield
UK

Editorial Office
MDPI
St. Alban-Anlage 66
4052 Basel, Switzerland

This is a reprint of articles from the Special Issue published online in the open access journal *Biomedicines* (ISSN 2227-9059) (available at: https://www.mdpi.com/journal/biomedicines/special_issues/anti-cancer_agents).

For citation purposes, cite each article independently as indicated on the article page online and as indicated below:

LastName, A.A.; LastName, B.B.; LastName, C.C. Article Title. *Journal Name* **Year**, *Volume Number*, Page Range.

ISBN 978-3-0365-5221-7 (Hbk)
ISBN 978-3-0365-5222-4 (PDF)

© 2022 by the authors. Articles in this book are Open Access and distributed under the Creative Commons Attribution (CC BY) license, which allows users to download, copy and build upon published articles, as long as the author and publisher are properly credited, which ensures maximum dissemination and a wider impact of our publications.
The book as a whole is distributed by MDPI under the terms and conditions of the Creative Commons license CC BY-NC-ND.

Contents

About the Editor

Simon J. Allison

Dr Simon J. Allison is a cancer biologist working at the interface of cancer cell biology and cancer pharmacology/drug discovery. He obtained his first degree in Natural Sciences (Biological) from the University of Cambridge, where he graduated with first class honours. This was followed by a PhD on the regulation of RNA polymerase III transcription at the University of Glasgow and postdoctoral cancer research positions at the University of York within the laboratory of Prof. Jo Milner. Following research spells at the University of Leeds and at the University of Bradford, Simon joined the University of Huddersfield in 2015 where he was appointed as a Reader in Cell Biology and Pharmacology in 2019. Simon has over 20 years of research experience in the cancer field, with a particular interest in cancer cell molecular and metabolic differences and their exploitation for the development of more cancer-selective efficacious therapies as well as those presented through phenotypic drug discovery activities and subsequent target deconvolution.

biomedicines

Editorial

Novel Anti-Cancer Agents and Cellular Targets and Their Mechanism(s) of Action

Simon J. Allison

School of Applied Sciences, University of Huddersfield, Huddersfield HD1 3DH, UK; s.allison@hud.ac.uk; Tel.: +44-(0)1484-256857

Whilst there have been some significant improvements in treatments and patient outcomes for some cancers, for other cancers there has been little change in survival rates for many years. Side effects resulting from the toxicity of agents towards normal cells and tissues and the development of drug resistance continue to limit the effectiveness of traditional and molecular-targeted anti-cancer agents. Other challenges include pharmacokinetics (PK) and delivering sufficient amounts of the active drug to tumour cells in vivo. There is an ongoing need for new anti-cancer agents with novel mechanism(s) of action and the identification of new putative cellular targets and therapeutic strategies that are both potent and selective towards cancer cells.

Addressing some of these therapeutic challenges, this book presents a collection of eight research articles and two reviews reproduced from the themed *Biomedicines* Special Issue 'Novel Anti-Cancer Agents and Cellular Targets and Their Mechanism(s) of Action'.

Traditional cytotoxic chemotherapeutic agents remain a vital part of treatments for many cancers, but their efficacy is severely restricted by their lack of selectivity and dose-limiting toxicity towards healthy cells and tissue. Affibody or antibody drug conjugates provide an opportunity to target or deliver cytotoxic agents specifically to cancer cells, exploiting a tumour-specific antigen specifically recognised by the conjugate for targeted delivery. Rinne et al. report on a human epidermal growth factor receptor 3 (HER3) affibody drug conjugate with a tubulin polymerisation inhibitor as its cytotoxic 'payload' for targeted delivery to HER3-overexpressing cancer cells [1]. The authors also include an albumin binding domain as part of the conjugate to improve its half-life and PK properties [1].

A different approach to using affibody drug conjugates to selectively target cancer cell microtubules is presented by Kwon and colleagues [2]. In triple-negative breast cancers (TNBC), which have the worst outcome of the different breast cancer subtypes, microtubules are found to be highly acetylated. Kwon et al. report on their identification of several compounds that disrupt microtubule acetylation in TNBC cells resulting in their apoptotic demise [2].

One cancer type with particularly poor outcome is glioblastoma (GBM), which has a five-year median patient survival rate of less than 5%. The invasion of cancer cells into healthy brain tissue makes tumour resection particularly challenging, resulting in high rates of tumour recurrence. This has stimulated interest in pharmacological inhibition of cell migration as part of the regime for treating glioblastomas. Ketchen et al. report on the migratory plasticity of GBM cells and how the pharmacological inhibition of mesenchymal migration, via the inhibition of cellular communication network factor 1 (CCN1), promotes cells to 'switch' to an alternative means of migration known as ameboid migration [3]. This work indicates the importance of simultaneously targeting multiple alternative modes of GBM cell migration.

With our increased understanding of cancer biology and of oncogene-driven molecular addictions, this has resulted in considerable research effort to exploit and target these addictions within specific cancers via molecular targeted anti-cancer drugs and through rational drug design approaches. However, it is also clear that whilst such agents can present

Citation: Allison, S.J. Novel Anti-Cancer Agents and Cellular Targets and Their Mechanism(s) of Action. *Biomedicines* **2022**, *10*, 1767. https://doi.org/10.3390/biomedicines10081767

Received: 14 July 2022
Accepted: 20 July 2022
Published: 22 July 2022

Publisher's Note: MDPI stays neutral with regard to jurisdictional claims in published maps and institutional affiliations.

Copyright: © 2022 by the author. Licensee MDPI, Basel, Switzerland. This article is an open access article distributed under the terms and conditions of the Creative Commons Attribution (CC BY) license (https://creativecommons.org/licenses/by/4.0/).

important advantages over traditional agents such as their increased cancer selectivity, other challenges remain such as the ability of cancer cells to adapt and develop resistance. This concept of some targeted agents being perhaps 'too targeted' has led to a resurgence of interest in the discovery of novel chemical entities with polypharmacological anti-cancer properties and in phenotypic drug discovery. The natural world continues to be a valuable source of novel chemicals with anti-cancer activity. Eldeeb et al. report on their testing of ten novel isatin sulfonamide derivatives for anti-cancer activity against hepatic cancer cell lines and associated in vitro mode of action and molecular docking studies [4].

A major challenge in phenotypic drug discovery and development is that of target deconvolution. Wu et al. report on an isoflavone derivative from fermented soybean, 8-hydroxydaidzein, which shows promising activity against chronic myeloid leukemia (CML) cells, with the authors reporting effects via multiple mechanisms of action [5]. Nikoleousakos et al. report on the synthesis of three novel β-lactam steroid alkylators and their in vitro evaluation against human ovarian cancer cell lines. As well as DNA damage induction, the inhibition of poly (ADP-ribose) polymerase (PARP) enzymatic activity is reported [6].

A major cause of treatment failure in vivo is the metabolic detoxification or inactivation of anti-cancer drugs by isoforms of the cytochrome P450 (CYP) family of enzymes. In other cases, the activity of specific CYP isoforms can result in the generation of more active, or cytotoxic, metabolites. Pors and colleagues review the role of CYP isoforms in breast cancers and how they influence different standard-of-care (SoC) treatments [7]. Importantly, they also discuss new therapeutic opportunities presented by the overexpression of particular CYP isoforms in different breast cancer subtypes, which is stimulating the development of CYP isoform-activated pro-drugs [7]. This tackles the challenge of developing drugs that are both potent and also tumour-selective, in contrast to the narrow therapeutic index of traditional chemotherapies.

For molecular targets in oncology that are considered 'undruggable', or perhaps more accurately 'difficult to drug', a potential solution to this druggability problem is RNA interference (RNAi)-based therapeutics. A key barrier to this latter approach, however, is the susceptibility of short interfering RNA (siRNA) to degradation in vivo by nucleases in serum. Approaches to counteract this and improve siRNA stability include chemical RNA modifications or siRNA extension with a 'protective' short DNA hairpin loop sequence. Another approach attracting considerable research interest is siRNA delivery within biodegradable nanoparticles which is discussed by Habib, Ariatti and Singh in their review article focused on RNAi-based nanotherapeutics in the context of the 'undruggable' *c-myc* oncogene [8].

A challenge with any anti-cancer agent is ensuring its sufficient delivery to the tumour in vivo. Poor in vivo biodistribution/PK often contributes to the in vivo failure of otherwise promising drugs. The ability to 'visualise' drug delivery in vivo provides the opportunity to assess whether this is an issue and for developing more effective drug delivery systems. Vepris et al. report on their development of novel poly(lactic-co-glycolic acid) (PGLA) fluorescent nanoparticles as delivery agents that enable in vivo molecular imaging through the phenomenon of photon triplet-triplet annihilation (TTA) upconversion [9].

The final article in this Special Issue book focuses on 5-fluorouracil (5-FU) drug resistance in colorectal cancers (CRC), where 5-FU is part of the standard-of-care, and pharmacogenomic analyses using publicly available resources to predict agents that can overcome 5-FU resistance depending on the particular CRC mutational signature [10].

In summary, this Special Issue collection of articles provides some examples of the many different approaches being used to tackle the multiple challenges that remain in the global battle against cancers. These range from the evaluation of novel compounds for anti-cancer activity and elucidation of their mechanism(s) of action to overcoming drug resistance and advances in targeted drug delivery and molecular imaging for the validation of effective delivery.

Funding: This research received no external funding.

Acknowledgments: I would like to thank all the contributing authors to this Special Issue.

Conflicts of Interest: The author declares no conflict of interest.

References

1. Rinne, S.S.; Yin, W.; Borras, A.M.; Abouzayed, A.; Leitao, C.D.; Vorobyeva, A.; Löfblom, J.; Ståhl, S.; Orlova, A.; Gräslund, T. Targeting Tumor Cells Overexpressing the Human Epidermal Growth Factor Receptor 3 with Potent Drug Conjugates Based on Affibody Molecules. *Biomedicines* **2022**, *10*, 1293. [CrossRef] [PubMed]
2. Kwon, A.; Lee, G.B.; Park, T.; Lee, J.H.; Ko, P.; You, E.; Ahn, J.H.; Eom, S.H.; Rhee, S.; Song, W.K. Potent Small-Molecule Inhibitors Targeting Acetylated Microtubules as Anticancer Agents against Triple-Negative Breast Cancer. *Biomedicines* **2020**, *8*, 338. [CrossRef] [PubMed]
3. Ketchen, S.E.; Gamboa-Esteves, F.O.; Lawler, S.E.; Nowicki, M.O.; Rohwedder, A.; Knipp, S.; Prior, S.; Short, S.C.; Ladbury, J.E.; Brüning-Richardson, A. Drug Resistance in Glioma Cells Induced by a Mesenchymal-Amoeboid Migratory Switch. *Biomedicines* **2021**, *10*, 9. [CrossRef] [PubMed]
4. Eldeeb, M.; Sanad, E.F.; Ragab, A.; Ammar, Y.A.; Mahmoud, K.; Ali, M.M.; Hamdy, N.M. Anticancer Effects with Molecular Docking Confirmation of Newly Synthesized Isatin Sulfonamide Molecular Hybrid Derivatives against Hepatic Cancer Cell Lines. *Biomedicines* **2022**, *10*, 722. [CrossRef] [PubMed]
5. Wu, P.S.; Yen, J.H.; Wang, C.Y.; Chen, P.Y.; Hung, J.H.; Wu, M.J. 8-Hydroxydaidzein, an Isoflavone from Fermented Soybean, Induces Autophagy, Apoptosis, Differentiation, and Degradation of Oncoprotein BCR-ABL in K562 Cells. *Biomedicines* **2020**, *8*, 506. [CrossRef] [PubMed]
6. Nikoleousakos, N.; Dalezis, P.; Polonifi, A.; Geromichalou, E.G.; Sagredou, S.; Alifieris, C.E.; Deligiorgi, M.V.; Sarli, V.; Trafalis, D.T. Cytocidal Antitumor Effects against Human Ovarian Cancer Cells Induced by B-Lactam Steroid Alkylators with Targeted Activity against Poly (ADP-Ribose) Polymerase (PARP) Enzymes in a Cell-Free Assay. *Biomedicines* **2021**, *9*, 1028. [CrossRef] [PubMed]
7. Sneha, S.; Baker, S.C.; Green, A.; Storr, S.; Aiyappa, R.; Martin, S.; Pors, K. Intratumoural Cytochrome P450 Expression in Breast Cancer: Impact on Standard of Care Treatment and New Efforts to Develop Tumour-Selective Therapies. *Biomedicines* **2021**, *9*, 290. [CrossRef] [PubMed]
8. Habib, S.; Ariatti, M.; Singh, M. Anti-c-myc RNAi-Based Onconanotherapeutics. *Biomedicines* **2020**, *8*, 612. [CrossRef] [PubMed]
9. Vepris, O.; Eich, C.; Feng, Y.; Fuentes, G.; Zhang, H.; Kaijzel, E.L.; Cruz, L.J. Optically Coupled PtOEP and DPA Molecules Encapsulated into PLGA-Nanoparticles for Cancer Bioimaging. *Biomedicines* **2022**, *10*, 1070. [CrossRef] [PubMed]
10. Huang, T.C.; Peng, K.C.; Kuo, T.T.; Lin, L.C.; Liu, B.C.; Ye, S.P.; Chu, C.C.; Hsia, S.M.; Chang, H.Y. Predicting Agents That Can Overcome 5-FU Resistance in Colorectal Cancers via Pharmacogenomic Analysis. *Biomedicines* **2021**, *9*, 882. [CrossRef] [PubMed]

biomedicines

Article

Targeting Tumor Cells Overexpressing the Human Epidermal Growth Factor Receptor 3 with Potent Drug Conjugates Based on Affibody Molecules

Sara S. Rinne [1,†], Wen Yin [2,†], Anna Mestre Borras [2], Ayman Abouzayed [1], Charles Dahlsson Leitao [2], Anzhelika Vorobyeva [3], John Löfblom [2], Stefan Ståhl [2], Anna Orlova [1,3,4,*] and Torbjörn Gräslund [2,*]

1 Department of Medicinal Chemistry, Uppsala University, 751 23 Uppsala, Sweden; sara.rinne@ilk.uu.se (S.S.R.); ayman.abouzayed@ilk.uu.se (A.A.)
2 Department of Protein Science, KTH Royal Institute of Technology, 114 17 Stockholm, Sweden; wenyin@kth.se (W.Y.); annamb@kth.se (A.M.B.); chdl@kth.se (C.D.L.); lofblom@kth.se (J.L.); ssta@kth.se (S.S.)
3 Department of Immunology, Genetics and Pathology, Uppsala University, 752 37 Uppsala, Sweden; anzhelika.vorobyeva@igp.uu.se
4 Science for Life Laboratory, Uppsala University, 752 37 Uppsala, Sweden
* Correspondence: anna.orlova@ilk.uu.se (A.O.); torbjorn@kth.se (T.G.); Tel.: +46-(0)73-992-2846 (A.O.); +46-(0)70-790-9627 (T.G.)
† These authors contributed equally to this work.

Abstract: Increasing evidence suggests that therapy targeting the human epidermal growth factor receptor 3 (HER3) could be a viable route for targeted cancer therapy. Here, we studied a novel drug conjugate, Z_{HER3}-ABD-mcDM1, consisting of a HER3-targeting affibody molecule, coupled to the cytotoxic tubulin polymerization inhibitor DM1, and an albumin-binding domain for in vivo half-life extension. Z_{HER3}-ABD-mcDM1 showed a strong affinity to the extracellular domain of HER3 (K_D 6 nM), and an even stronger affinity (K_D 0.2 nM) to the HER3-overexpressing pancreatic carcinoma cell line, BxPC-3. The drug conjugate showed a potent cytotoxic effect on BxPC-3 cells with an IC_{50} value of 7 nM. Evaluation of a radiolabeled version, [^{99m}Tc]Tc-Z_{HER3}-ABD-mcDM1, showed a relatively high rate of internalization, with a 27% internalized fraction after 8 h. Further in vivo evaluation showed that it could target BxPC-3 (pancreatic carcinoma) and DU145 (prostate carcinoma) xenografts in mice, with an uptake peaking at 6.3 ± 0.4% IA/g at 6 h post-injection for the BxPC-3 xenografts. The general biodistribution showed uptake in the liver, lung, salivary gland, stomach, and small intestine, organs known to express murine ErbB3 naturally. The results from the study show that Z_{HER3}-ABD-mcDM1 is a highly potent and selective drug conjugate with the ability to specifically target HER3 overexpressing cells. Further pre-clinical and clinical development is discussed.

Keywords: affibody molecule; human epidermal growth factor receptor 3 (HER3); BxPC-3; emtansine; DM1; albumin binding domain; affibody drug conjugate (AffiDC)

Citation: Rinne, S.S.; Yin, W.; Borras, A.M.; Abouzayed, A.; Leitao, C.D.; Vorobyeva, A.; Löfblom, J.; Ståhl, S.; Orlova, A.; Gräslund, T. Targeting Tumor Cells Overexpressing the Human Epidermal Growth Factor Receptor 3 with Potent Drug Conjugates Based on Affibody Molecules. *Biomedicines* **2022**, *10*, 1293. https://doi.org/10.3390/biomedicines10061293

Academic Editor: Simon J Allison

Received: 19 April 2022
Accepted: 28 May 2022
Published: 31 May 2022

Publisher's Note: MDPI stays neutral with regard to jurisdictional claims in published maps and institutional affiliations.

Copyright: © 2022 by the authors. Licensee MDPI, Basel, Switzerland. This article is an open access article distributed under the terms and conditions of the Creative Commons Attribution (CC BY) license (https://creativecommons.org/licenses/by/4.0/).

1. Introduction

Approaches to specifically target receptors that are abnormally expressed on cancer cells have become a viable therapeutic strategy and several drugs are in clinical trials or have been approved for clinical use. One of the most common targeting moieties are monoclonal antibodies (mAbs) [1]. However, pre-clinical as well as clinical trials have also shown that fragments of mAbs [2] or engineered affinity scaffold proteins (EAPs) [3], may be used.

A well-studied group of receptors with relevance to cancer is the epidermal growth factor receptor (ERBB) family, for which many targeted therapies have been investigated. The members of the ERBB family are tyrosine kinase receptors, which are involved in driving tumor progression and in the acquisition of resistance to therapy [4]. One of the family members, the human epidermal growth factor receptor 3 (HER3, human ErbB3), is a transmembrane receptor with impaired tyrosine kinase activity. It has been observed that the expression level of HER3 is associated with malignancy for several different cancer types, such as non-small cell lung carcinoma (NSCLC) [5], pancreatic carcinoma [6], melanoma [7], prostate carcinoma [8], and breast carcinoma [9]. The two main ligands of HER3 are heregulin (HRG) and neuregulin 2 (NRG2). Upon binding, HRG and NRG2 activate the receptor, which in turn can form heterodimers with the other members of the ERBB family, followed by the activation of downstream signaling [10]. HRG and NRG2 expression is sometimes upregulated, which over-activates HER3, which in turn drives tumor growth [11].

Due to HER3's upregulated expression in different cancers, therapies specifically targeting HER3 are of increasing interest. Seribantumab is a mAb which has shown promise in combination with other drugs in clinical phase I or II studies for several different indications, including breast carcinoma, NSCLC, colorectal carcinoma, and ovarian carcinoma [4]. Other HER3-targeting mAbs that have shown promise in early clinical trials include patritumab, lumretuzumab, and elgemtumab.

A more recent approach to HER3-targeted therapy is to link the mAb to a potent cytotoxic drug, resulting in an antibody drug conjugate (ADC) [12,13]. These types of targeted drugs often have several modes of action: (i) the mAb part may prevent the transduction of signaling by the targeted receptor, and/or induce an antibody-dependent cellular cytotoxicity (ADCC) response, (ii) the linked drug delivers a cytotoxic effect leading to cell death. Until now, eleven ADCs have been approved for clinical use for different cancer indications by the regulatory authorities in Europe and/or the USA. Data from studies of these compounds suggest that they are often well-tolerated and efficient therapy options. However, severe side-effects occur for some patients; most commonly liver damage from off-target uptake of the ADC, and a low blood count from premature release of the cytotoxic drug from the mAb [14]. Patritumab deruxtecan is a HER3-targeting ADC in phase I/II clinical trials for patients suffering from NSCLC and breast carcinoma, and has shown promising results [15]. Compared to other ADCs, patritumab deruxtecan has a relatively weak affinity for its receptor and a relatively high rate of internalization [16].

Another drug candidate under clinical development is zenocutuzumab, a bi-specific mAb targeting HER2 (human epidermal growth factor receptor 2) and HER3. Zenocutuzumab prevents HRG binding to HER3 and has been engineered for an enhanced ADCC response. It was granted fast track designation by the U.S. Food and Drug Administration (FDA) in early 2021 for the clinical testing of metastatic, HRG-positive solid tumors. Three patients, one with NSCLC and two with pancreatic carcinoma receiving zenocutuzumab all responded with tumor shrinkage [17].

The early clinical evaluations of HER3-targeted therapies on small cohorts have shown promise. However, the few clinical phase III studies performed on mAbs have thus far not been successful [4]. Furthermore, clinical phase III studies on HER3-targeting ADCs have not yet been undertaken. While HER3 targeting may be a viable strategy for cancer treatment, based on the clinical evaluations presented above, it is not evident how to best design a HER3-targeting drug. Compared to many other plasma membrane-anchored receptors, for which targeted therapies have been developed, HER3 has a substantial expression on normal tissues, and its overexpression in tumors is only modest for most patients, up to approximately 50,000 receptors/tumor cells [18], which makes the development of targeted therapies challenging.

Affibody molecules are engineered affinity scaffold proteins (EAPs) derived from the B-domain of protein A from *Staphylococcus aureus*, normally folding into an anti-parallel triple helical structure [19]. They consist of 58 amino acids (Mw 7 kDa) and the scaffold is devoid of cysteine amino acids. The scaffold has a natural affinity for some IgGs. By randomizing 13 amino acids in the two helices directly involved in IgG binding or in close vicinity to the binding surface, combinatorial libraries have been generated. From these libraries, variants that no longer bind to IgG but to desired targets have been selected by e.g., phage or cell display techniques [20].

Affibody molecules, specifically interacting with HER3 have been described [21,22]. These binders have been radiolabeled and investigated as radionuclide molecular imaging agents in pre-clinical mouse models, where they were able to visualize HER3-expressing tumors and to discriminate between high and low receptor expression [23–29]. The most promising variant for molecular imaging was $Z_{HER3:08698}$ with specific and strong interaction with HER3 (equilibrium dissociation constant, K_D 50 pM) [22]. When evaluated for imaging, a tag with the amino acid sequence His-Glu-His-Glu-His-Glu, a $(HE)_3$-tag, was placed in the N-terminus. It was used for purification by immobilized metal-ion affinity chromatography (IMAC), and for radiolabeling with $[^{99m}Tc]Tc$. It has also been shown to minimize unspecific uptake in the liver compared to the more commonly used hexahistidine tag [30–32]. We hypothesized that affibody molecules binding to HER3 could be used to create drug conjugates, similar to an antibody drug conjugate.

To create an affibody drug conjugate (AffiDC), a suitable payload has to be conjugated with the affibody carrier. DM1 is a highly toxic tubulin polymerization inhibitor derived from maytansine [33]. It is for example used in trastuzumab emtansine (T-DM1), an FDA-approved ADC for the treatment of HER2-positive, metastatic breast cancer [34]. In T-DM1, the payload is linked via a non-cleavable linker. These types of ADCs bind to their intended receptor on the tumor cells, become internalized, and are transported to the lysosome where the protein part is degraded. When the hydrophobic drug is separated from the protein part, it can enter the cytosol by diffusion through the lysosomal membrane and can inhibit tubulin polymerization, leading to cell death.

Affibody molecules are small and are quickly excreted from circulation due to renal filtration. Different methods can be employed to extend the in vivo half-life [35]. One strategy often used for affibody molecules is head-to-tail fusion with an albumin-binding domain (ABD), which upon entering the blood stream associates with albumin, leading to an increase in the molecular weight of the complex by the molecular weight of albumin (67 kDa). These complexes become larger than the cut-off of renal filtration, extending the plasma half-life [36]. In particular, ABD_{035}, with sub-picomolar affinity (K_D) has been found to be suitable as a fusion partner to affibody molecules [37]. A potential drawback with the use of an albumin-binding domain in cancer drugs is that some cancer patients develop hypoalbuminemia. It occurs, for example, in approximately 40% of all patients with pancreatic carcinoma [38], in which cases the bioavailabilty of ABD-containing drugs may be lower than for patients with a normal albumin level.

We have previously investigated AffiDCs targeting HER2, a receptor belonging to the same family as HER3 [36,39–41]. From these studies it was found that an architecture consisting of a monovalent affibody-targeting domain followed by an ABD led to the best performing AffiDC. Furthermore, a C-terminal extension after the ABD, with the amino acid sequence Glu-Glu-Glu-Cys, where the cytotoxic DM1 was attached to the cysteine, was found to minimize unwanted unspecific uptake in the liver [39,40].

In this study, we generated a novel drug conjugate targeting HER3, based on the affibody molecule ($Z_{HER3:08698}$) carrying the drug DM1. We utilized an analogous architecture to the best performing HER2-targeting AffiDC [40]. The drug conjugate was characterized biochemically, and its cytotoxic potential was evaluated. After radiolabeling with $[^{99m}Tc]Tc$, the specificity of binding and rate of internalization was determined, and the distribution in mice bearing HER3-positive xenografts was quantified.

2. Materials and Methods

2.1. General

Unless stated otherwise, all chemicals were from Merck (Darmstadt, Germany) or Sigma-Aldrich (St. Louis, MO, USA). Restriction digestion enzymes were from New England Biolabs (Ipswitch, MA, USA). Statistical significance ($p < 0.05$) was determined using Prism, version 9.3.1 (Graphpad Software, San Diego, CA, USA). For comparison of two groups, a Mann–Whitney test was used. For comparison of multiple groups, a Kruskal–Wallis test with a post-hoc Dunn's test was used.

2.2. Gene Construction

The HER3-targeting affibody molecule used in this study was $Z_{HER3:08698}$ [22], herein abbreviated as Z_{HER3}. The albumin-binding domain used for in vivo half-life extension was ABD_{035} [37], herein abbreviated as ABD. Starting from a plasmid-encoding Z_{HER3}-ABD [29], the affibody coding gene was sub-cloned to the pET-21a(+) plasmid (Novagen, Madison, WI, USA) by in-fusion cloning according to the manufacturer's recommendations (Takara, Kusatsu, Japan). The sequence consisted of codons encoding a Met-His-Glu-His-Glu-His-Glu tag in the N-terminus, followed by the Z_{HER3} affibody sequence, the sequence encoding the ABD domain, and codons encoding a Glu-Glu-Glu-Cys extension in the C-terminus. The plasmid was sequenced by sanger sequencing (Eurofins genomics, Ebersberg, Germany).

2.3. Protein Expression and Purification

The plasmid-encoding Z_{HER3}-ABD was transformed to *Escherichia coli* BL21 (DE3) cells (Thermo Fisher Scientific, Waltham, MA, USA) for protein production. Transformed cells were grown in 1 L cultures in TSB-medium (30 g/L tryptic soy broth, 5 g/L yeast extract and 50 µg/L Kanamycin), in 5 L shake flasks, at 200 rpm (37 °C). When the OD_{600} was approximately 0.9, isopropyl-β-D-1-thiogalactopyranoside (Appolo Scientific, Stockport, UK) was supplemented (1 mM final concentration) to induce protein expression. The cells were then grown for 16 h at 25 °C, after which they were collected by centrifugation. The cells were lysed by sonication and the proteins were purified by affinity chromatography using the ligand human serum albumin (HSA) immobilized on a sepharose column on an ÄKTA pure system (GE Healthcare, Uppsala, Sweden). TST (25 mM Tris-HCl, 1 mM EDTA, 200 mM NaCl, 0.05% Tween, pH 8.0) was used as a running buffer, ammonium acetate (5 mM, pH 5.5) was used for washing, and acetic acid (0.5 M, pH 2.8) was used for elution. The protein were subsequently lyophilized.

2.4. Conjugation with DM1

The lyophilized protein was reconstituted in PBS (pH 6.7) to 0.1 mM as the final concentration. To reduce potentially oxidized cysteines, freshly prepared tris (2-carboxyethyl) phosphine (TCEP, pH 6.7) was added to 5 mM as the final concentration, followed by incubation at 37 °C for 1 h. Maleimidocaproyl-DM1 (mcDM1) (Levena Biopharma, San Diego, CA, USA) was added to 0.3 mM as the final concentration, followed by overnight incubation at room temperature. Unreacted mcDM1 was removed by passage through a NAP-5 desalting column.

Reversed-phase high-performance liquid chromatography (RP-HPLC) was used for further purification. Eluted material from the NAP-5 column was loaded on a Zorbax SB-C18 semi-preparative column (Agilent, Santa Clara, CA, USA), followed by elution with a gradient from 25% to 65% acetonitrile in water, supplemented with 0.1% trifluoroacetic acid (TFA), for 40 min. The fractions containing Z_{HER3}-ABD-mcDM1 were pooled and lyophilized.

A non-toxic control construct, Z_{HER3}-ABD-AA, was also created, where the C-terminal cysteine was alkylated. Lyophilized Z_{HER3}-ABD was reconstituted in a 200 mM ammonium bicarbonate buffer (pH 8.0) supplemented with 2% sodium dodecyl sulfate. Freshly dissolved TCEP was added to 10 mM as the final concentration, followed by incubation for 1 h at 55 °C. Freshly prepared 2-iodoacetamide was added to 20 mM as the final concentration, followed by incubation for 30 min in the dark. Z_{HER3}-ABD-AA was purified by RP-HPLC and lyophilized in the same way as Z_{HER3}-ABD-mcDM1.

2.5. Biochemical Characterization

The two constructs, Z_{HER3}-ABD-mcDM1 and Z_{HER3}-ABD-AA, were reconstituted in PBS, and were characterized biochemically by sodium dodecyl sulphate-polyacrylamide gel electrophoresis (SDS-PAGE), analytical size-exclusion chromatography (SEC), analytical RP-HPLC, and circular dichroism spectroscopy (CD).

For the SDS-PAGE analysis, 10 µg of each construct was loaded on a NuPAGE Bis-Tris protein gel. After 1 h electrophoresis at 200 V, the gel was stained with Gelcode blue safe protein stain for 1 h. Then, the gel was destained with water and photographed.

For analytical SEC, 5 µg protein was analyzed on a prepacked Superdex-75 5/150 column (GE Healthcare). The column was equilibrated with PBS and the samples were loaded followed by elution with PBS at a flow rate of 0.45 mL/min.

For analytical RP-HPLC, the constructs were loaded on a Zorbax 300SB-C18 column (Agilent), followed by elution by a 25 min linear gradient from 30% to 60% acetonitrile in water supplemented with 0.1% TFA.

Circular dichroism (CD) measurements were carried out on a Chirascan spectropolarimeter (Applied Photophysics, Leatherhead, UK). A cuvette with an optical path length of 1 mm was used. To determine thermal stability, a sample (0.4 mg/mL) of each construct was heated from 20 to 95 °C (5 °C/min). During the heating process, the ellipticity at 221 nm was measured. Since both Z_{HER3} and ABD have a high alpha-helical content, the ellipticity at 221 nm is an estimate of helical unfolding and can be used for melting temperature (T_m) estimation. Spectra before and after heating to 95 °C were recorded between 195 and 260 nm.

2.6. Affinity Determination

The affinities of the affibody constructs towards HER3 and murine ErbB3 were measured by surface plasmon resonance analysis using a capture setup on a Biacore 8K instrument (GE Healthcare). Human serum albumin (HSA) was immobilized on a CM5 chip using an amine coupling kit, according to the manufacturer's protocol (GE Healthcare). The level of HSA immobilization in each channel was approximately 1000 RU. The affibody constructs were captured on the surface by a 30 s injection of a 100 nM solution. Finally, a multi-cycle kinetic analysis with a dilution series of HER3 or murine ErbB3 (6.3 nM, 12.5 nM, 25 nM, 50 nM, and 100 nM) was performed at two different temperatures, 25 °C and 37 °C. Samples heated to 95 °C and then cooled down to 25 °C during CD measurements were also analyzed in the same way. The flow rate during the experiments was 30 µL/min. The chip surfaces were regenerated between the runs by injecting 10 mM HCl during 30 s. The sensorgrams were analyzed using a Langmuir 1:1 kinetic model.

2.7. Cell Lines

The BxPC-3 (pancreatic cancer) and DU145 (prostate cancer) cell lines were purchased from ATCC (Manassas, VA, USA) and were maintained in an RPMI 1640 growth medium supplemented with 10% fetal bovine serum (FBS) (Sigma-Aldrich, St. Louis, MO, USA). A Trypsin-EDTA solution (Sigma-Aldrich) was used for cell detachment. The cells were grown in a humidified incubator at 37 °C in a 5% CO_2 atmosphere.

2.8. Determination of Cytotoxicity

To investigate the cytotoxic potential of Z_{HER3}-ABD-mcDM1, 5000 BxPC-3 cells were seeded per well in 96-well plates and were allowed to attach for 24 h. Then, the medium was aspirated, and a new medium was added containing a series of concentrations of Z_{HER3}-ABD-mcDM1 or Z_{HER3}-ABD-AA. The cells were incubated for 72 h, and then cell viability was determined with the Cell Counting Kit-8 (Sigma-Aldrich).

2.9. Radiolabeling with [^{99m}Tc]Tc and Analysis of the Stability of the Labeled Constructs

Both Z_{HER3}-ABD-mcDM1 and Z_{HER3}-ABD-AA were radiolabeled with [^{99m}Tc]Tc$(CO)_3$ on the N-terminal amino acid sequence His-Glu-His-Glu-His-Glu according to a previously published protocol [25,42]. Briefly, [^{99m}Tc]NaTcO$_4$ eluate (400–500 µL, ~4 GBq) was added to a glass vial containing a CRS kit (PSI, Villigen, Switzerland) and was incubated at 100 °C for 30 min. Thereafter, 40 µL (340–400 MBq) of the obtained [^{99m}Tc]Tc(CO)$_3$ solution was added to 50 µg of the affibody constructs in 40 µL PBS. The mixture was incubated at 50 °C for 60 min.

The radiochemical yield was determined by Instant Thin Layered Chromatography (ITLC). A sample of the reaction mixture was applied to a Silica Gel impregnated Chromatography strip (Agilent) and was eluted with PBS. To check for the presence of reduced hydrolyzed technetium, a sample was applied to a second strip, and was eluted with pyridine:acetic acid:water (5:3:1.5).

The radiolabeled conjugates were purified using NAP-5 size-exclusion columns (GE Healthcare), pre-equilibrated with 1% BSA in PBS, and the purity was determined by ITLC. To test the stability of the technetium label, 1 µg each of [^{99m}Tc]Tc-Z$_{HER3}$-ABD-mcDM1 and [^{99m}Tc]Tc-Z$_{HER3}$-ABD-AA were incubated with a 500-fold molar excess of histidine at room temperature and at 37 °C. Control samples in PBS were also incubated at room temperature for up to 4 h. The release of the radiolabel was determined by ITLC.

A control for the biodistribution experiment, (HE)$_3$-Z$_{HER3}$, was similarly labeled according to a previously published protocol [25].

2.10. Binding Specificity, Internalization, and Retention

To confirm the binding specificity of [^{99m}Tc]Tc-Z$_{HER3}$-ABD-mcDM1 and [^{99m}Tc]Tc-Z$_{HER3}$-ABD-AA, HER3-expressing BxPC-3 and DU145 cells were incubated with 0.1 nM of the radiolabeled constructs for 1 h at 37 °C. Prior to incubation, HER3 receptors were pre-saturated by the addition of 50 nM of the Z$_{HER3:0698}$ affibody molecule. After incubation, the cells were detached, collected, and measured for radioactivity content.

Internalization of [^{99m}Tc]Tc-Z$_{HER3}$-ABD-mcDM1 was studied in BxPC-3 and DU145 cells using the 'acid wash' method [24]. In brief, the cells were incubated with 0.1 nM [^{99m}Tc]Tc-Z$_{HER3}$-ABD-mcDM1 at 37 °C for 1, 2, 4, and 8 h. At each time point a set of dishes was removed from the incubator and cells were incubated with 0.2 M glycine buffer (with 0.15 M NaCl, 4 M Urea, pH 2.0) for 5 min on ice. The solution was collected and considered membrane-bound activity. Thereafter the cells were incubated with 1 M NaOH for 30 min at 37 °C, scraped, and collected. The collected activity from these samples was considered to be the internalized fraction.

2.11. Cell Binding Analysis

The binding kinetics of [^{99m}Tc]Tc-Z$_{HER3}$-ABD-mcDM1, [^{99m}Tc]Tc-Z$_{HER3}$-ABD-AA and [^{99m}Tc]Tc-(HE)$_3$-Z$_{HER3}$ to BxPC-3 cells were measured in real-time using a LigandTracer Yellow (Ridgeview Instruments, Uppsala, Sweden). Cells were plated in a dedicated area of a 10 cm petri dish 1 d before the experiment. The petri dish was placed in a rotating holder and three concentrations of radiolabeled construct were added stepwise to the dish (0.3 nM, 1 nM, 3 nM), starting from the lowest concentration. The next concentration was added when the signal had reached equilibrium. The radioactive solution was replaced by fresh culture medium to record the dissociation from the cells. Data were analyzed and k_a, k_d, and K_D were determined by TraceDrawer Software (Version 1.9.2, Ridgeview Instruments, Uppsala, Sweden). The measurements were performed in triplicate at room temperature.

2.12. Animal Studies

Female BALB/c nu/nu mice were inoculated with HER3-expressing BxPC-3 (n = 28) or DU145 cells (n = 8), or HER3-negative RAMOS cells (n = 8) 2–3 weeks before the biodistribution experiment. At the start of the experiment, the average animal weight was 18 ± 1 g, and the average tumor weights were 0.08 ± 0.06 g (BxPC-3), 0.07 ± 0.05 g (DU145), and 0.09 ± 0.04 g (RAMOS). For each data point, a group of four animals was used.

The biodistribution of [^{99m}Tc]Tc-Z$_{HER3}$-ABD-mcDM1 and [^{99m}Tc]Tc-Z$_{HER3}$-ABD-AA was studied in BxPC-3 xenografted mice, 1, 6, and 24 h pi and in DU145-xenografted mice,

24 h pi. Mice were injected intravenously with 20 μg of the affibody constructs, the injected activity was adjusted to be 40 kBq at the time of sample collection, and the protein dose was adjusted with non-labeled construct. The biodistribution of $[^{99m}Tc]Tc-(HE)_3-Z_{HER3}$ was included for comparison. For this control, one group of BxPC-3-xenografted mice were injected with 2 μg (40 kBq) of $[^{99m}Tc]Tc-(HE)_3-Z_{HER3}$ and were sacrificed 1 h pi. All animals were sacrificed by ip injection of a ketamine (250 mg/kg) and xylazine (25 mg/kg) solution followed by heart puncture. Samples from the blood, salivary gland, lung, liver, spleen, stomach (without content), small intestine (without content), kidney, tumor, muscle, and bone were collected, weighed, and measured in an automated gamma counter. The gastrointestinal tract and carcass were collected and measured for activity content.

For the in vivo specificity test, mice with HER3-negative RAMOS xenografts were injected with 20 μg (640 kBq) of $[^{99m}Tc]Tc-Z_{HER3}$-ABD-mcDM1 or $[^{99m}Tc]Tc-Z_{HER3}$-ABD-AA and were sacrificed 24 h pi according to the protocol described above.

3. Results

3.1. Drug Conjugate Design

To study the potential of targeting HER3-expressing tumors with an affibody-based drug conjugate, Z_{HER3}-ABD-mcDM1 was generated. It consisted of Z_{HER3}, an affibody molecule with a strong and specific affinity for HER3 linked via the C-terminus to an albumin-binding domain (ABD) for in vivo half-life extension. Flanking the fusion protein was an N-terminal tag for radiolabeling with $[^{99m}Tc]Tc$, with the amino acid sequence, His-Glu-His-Glu-His-Glu and a C-terminal ending of Glu-Glu-Glu-Cys. The cysteine was used to site-specifically attach DM1 via a non-cleavable malemidocaproyl (mc) linker (Figure 1A).

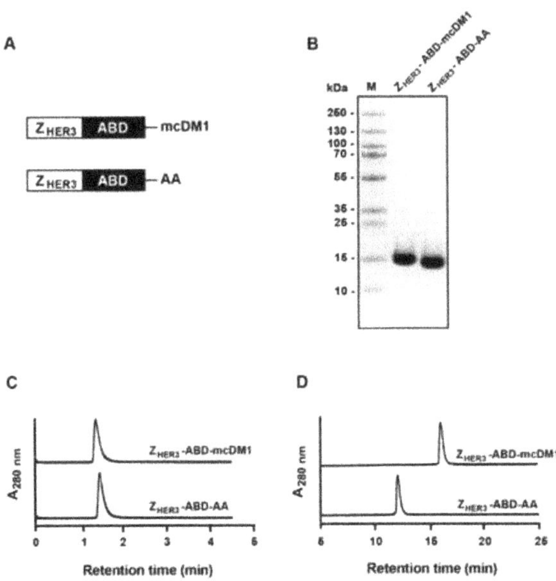

Figure 1. Schematic description and biochemical characterization of the drug conjugate Z_{HER3}-ABD-mcDM1 and the non-toxic control Z_{HER3}-ABD-AA. (**A**) The schematic figure shows the two constructs, Z_{HER3}-ABD-mcDM1 and Z_{HER3}-ABD-AA. (**B**) Samples (10 μg each) of the two constructs were analyzed by sodium dodecyl sulfate polyacrylamide gel electrophoresis (SDS-PAGE) under reducing conditions. Lane M corresponds to the electrophoretic separation of marker proteins, with

the indicated molecular weights, 10 to 250 kDa, to the left of the gel. (**C**) Size-exclusion chromatography analysis of the two constructs. (**D**) RP-HPLC analysis of Z_{HER3}-ABD-mcDM1 and Z_{HER3}-ABD-AA. The constructs were eluted with a 25 min linear gradient from 30% to 60% acetonitrile in water supplemented with 0.1% trifluoroacetic acid.

3.2. Protein Expression, DM1 Conjugation, and Biochemical Characterization

The fusion protein Z_{HER3}-ABD was recombinantly expressed in *Escherichia coli* in a soluble form. It was purified by affinity chromatography using immobilized human serum albumin (HSA) as the ligand. After purification, mcDM1 was conjugated with the C-terminal cysteine yielding Z_{HER3}-ABD-mcDM1. A non-toxic control was also created where the cysteine was alkylated, yielding Z_{HER3}-ABD-AA (Figure 1A). Both constructs were purified by reversed-phase high-performance liquid chromatography (RP-HPLC). Samples of Z_{HER3}-ABD-mcDM1 and Z_{HER3}-ABD-AA were analyzed by SDS-PAGE (Figure 1B). According to the gel, both constructs were of essentially the correct molecular weight with no visible extra bands, indicating a high purity. The constructs were further analyzed by size-exclusion chromatography under native conditions (Figure 1C). Both constructs were eluted as single, symmetrical peaks, suggesting that they were in a monomeric state, and that no oligo- or multimers were formed. Both Z_{HER3}-ABD-mcDM1 and Z_{HER3}-ABD-AA were analyzed by analytical RP-HPLC and were eluted essentially as single peaks (Figure 1D). Calculation of the area under curve (AUC), showed that they were of more than 95% purity, close to 100%. In Figure 1D it is evident that Z_{HER3}-ABD-mcDM1 was eluted later than the non-toxic control. This was not surprising since DM1 is relatively hydrophobic.

The constructs were further subjected to circular dichroism analysis. The spectra of both Z_{HER3}-ABD-mcDM1 and Z_{HER3}-ABD-AA showed a high level of alpha-helicity, characteristic of affibody molecules and the ABD (Figure 2A,C). The thermal stability of the constructs was also investigated by measuring the ellipticity at 221 nm during heating (Figure 2B,D). At this wavelength, the highly alpha-helical constructs have a large negative ellipticity when folded. The negative ellipticity is lost upon heating, which indicates the unfolding of the conjugates. After heat-induced denaturation, the samples were cooled to 25 °C, and new spectra were recorded and overlayed with the spectra recorded before heating (Figure 2A,C). For both constructs, the spectra before and after heating were essentially identical, suggesting efficient refolding after heat-induced denaturation.

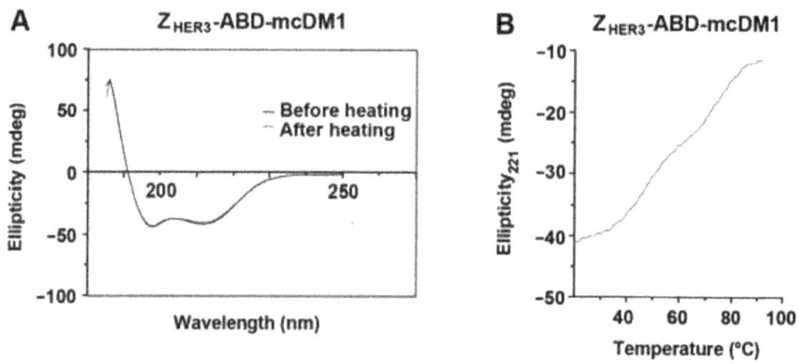

Figure 2. *Cont.*

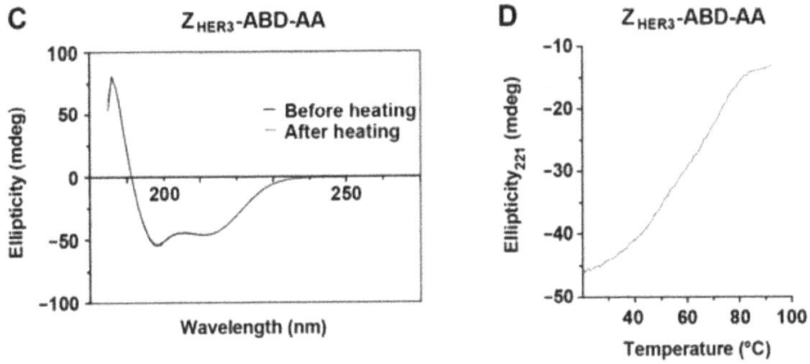

Figure 2. Circular dichroism measurements. Panels (**A**,**C**) show the measured ellipticity (y-axis) at different wavelengths (x-axis). In both panels the spectra obtained before and after heat-induced denaturation followed by refolding are included. The ellipticity at 221 nM was also measured as a function of temperature (Panel (**B**,**D**)). The secondary structure of the constructs is to a large extent alphahelical and the ellipticity at 221 nm is therefore a measurement of the folded state of the constructs.

3.3. *Determination of Affinity to HER3 and Murine ErbB3*

The affinities of Z_{HER3}-ABD-mcDM1 and Z_{HER3}-ABD-AA to HER3 and murine ErbB3 (mErbB3) were determined by surface plasmon resonance analysis, and the results are displayed in Table 1.

Table 1. Affinity between the constructs and HER3 and mErbB3, analyzed by a real-time biosensor.

Ligand	Temperature (°C)	Analyte	k_a (1/Ms)	k_d (1/s)	K_D (nM)
Z_{HER3}-ABD-mcDM1	25	HER3	4.3×10^4	2.5×10^{-4}	6
Z_{HER3}-ABD-AA	25	HER3	1.4×10^5	2.5×10^{-4}	2
Z_{HER3}-ABD-mcDM1	25	mErbB3	1.1×10^5	5.7×10^{-4}	5
Z_{HER3}-ABD-AA	25	mErbB3	1.0×10^5	4.7×10^{-4}	5
Z_{HER3}-ABD-mcDM1 *	25	HER3	6.0×10^4	2.2×10^{-4}	4
Z_{HER3}-ABD-AA *	25	HER3	1.5×10^5	2.3×10^{-4}	2
Z_{HER3}-ABD-mcDM1 *	25	mErbB3	5.6×10^4	4.3×10^{-4}	8
Z_{HER3}-ABD-AA *	25	mErbB3	1.0×10^5	4.7×10^{-4}	5
Z_{HER3}-ABD-mcDM1	37	HER3	1.3×10^5	1.5×10^{-3}	10
Z_{HER3}-ABD-AA	37	HER3	2.2×10^5	1.2×10^{-3}	5
Z_{HER3}-ABD-mcDM1	37	mErbB3	2.7×10^4	2.7×10^{-3}	100
Z_{HER3}-ABD-AA	37	mErbB3	1.2×10^5	2.7×10^{-3}	20
Z_{HER3}-ABD-mcDM1 *	37	HER3	1.6×10^5	1.8×10^{-3}	10
Z_{HER3}-ABD-AA *	37	HER3	2.4×10^5	1.2×10^{-3}	5
Z_{HER3}-ABD-mcDM1 *	37	mErbB3	6.6×10^4	3.7×10^{-3}	60
Z_{HER3}-ABD-AA *	37	mErbB3	9.2×10^4	2.4×10^{-3}	30

* Measurement after heat-induced unfolding and refolding.

The affinities of Z_{HER3}-ABD-mcDM1 to HER3 and mErbB3 at 25 °C were similar with equilibrium dissociation constants (K_D values) of 6 nM and 5 nM, respectively. The affinity measurements were also performed at 37 °C to have a milieu closer to the in vivo situation. The affinities were weaker at this temperature, particularly to mErbB3, with K_D values of 12 nM (HER3) and 100 nM (mErbB3). At both 25 °C and 37 °C, the affinities obtained after heat-induced denaturation were essentially the same, further showing the efficient and functional refolding of the drug conjugate. For Z_{HER3}-ABD-AA, a similar pattern was found with similar K_D-values for HER3 and mErbB3 at 25 °C, but with a weaker affinity for the murine variant at 37 °C. Moreover, this construct had similar affinities before and after heat-induced denaturation, showing efficient refolding. Generally, the affinities measured for Z_{HER3}-ABD-AA were slightly stronger than the affinities measured for Z_{HER3}-ABD-mcDM1.

The affinity measurement was set up as a capture assay, where HSA was immobilized on the sensor chip, followed by the injection of a construct and then HER3 or mErbB3. The results thus show that in all cases, the constructs were able to bind to HSA and HER3 or mErbB3 simultaneously.

3.4. Determination of Cytotoxic Potential

An essential characteristic of drug conjugates is their ability to target the relevant cells and deliver a cytotoxic effect. To investigate the cytotoxic effect of Z_{HER3}-ABD-mcDM1 to HER3-overexpressing cells, a pancreatic carcinoma model was chosen. This is one of the cancer types where HER3 overexpression has been correlated to a more dismal prognosis for the patients [6]. The cytotoxic effect was investigated by treating BxPC-3 cells with different concentrations of Z_{HER3}-ABD-mcDM1, followed by the determination of cell viability (Figure 3). Z_{HER3}-ABD-mcDM1 showed a dose-dependent cytotoxic effect with an IC_{50} value of 7 nM. Treatment of the cells with the non-toxic control Z_{HER3}-ABD-AA showed some loss in viability at higher concentrations.

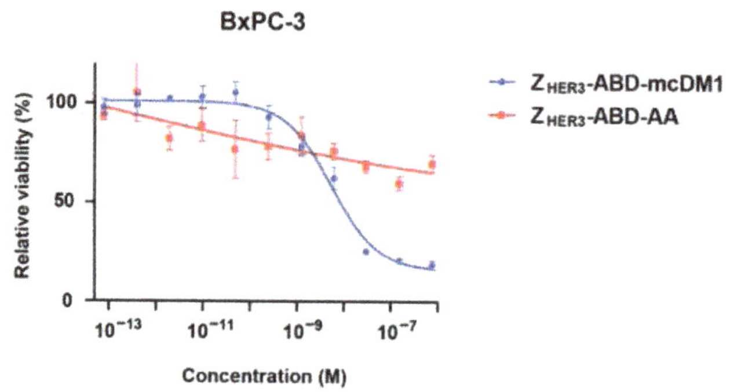

Figure 3. Effect on viability on the human pancreatic cancer cell line BxPC-3 by Z_{HER3}-ABD-mcDM1 and Z_{HER3}-ABD-AA. The cells were incubated with different concentrations of the constructs (*x*-axis), followed by the measurement of cell viability (*y*-axis). The viability of cells grown without the addition of any construct was set to 100%. Each data point is the average of four experiments. The error bars correspond to 1 SD.

3.5. Radiolabeling with [^{99m}Tc]Tc

To allow further analysis in vitro and in vivo, Z_{HER3}-ABD-mcDM1 and Z_{HER3}-ABD-AA were radiolabeled with the residualizing nuclide, [^{99m}Tc]Tc. Specifically, the constructs were labeled with [^{99m}Tc][$Tc(CO)_3(H_2O)_3$]$^+$ on the N-terminal $(HE)_3$-tag. The radiochemical yields were 93.0 ± 2% and 90.9 ± 0.4% for [^{99m}Tc]Tc-Z_{HER3}-ABD-mcDM1 and [^{99m}Tc]Tc-Z_{HER3}-ABD-AA, respectively. After purification by size exclusion chromatography, the purity for both constructs was greater than 99.9%. A control construct to be used in vivo, $(HE)_3$-Z_{HER3}, was also radiolabeled and purified. The resulting purity after size-exclusion chromatography was >98%.

To test the stability of the [^{99m}Tc]Tc-label, the constructs were challenged with a high concentration of histidine. The results are shown in Table 2. Only a minor release of the label, up to 2.3% for [^{99m}Tc]Tc-Z_{HER3}-ABD-mcDM1, was observed at room temperature at both 1 h and 4 h. The release was higher for both constructs at 37 °C with 8.8 ± 0.6%, and 5.7 ± 0.2% released within 4 h from [^{99m}Tc]Tc-Z_{HER3}-ABD-mcDM1 and [^{99m}Tc]Tc-Z_{HER3}-ABD-AA, respectively. Only a minor release was observed in the control experiment (PBS), which was performed in a PBS buffer without histidine.

Table 2. Radiolabel stability of [^{99m}Tc]Tc-Z_{HER3}-ABD-mcDM1 and [^{99m}Tc]Tc-Z_{HER3}-ABD-AA.

Condition	[^{99m}Tc]Tc-Z_{HER3}-ABD-mcDM1		[^{99m}Tc]Tc-Z_{HER3}-ABD-AA	
	1 h	4 h	1 h	4 h
500× histidine, 22 °C	1.4 ± 0.5 *	2.3 ± 0.1	1.3 ± 0.1	2.0 ± 0.4
500× histidine, 37 °C	4.0 ± 0.5	8.8 ± 0.6	4.3 ± 0.5	5.7 ± 0.2
PBS, 22 °C	1.3 ± 0.9	0.9 ± 0.5	1.0 ± 0.6	1.0 ± 0.5

* The values in the table are given as a percent of the released radioactivity from the constructs ± 1 SD. Each value corresponds to the average of three replicate experiments.

3.6. Cell-Binding Specificity and Rate of Internalization

The interaction of the radiolabeled constructs with the pancreatic carcinoma cell line BxPC-3 and the prostate carcinoma cell line DU145, was further characterized. The specificity in the interaction was determined by incubating the cells with [^{99m}Tc]Tc-Z_{HER3}-ABD-mcDM1 or [^{99m}Tc]Tc-Z_{HER3}-ABD-AA, and comparing the result with cells where available HER3 receptors had been pre-blocked with a large excess of the non-ABD conjugated, non-radiolabeled Z_{HER3} affibody. The experiment showed that the blocking of available HER3 receptors resulted in a significant decrease in the uptake of [^{99m}Tc]Tc-Z_{HER3}-ABD-mcDM1 and [^{99m}Tc]Tc-Z_{HER3}-ABD-AA in both cell lines (Figure 4), suggesting a HER3-specific uptake of the constructs.

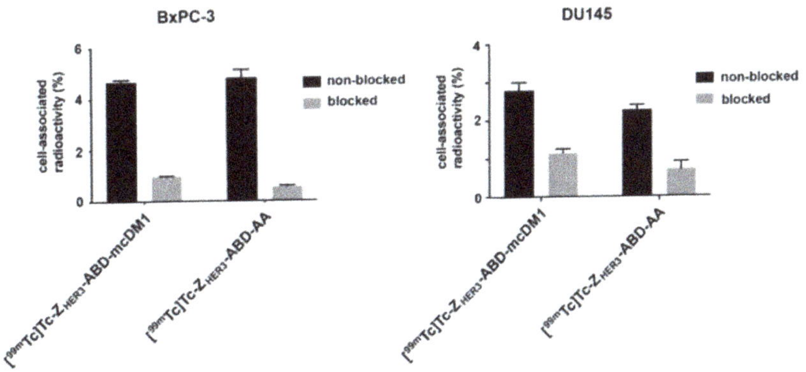

Figure 4. Cell-binding specificity of [^{99m}Tc]Tc-Z_{HER3}-ABD-mcDM1 and [^{99m}Tc]Tc-Z_{HER3}-ABD-AA. BxPC-3 or DU145 cells were pre-incubated (blocked) with $Z_{HER3:08698}$ or not pre-incubated (non-blocked). The concentration of the construct for pre-blocking was 50 nM. Subsequently, the cells were incubated with [^{99m}Tc]Tc-Z_{HER3}-ABD-mcDM1 or [^{99m}Tc]Tc-Z_{HER3}-ABD-AA (0.1 nM). Both panels show the activity retained on the cells after incubation, as the percentage of the activity initially added. The results are shown as the average of three samples with the error bars corresponding to 1 SD.

Next, the rates of association and internalization of [^{99m}Tc]Tc-Z_{HER3}-ABD-mcDM1 were investigated. The association of [^{99m}Tc]Tc-Z_{HER3}-ABD-mcDM1 to both cell lines was quick with more than 75% of the total cell-associated activity bound after 1 h of incubation (Figure 5). The level of internalized radioactivity increased with time, and after 8 h it was 27% of the cell-associated activity for both cell lines. The difference between BxPC-3 and DU145 was not significant.

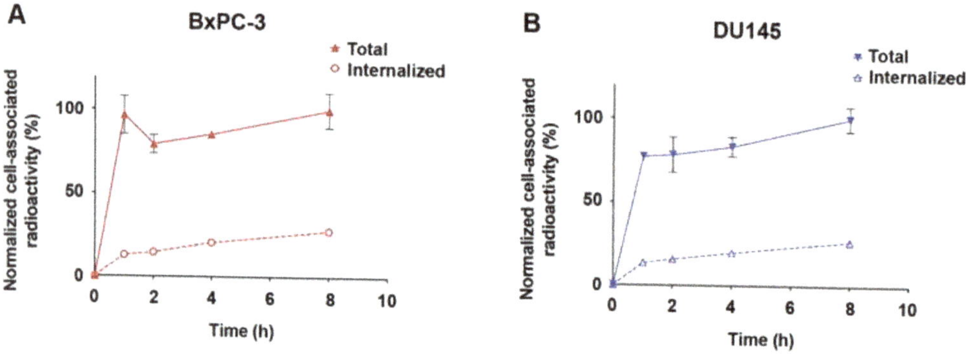

Figure 5. Cellular association and internalization of [^{99m}Tc]Tc-Z_{HER3}-ABD-mcDM1 in BxPC-3 (**A**) and DU145 (**B**) cells. The cells, BxPC-3 and DU145, were subjected to continuous incubation with 0.1 nM [^{99m}Tc]Tc-Z_{HER3}-mcDM1 at 37 °C. The percent of total cell-associated radioactivity and internalized radioactivity was measured and is plotted as a function of time. All values were normalized to the highest value observed (which was set to 100%). The data points represent the average values measured for three samples with error bars corresponding to 1 standard deviation. In some cases, the error bars are smaller than the symbols, and are therefore not visible.

3.7. Affinity to BxPC-3 Cells

The affinities (equilibrium dissociation constants, K_D values) of [^{99m}Tc]Tc-Z_{HER3}-ABD-mcDM1 and [^{99m}Tc]Tc-Z_{HER3}-ABD-AA to HER3-expressing BxPC-3 cells were measured, and the kinetic parameters of the interactions are shown in Table 3. The K_D values of [^{99m}Tc]Tc-Z_{HER3}-ABD-mcDM1 and [^{99m}Tc]Tc-Z_{HER3}-ABD-AA were in the sub-nanomolar range. The control construct used in the in vivo experiments, [^{99m}Tc]Tc-(HE)$_3$-Z_{HER3}, was also characterized. The K_D value of that construct for BxPC-3 cells was stronger (0.04 nM) compared to the values for [^{99m}Tc]Tc-Z_{HER3}-ABD-mcDM1 and [^{99m}Tc]Tc-Z_{HER3}-ABD-AA. The interactions of the constructs and the HER3-binding mAb seribantumab with BxPC-3 cells was also investigated by flow cytometry. Both constructs and seribantumab were found to bind to BxPC-3 cells (Figure S1).

Table 3. Kinetic parameters and equilibrium dissociation constants for the interaction between the constructs and BxPC-3 cells.

Construct	k_a (1/Ms) [a]	k_d (1/s)	K_D (nM)
[^{99m}Tc]Tc-Z_{HER3}-ABD-mcDM1	$1.4 \times 10^5 \pm 0.1 \times 10^5$	$3.0 \times 10^{-5} \pm 2.0 \times 10^{-5}$	0.2 ± 0.1
[^{99m}Tc]Tc-Z_{HER3}-ABD-AA	$1.8 \times 10^5 \pm 0.1 \times 10^5$	$4.0 \times 10^{-5} \pm 3.0 \times 10^{-5}$	0.2 ± 0.1
[^{99m}Tc]Tc-(HE)$_3$-Z_{HER3}	$3.0 \times 10^5 \pm 0.4 \times 10^5$	$1.1 \times 10^{-5} \pm 0.1 \times 10^{-5}$	0.04 ± 0.01

[a] Each value in the table is the average of three ([^{99m}Tc]Tc-Z_{HER3}-ABD-mcDM1 and [^{99m}Tc]Tc-Z_{HER3}-ABD-AA) or two ([^{99m}Tc]Tc-(HE)$_3$-Z_{HER3}) independent experiments. The error corresponds to 1 standard deviation.

3.8. Biodistribution

The biodistribution of [^{99m}Tc]Tc-Z_{HER3}-ABD-mcDM1 and [^{99m}Tc]Tc-Z_{HER3}-ABD-AA was studied in animals with xenografts derived from the BxPC-3 or DU145 cell lines, both expressing HER3. As a control, xenografts derived from the RAMOS cell line were used, which do not express HER3. The uptake of [^{99m}Tc]Tc-Z_{HER3}-ABD-mcDM1 and [^{99m}Tc]Tc-Z_{HER3}-ABD-AA was significantly higher in the BxPC-3 xenografts than in the RAMOS-derived xenografts at 24 h, indicating a specific binding of the constructs to HER3 in vivo (Figure 6). The uptake in the four DU145 xenografts at 24 h was higher than the uptake in the four RAMOS xenografts, but the difference was not significant. Furthermore, there was no significant difference in tumor uptake between [^{99m}Tc]Tc-Z_{HER3}-ABD-mcDM1 and [^{99m}Tc]Tc-Z_{HER3}-ABD-AA.

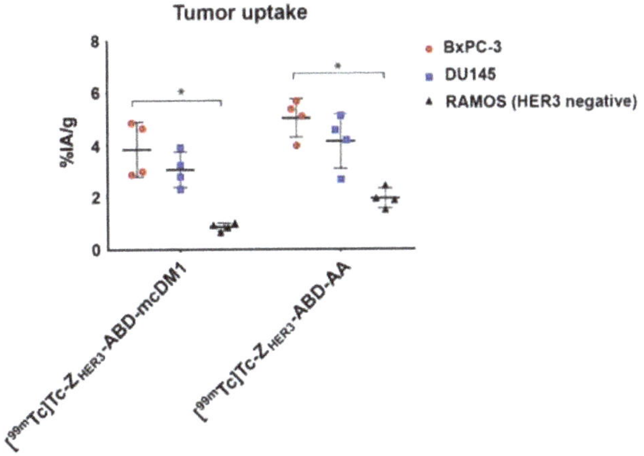

Figure 6. In vivo tumor targeting. The uptake (expressed as %IA/g) of $[^{99m}Tc]Tc$-Z_{HER3}-ABD-mcDM1 and $[^{99m}Tc]Tc$-Z_{HER3}-ABD-AA was studied in mice carrying HER3-expressing BxPC-3 or DU145-derived xenografts and in HER3-negative RAMOS-derived xenografts at 24 h pi. The mice (n = 4 animals/group) were injected with 20 µg of $[^{99m}Tc]Tc$-labeled constructs. * corresponds to significant differences ($p < 0.05$).

The general biodistribution of $[^{99m}Tc]Tc$-Z_{HER3}-ABD-mcDM1 and $[^{99m}Tc]Tc$-Z_{HER3}-ABD-AA in mice carrying BxPC-3 xenografts was also studied as a function of time (Figure 7, Table S1). The results demonstrated an uptake in organs with the expression of mErbB3 (liver, lung, salivary gland, stomach, and small intestine) and an additionally elevated uptake in the spleen. Considering all organs and the tumors, the uptake of $[^{99m}Tc]Tc$-Z_{HER3}-ABD-mcDM1 was significantly lower at 24 h pi compared with the uptake at 1 h pi with the exceptions of the salivary gland, tumor, and GI tract. A similar pattern was observed for animals injected with $[^{99m}Tc]Tc$-Z_{HER3}-ABD-AA with the exception of salivary gland, tumor, muscle, and GI tract. A comparison of the uptake of $[^{99m}Tc]Tc$-Z_{HER3}-ABD-mcDM1 and $[^{99m}Tc]Tc$-Z_{HER3}-ABD-AA showed some differences. At both 1 h and 6 h pi, the uptake in the liver of $[^{99m}Tc]Tc$-Z_{HER3}-ABD-mcDM1 was significantly higher compared with the hepatic uptake of $[^{99m}Tc]Tc$-Z_{HER3}-ABD-AA. Moreover, at 1 and 6 h, the uptake of $[^{99m}Tc]Tc$-Z_{HER3}-ABD-mcDM1 in the small intestine was higher than the uptake of $[^{99m}Tc]Tc$-Z_{HER3}-ABD-AA. For the tumors, the uptake increased significantly from 1 h to 6 h pi, where it peaked with 6.3 ± 0.4% IA/g and 5.9 ± 0.2% IA/g, for $[^{99m}Tc]Tc$-Z_{HER3}-ABD-mcDM1 and $[^{99m}Tc]Tc$-Z_{HER3}-ABD-AA, respectively.

As a control, the biodistribution of $[^{99m}Tc]Tc$-$(HE)_3$-Z_{HER3} was studied at 1 h pi (Table S2) This construct lacked both the half-life extending ABD domain, and the cytotoxic DM1 molecule. The uptake of $[^{99m}Tc]Tc$-$(HE)_3$-Z_{HER3} was significantly lower than the uptake of both $[^{99m}Tc]Tc$-Z_{HER3}-ABD-mcDM1 and $[^{99m}Tc]Tc$-Z_{HER3}-ABD-AA in most organs, with the exception of the kidneys where the uptake was significantly higher. Compared with $[^{99m}Tc]Tc$-$(HE)_3$-Z_{HER3}, both $[^{99m}Tc]Tc$-Z_{HER3}-ABD-mcDM1 and $[^{99m}Tc]Tc$-Z_{HER3}-ABD-AA had a significant, 10-fold, higher concentration in blood at 1 h pi, showing the impact of including the ABD in the constructs for an increased in vivo half-life. For both ABD-containing constructs, a significant decrease in activity in the blood was observed between the 1 h and 24 h time points. At 24 h the blood concentration of $[^{99m}Tc]Tc$-Z_{HER3}-ABD-mcDM1 was comparable to the concentration of $[^{99m}Tc]Tc$-$(HE)_3$-Z_{HER3} at 1 h pi, and was significantly lower than the blood concentration of $[^{99m}Tc]Tc$-Z_{HER3}-ABD-AA, showing a more rapid clearance of the construct with DM1.

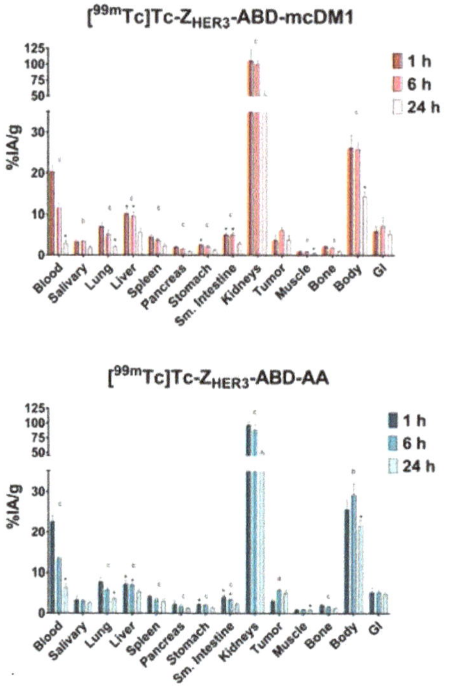

Figure 7. General biodistribution of [^{99m}Tc]Tc-Z$_{HER3}$-ABD-mcDM1 and [^{99m}Tc]Tc-Z$_{HER3}$-ABD-AA. Analyses were performed at 1 h, 6 h, and 24 h pi in mice bearing BxPC-3-derived xenografts. The bars represent the average of four mice. Data are presented as the uptake in different organs as a percentage of the injected activity divided by the weight of the organ or tumor (%IA/g). The data for the GI tract and body are presented as %IA per whole sample. Each mouse received an injection of 20 µg of the [^{99m}Tc]Tc-labeled conjugates. * corresponds to statistically significant differences ($p < 0.05$) between [^{99m}Tc]Tc-Z$_{HER3}$-ABD-mcDM1 and [^{99m}Tc]Tc-Z$_{HER3}$-ABD-AA for the same organ at the same time-point. [a] corresponds to significant differences ($p < 0.05$) between the 1 and 6 h time points. [b] corresponds to significant differences ($p < 0.05$) between the 6 h and 24 h time points. [c] corresponds to significant differences ($p < 0.05$) between the 1 h and 24 h time points.

Furthermore, the biodistribution of [^{99m}Tc]Tc-Z$_{HER3}$-ABD-mcDM1 and [^{99m}Tc]Tc-Z$_{HER3}$-ABD-AA was studied in mice carrying DU145-derived xenografts at 24 h pi (Table S3). There was no significant difference in tumor uptake compared to the BxPC-3 model. Moreover, as expected, no difference in general biodistribution in normal organs was observed between the two xenografts models.

4. Discussion

Targeted therapy directed to the human epidermal growth factor receptor 3 (HER3) appears challenging due to its expression on several normal organs and relatively low over-expression on tumor cells [18]. HER3-targeted therapies have been investigated in both clinical and pre-clinical settings in several tumor types, but so far, no drugs have been approved for clinical use. It is not clear if any of the currently investigated modalities will reach the clinic and there is thus a need to also investigate other drug formats.

Here, a novel type of HER3-targeted drug candidate was constructed and studied, Z$_{HER3}$-ABD-mcDM1, an affibody–drug conjugate (AffiDC). It is a well-defined and relatively small compound (Mw 14 kDa), consisting of a HER3-targeting affibody molecule, an albumin-binding domain for in vivo half-life extension, and the highly potent cytotoxic drug DM1. Similar to several monoclonal antibodies (mAbs) directed against HER3, Z_{HER3} by itself has been found to be able to block the activation of the receptor by HRG. In an earlier study by our group, experimental therapy with the Z_{HER3}-ABD fusion protein led to a cytostatic effect, delayed tumor growth, and prolonged survival in xenograft-bearing mice, comparable to the therapeutic effect of the mAb seribantumab [43]. In the present study, we found a decreased viability of the HER3-expressing pancreatic cancer cell line BxPC-3 in the presence of increasing concentrations on the non-toxic control construct Z_{HER3}-ABD-AA (Figure 3), which may have been the result of the cytostatic effect observed earlier. However, from comparing the effect of Z_{HER3}-ABD-mcDM1 and Z_{HER3}-ABD-AA in Figure 3, it was evident that the main inhibitory effect on cell viability was delivered through the cytotoxic drug DM1. It can therefore be concluded that Z_{HER3}-ABD-mcDM1 has two modes of action, blocking the binding site of HRG and poisoning by DM1.

Monoclonal antibodies under pre-clinical and clinical development targeting HER3 usually deliver an antibody-dependent cell-mediated cytotoxicity (ADCC) effect, sometimes coupled with a blocking of the HRG-binding site and/or locking of the receptor in a conformation that prevents its dimerization and signaling, and/or triggering internalization of HER3. A more recent strategy for HER3-targeted cancer therapy, which inspired us to investigate the AffiDCs, is the antibody–drug conjugates (ADCs). These compounds take advantage of the targeting ability of the mAb, as well as the therapeutic effects described above for mAbs, and add the cytotoxic action of the attached drug. There is currently one HER3-targeting ADC in clinical development, patritumab deruxtecan, and it has shown promising results in a phase I/II trial [15]. Compared to patritumab deruxtecan, Z_{HER3}-ABD-mcDM1 is considerably smaller, and should thus penetrate solid tumors more efficiently, possibly resulting in more efficient therapy [44,45]. These two molecules also contain drugs with different toxic effects; patritumab deruxtecan inhibits topoisomerase 1 activity, and Z_{HER3}-ABD-mcDM1 prevents tubulin polymerization. Therefore, it is possible that the two compounds could be used concomitantly for a synergistic therapeutic effect with limited toxicity to healthy tissue, since they likely have differences in their general biodistributions.

Engineered alternative scaffold proteins (ESPs) have been gaining attention as delivery vehicles for different payloads, including cytotoxic drugs and radionuclides [20,46,47]. In addition to their smaller size, leading to a higher rate of tumor penetration [48], the use of affibody molecules also allows for production in procaryotic hosts, resulting in a comparatively low cost-of-goods compared to more advanced host cells [49]. Affibody molecules have even been produced by peptide synthesis [50], which may further reduce the cost in an industrial-scale manufacturing process.

During the biochemical evaluation of our novel AffiDC, the equilibrium dissociation constants (K_D values) for HER3 and murine ErbB3 were evaluated in a Biacore biosensor and were found to differ by 10-fold at 37 °C, 11 nM, and 100 nM, respectively. It is therefore possible that the on-target uptake in normal organs will be higher for a human subject than for the mice investigated in this study. However, the conditions during the experiment could only mimic the in vivo conditions. It is also interesting to note that the K_D values obtained at 25 °C for the two receptors were similar, with 6 nM and 5 nM for the human and murine version, respectively. The results highlight the importance of performing experiments as close to the relevant temperature as possible, 37 °C in this case.

Furthermore, the affinity of Z_{HER3}-ABD-mcDM1 to the HER3-expressing BxPC-3 cells was stronger (K_D 0.2 nM) than the measured affinity to the human receptor with the biosensor (6 nM). It highlights the importance of studying interactions with several methods to be able to draw robust conclusions about their performance. A comparison of the affinity of Z_{HER3}-ABD-mcDM1 and the non-toxic control Z_{HER3}-ABD-AA to BxPC-3 cells showed the same strong affinity (K_D 0.2 nM) in both cases. This strongly suggests that the addition of DM1 to the C-terminus of the affibody carrier does not affect its binding properties to HER3. This finding was corroborated by the biodistribution experiment where the uptake of both constructs by the BxPC-3 and DU145 xenografts was similar.

The biodistribution of $[^{99m}Tc]Tc$-Z_{HER3}-ABD-mcDM1 and $[^{99m}Tc]Tc$-Z_{HER3}-ABD-AA was studied in tumor-bearing mice, with the measurement of the uptake in organs and tissues as a function of time. Since HER3 is expressed on some normal organs and tissues, it is important to note that Z_{HER3} is cross-reactive, and thus allows for an investigation of both the on-target and off-target uptake, to be able to determine if future pre-clinical and clinical testing is feasible. The results showed a striking difference in uptake by HER3-positive and negative tumors. The uptake of $[^{99m}Tc]Tc$-Z_{HER3}-ABD-mcDM1 was 10-fold higher in the HER3-positive xenografts derived from BxPC-3 cells, compared to the HER3-negative xenografts derived from RAMOS cells. Moreover, the relatively high kidney uptake of $[^{99m}Tc]Tc$-Z_{HER3}-ABD-mcDM1, in combination with a relatively low uptake in the liver suggests elimination mainly by the kidneys. The biodistribution pattern for $[^{99m}Tc]Tc$-Z_{HER3}-ABD-mcDM1 over time showed a significant decrease in uptake in all normal organs, which in general terms, followed the activity concentration in blood (Figure 7). There was a notable delay in the excretion of the activity for the kidney and liver that could be explained by the degradation of the labeled construct and the elimination of radiocatabolites in these organs, since the $[^{99m}Tc]Tc(CO)_3$ label had residualizing properties. This is in agreement with the constant activity uptake in the GI tract that was collected together with its content, and overall reflects hepatobiliary excretion to some extent. For the tumors, the activity instead accumulated over time, suggesting binding followed by internalization. Such a pattern of tumor activity uptake is in a good agreement with the in vitro data on the association and internalization of the constructs (Figure 5).

The uptake in the liver of $[^{99m}Tc]Tc$-Z_{HER3}-ABD-mcDM1 was significantly higher than the uptake of $[^{99m}Tc]Tc$-Z_{HER3}-ABD-AA, at 1 h and 6 h, possibly due to the hydrophobic character of the DM1 payload. An earlier study on an affibody–drug conjugate targeting HER2 showed that the addition of three hydrophilic and negatively charged glutamic acids next to the cysteine where DM1 was attached efficiently lowered liver accumulation [39]. In the present study, three glutamic acid residues were added in the same way to Z_{HER3}-ABD-mcDM1. It is possible that the uptake of Z_{HER3}-ABD-mcDM1 in the liver would have been higher without the glutamic acid residues.

Affibody-based drug conjugates targeting the HER2 receptor have been studied [36,39,40,51]. Similar to those studies, the drug conjugate in this study was able to deliver a strong cytotoxic effect to cells overexpressing the receptor, with an IC_{50} value of 7 nM to BxPC-3 cells. However, it was evident from the study on Z_{HER2}-ABD-mcDM1 that the cytotoxic potential varied among different cell lines, all having a very high expression level of HER2 (AU565, SKBR3 and SKOV3), and that the strength of cell binding can only be partially correlated to the cytotoxic effect. The IC_{50} value of Z_{HER2}-ABD-mcDM1 towards SKOV-3 cells was 33 nM, which was weaker than the IC_{50} value of Z_{HER3}-ABD-mcDM1 towards BxPC-3 cells (7 nM). Since Z_{HER2}-ABD-mcDM1 could efficiently eradicate SKOV3-derived xenografts in mice, it is likely that Z_{HER3}-ABD-mcDM1 might be effective in future pre-clinical development, if the animals can tolerate a similar dosing scheme. The liver is often a sensitive organ and may limit the maximum tolerated dose. Both constructs were taken up in the liver to a relatively low extent, with 6% IA/g, 4 h after injection of 6 µg Z_{HER2}-ABD-mcDM1, compared to 9.7% IA/g, 6 h after the injection of 20 µg Z_{HER3}-ABD-mcDM1 (Figure 7, Table S1). The uptake in the tumors was similar with 4.2% IA/g and 6.3% IA/g for Z_{HER2}-ABD-mcDM1 and Z_{HER3}-ABD-mcDM1, respectively.

In conclusion, we generated a HER3-targeting affibody-based drug conjugate, Z_{HER3}-ABD-mcDM1, and extensively investigated its in vitro properties as well as determined its biodistribution in tumor-bearing mice. The conjugate is a well-defined and robust molecule, with potent cytotoxicity to HER3-overexpressing cells in vitro and with the ability to target HER3-expressing human tumors in mice. The results hold promise for further pre-clinical and clinical development.

Supplementary Materials: The following supporting information can be downloaded at: https://www.mdpi.com/xxx/s1, Figure S1: Flow cytometry analysis of binding of Z_{HER3}-ABD-mcDM1, Z_{HER3}-ABD-AA, and seribantumab to BxPC-3 55 cells; Table S1: Biodistribution of $[^{99m}Tc]Tc$-Z_{HER3}-

ABD-mcDM1, and $[^{99m}Tc]Tc-Z_{HER3}$-ABD-AA in BxPC-3 xenograft bearing mice at 1 h, 6 h, and 24 h pi; Table S2: Biodistribution of $[^{99m}Tc]Tc-(HE)_3-Z_{HER3}$ in BxPC-3 xenograft bearing mice at 1 h pi; Table S3: Biodistribution of $[^{99m}Tc]Tc-Z_{HER3}$-ABD-mcDM1 and $[^{99m}Tc]Tc-Z_{HER3}$-ABD-AA in DU145 xenograft bearing mice at 24 h pi.

Author Contributions: Conceptualization, A.V., J.L., S.S., A.O. and T.G.; methodology, J.L., A.O. and T.G.; validation, S.S.R., W.Y., A.V. and T.G.; formal analysis, S.S.R., W.Y., A.M.B. and T.G.; investigation, S.S.R., W.Y., A.M.B., A.A. and C.D.L.; resources, A.V., J.L., S.S., A.O. and T.G.; data curation, S.S.R., W.Y., A.M.B., A.A., C.D.L. and T.G.; writing—original draft preparation, S.S.R. and T.G.; writing—review and editing, S.S.R., A.M.B., J.L., S.S., A.O. and T.G.; supervision, A.V., J.L., S.S., A.O. and T.G.; project administration, J.L., A.O. and T.G.; funding acquisition, A.V., J.L., S.S., A.O. and T.G. All authors have read and agreed to the published version of the manuscript.

Funding: This research was funded by the Swedish Agency for Innovation VINNOVA (2019/00104) and the Swedish Cancer Society (Cancerfonden) (grant number 21 1861 Pj (T.G.); 2020/181 (A.V.); 20 0893 Pj (A.V.); 20 0815 PjF (A.O.); 20 1090 PjF (J.L.); 190101Pj01H (S.S.)), and the Swedish Research Council (Vetenskapsrådet) (grant number 2019-00986 (A.O); 2019-05115 (J.L.)), and the Knut and Alice Wallenberg Foundation through the Wallenberg Center for Protein Technology (grant number 2019.0341 (S.S)). Support for Wen Yin was partly covered by a grant from the China Scholarship Council (CSC).

Institutional Review Board Statement: Animal experiments were carried out in agreement with the Swedish legislation for animal welfare and were approved by the Animal Research Committee at Uppsala University (ethical permit number C 5/16, approved 26 February 2016).

Data Availability Statement: The data generated in the study are available upon reasonable request through the corresponding authors.

Conflicts of Interest: S.S.R., W.Y., A.M.B., A.A., C.D.L., A.V., J.L. and T.G. declare no conflict of interest. S.S. and A.O. are minority shareholders in Affibody AB. The funders/company had no role in the design of the study; in the collection, analyses, or interpretation of data; in the writing of the manuscript, or in the decision to publish the results.

References

1. Zahavi, D.; Weiner, L. Monoclonal antibodies in cancer therapy. *Antibodies* **2020**, *9*, 34. [CrossRef] [PubMed]
2. Xenaki, K.T.; Oliveira, S.; van Bergen en Henegouwen, P.M.P. Antibody or antibody fragments: Implications for molecular imaging and targeted therapy of solid tumors. *Front. Immunol.* **2017**, *8*, 1287. [CrossRef] [PubMed]
3. Gebauer, M.; Skerra, A. Engineered Protein Scaffolds as Next-Generation Therapeutics. *Annu. Rev. Pharm. Toxicol.* **2020**, *6*, 391–415. [CrossRef] [PubMed]
4. Mishra, R.; Patel, H.; Alanazi, S.; Yuan, L.; Garrett, J.T. HER3 signaling and targeted therapy in cancer. *Oncol. Rev.* **2018**, *12*, 45–62. [CrossRef]
5. Siegfried, J.M.; Lin, Y.; Diergaarde, B.; Lin, H.M.; Dacic, S.; Pennathur, A.; Weissfeld, J.L.; Romkes, M.; Nukui, T.; Stabile, L.P. Expression of PAM50 Genes in Lung Cancer: Evidence that Interactions between Hormone Receptors and HER2/HER3 Contribute to Poor Outcome. *Neoplasia* **2015**, *17*, 817–825. [CrossRef]
6. Liles, J.S.; Arnoletti, J.P.; Tzeng, C.W.D.; Howard, J.H.; Kossenkov, A.V.; Kulesza, P.; Heslin, M.J.; Frolov, A. ErbB3 expression promotes tumorigenesis in pancreatic adenocarcinoma. *Cancer Biol. Ther.* **2010**, *10*, 555–563. [CrossRef]
7. Reschke, M.; Mihic-Probst, D.; Van Horst, E.H.; Der Knyazev, P.; Wild, P.J.; Hutterer, M.; Meyer, S.; Dummer, R.; Moch, H.; Ullrich, A. HER3 Is a determinant for poor prognosis in melanoma. *Clin. Cancer Res.* **2008**, *14*, 5188–5197. [CrossRef]
8. Koumakpayi, I.H.; Diallo, J.S.; Le Page, C.; Lessard, L.; Gleave, M.; Bégin, L.R.; Mes-Masson, A.M.; Saad, F. Expression and nuclear localization of ErbB3 in prostate cancer. *Clin. Cancer Res.* **2006**, *12*, 2730–2737. [CrossRef]
9. Lipton, A.; Goodman, L.; Leitzel, K.; Cook, J.; Sperinde, J.; Haddad, M.; Köstler, W.J.; Huang, W.; Weidler, J.M.; Ali, S.; et al. HER3, p95HER2, and HER2 protein expression levels define multiple subtypes of HER2-positive metastatic breast cancer. *Breast Cancer Res. Treat.* **2013**, *141*, 43–53. [CrossRef]
10. Lemmon, M.A.; Schlessinger, J. Cell signaling by receptor tyrosine kinases. *Cell* **2010**, *141*, 1117–1134. [CrossRef]
11. Laskin, J.; Liu, S.V.; Tolba, K.; Heining, C.; Schlenk, R.F.; Cheema, P.; Cadranel, J.; Jones, M.R.; Drilon, A.; Cseh, A.; et al. NRG1 fusion-driven tumors: Biology, detection, and the therapeutic role of afatinib and other ErbB-targeting agents. *Ann. Oncol.* **2020**, *31*, 1693–1703. [CrossRef] [PubMed]
12. Khongorzul, P.; Ling, C.J.; Khan, F.U.; Ihsan, A.U.; Zhang, J. Antibody-drug conjugates: A comprehensive review. *Mol. Cancer Res.* **2020**, *18*, 3–19. [CrossRef] [PubMed]
13. Joubert, N.; Beck, A.; Dumontet, C.; Denevault-sabourin, C. Antibody—Drug Conjugates: The Last Decade. *Pharmaceuticals* **2020**, *13*, 245. [CrossRef] [PubMed]

14. Wolska-Washer, A.; Robak, T. Safety and Tolerability of Antibody-Drug Conjugates in Cancer. *Drug Saf.* **2019**, *42*, 295–314. [CrossRef] [PubMed]
15. Jänne, P.A.; Baik, C.; Su, W.-C.; Johnson, M.L.; Hayashi, H.; Nishio, M.; Kim, D.-W.; Koczywas, M.; Gold, K.A.; Steuer, C.E.; et al. Efficacy and Safety of Patritumab Deruxtecan (HER3-DXd) in EGFR Inhibitor–Resistant, EGFR-Mutated Non–Small Cell Lung Cancer. *Cancer Discov.* **2021**, *12*, 74–89. [CrossRef]
16. Hashimoto, Y.; Koyama, K.; Kamai, Y.; Hirotani, K.; Ogitani, Y.; Zembutsu, A.; Abe, M.; Kaneda, Y.; Maeda, N.; Shiose, Y.; et al. A novel HER3-targeting antibody–drug conjugate, U3-1402, exhibits potent therapeutic efficacy through the delivery of cytotoxic payload by efficient internalization. *Clin. Cancer Res.* **2019**, *25*, 7151–7161. [CrossRef]
17. Lee, S.; Larson, R. MCLA-128 Fights NRG1 Fusion-Positive Cancers. *Cancer Discov.* **2019**, *9*, 1636.
18. Robinson, M.K.; Hodge, K.M.; Horak, E.; Sundberg, Å.L.; Russeva, M.; Shaller, C.C.; Von Mehren, M.; Shchaveleva, I.; Simmons, H.H.; Marks, J.D.; et al. Targeting ErbB2 and ErbB3 with a bispecific single-chain Fv enhances targeting selectivity and induces a therapeutic effect in vitro. *Br. J. Cancer* **2008**, *99*, 1415–1425. [CrossRef]
19. Löfblom, J.; Feldwisch, J.; Tolmachev, V.; Carlsson, J.; Ståhl, S.; Frejd, F.Y. Affibody molecules: Engineered proteins for therapeutic, diagnostic and biotechnological applications. *FEBS Lett.* **2010**, *584*, 2670–2680. [CrossRef]
20. Ståhl, S.; Gräslund, T.; Eriksson Karlström, A.; Frejd, F.Y.; Nygren, P.Å.; Löfblom, J. Affibody Molecules in Biotechnological and Medical Applications. *Trends Biotechnol.* **2017**, *35*, 691–712. [CrossRef]
21. Kronqvist, N.; Malm, M.; Göstring, L.; Gunneriusson, E.; Nilsson, M.; Höidén Guthenberg, I.; Gedda, L.; Frejd, F.Y.; Ståhl, S.; Löfblom, J. Combining phage and staphylococcal surface display for generation of ErbB3-specific Affibody molecules. *Protein Eng. Des. Sel.* **2011**, *24*, 385–396. [CrossRef] [PubMed]
22. Malm, M.; Kronqvist, N.; Lindberg, H.; Gudmundsdotter, L.; Bass, T.; Frejd, F.Y.; Höidén-Guthenberg, I.; Varasteh, Z.; Orlova, A.; Tolmachev, V.; et al. Inhibiting HER3-Mediated Tumor Cell Growth with Affibody Molecules Engineered to Low Picomolar Affinity by Position-Directed Error-Prone PCR-Like Diversification. *PLoS ONE* **2013**, *8*, e62791. [CrossRef]
23. Da Pieve, C.; Allott, L.; Martins, C.D.; Vardon, A.; Ciobota, D.M.; Kramer-Marek, G.; Smith, G. Efficient [18F]AlF Radiolabeling of ZHER3:8698 Affibody Molecule for Imaging of HER3 Positive Tumors. *Bioconjug. Chem.* **2016**, *27*, 1839–1849. [CrossRef] [PubMed]
24. Rinne, S.S.; Leitao, C.D.; Abouzayed, A.; Vorobyeva, A.; Tolmachev, V.; Ståhl, S.; Löfblom, J.; Orlova, A. HER3 pet imaging:^{68}Ga-labeled affibody molecules provide superior HER3 contrast to ^{89}Zr-labeled antibody and antibody-fragment-based tracers. *Cancers* **2021**, *13*, 4791. [CrossRef] [PubMed]
25. Orlova, A.; Malm, M.; Rosestedt, M.; Varasteh, Z.; Andersson, K.; Selvaraju, R.K.; Altai, M.; Honarvar, H.; Strand, J.; Ståhl, S.; et al. Imaging of HER3-expressing xenografts in mice using a 99mTc(CO) 3-HEHEHE-ZHER3:08699 affibody molecule. *Eur. J. Nucl. Med. Mol. Imaging* **2014**, *41*, 1450–1459. [CrossRef]
26. Leitao, C.D.; Rinne, S.S.; Mitran, B.; Vorobyeva, A.; Andersson, K.G.; Tolmachev, V.; Ståhl, S.; Löfblom, J.; Orlova, A. Molecular design of HER3-targeting affibody molecules: Influence of chelator and presence of HEHEHE-Tag on biodistribution of 68 Ga-labeled tracers. *Int. J. Mol. Sci.* **2019**, *20*, 1080. [CrossRef]
27. Rinne, S.S.; Leitao, C.D.; Saleh-Nihad, Z.; Mitran, B.; Tolmachev, V.; Ståhl, S.; Löfblom, J.; Orlova, A. Benefit of later-time-point PET imaging of her3 expression using optimized radiocobalt-labeled affibody molecules. *Int. J. Mol. Sci.* **2020**, *21*, 1972. [CrossRef]
28. Rinne, S.S.; Orlova, A.; Tolmachev, V. Pet and spect imaging of the egfr family (Rtk class i) in oncology. *Int. J. Mol. Sci.* **2021**, *22*, 3663. [CrossRef]
29. Rosestedt, M.; Andersson, K.G.; Mitran, B.; Tolmachev, V.; Löfblom, J.; Orlova, A.; Ståhl, S. Affibody-mediated PET imaging of HER3 expression in malignant tumours. *Sci. Rep.* **2015**, *5*, 15226. [CrossRef]
30. Tolmachev, V.; Hofström, C.; Malmberg, J.; Ahlgren, S.; Hosseinimehr, S.J.; Sandström, M.; Abrahmsén, L.; Orlova, A.; Gräslund, T. HEHEHE-tagged affibody molecule may be purified by IMAC, is conveniently labeled with [^{99m}Tc(CO)$_3$]+, and shows improved biodistribution with reduced hepatic radioactivity accumulation. *Bioconjug. Chem.* **2010**, *21*, 2013–2022. [CrossRef]
31. Hofström, C.; Altai, M.; Honarvar, H.; Strand, J.; Malmberg, J.; Hosseinimehr, S.J.; Orlova, A.; Gräslund, T.; Tolmachev, V. HAHAHA, HEHEHE, HIHIHI, or HKHKHK: Influence of position and composition of histidine containing tags on biodistribution of [^{99m}Tc(CO)$_3$]+-labeled affibody molecules. *J. Med. Chem.* **2013**, *56*, 4966–4974. [CrossRef] [PubMed]
32. Hofström, C.; Orlova, A.; Altai, M.; Wangsell, F.; Gräslund, T.; Tolmachev, V. Use of a HEHEHE purification tag instead of a hexahistidine tag improves biodistribution of affibody molecules site-specifically labeled with ^{99m}Tc, ^{111}In, and ^{125}I. *J. Med. Chem.* **2011**, *54*, 3817–3826. [CrossRef] [PubMed]
33. Wong, D.J.L.; Hurvitz, S.A. Recent advances in the development of anti-HER2 antibodies and antibody-drug conjugates. *Ann. Transl. Med.* **2014**, *2*, 122–136. [PubMed]
34. Verma, S.; Miles, D.; Gianni, L.; Krop, I.E.; Welslau, M.; Baselga, J.; Pegram, M.; Oh, D.-Y.; Diéras, V.; Guardino, E.; et al. Trastuzumab emtansine for HER2-positive advanced breast cancer. *N. Engl. J. Med.* **2012**, *367*, 1783–1791. [CrossRef] [PubMed]
35. Kontermann, R.E. Strategies for extended serum half-life of protein therapeutics. *Curr. Opin. Biotechnol.* **2011**, *22*, 868–876. [CrossRef] [PubMed]
36. Altai, M.; Liu, H.; Ding, H.; Mitran, B.; Edqvist, P.-H.; Tolmachev, V.; Orlova, A.; Gräslund, T. Affibody-derived drug conjugates: Potent cytotoxic molecules for treatment of HER2 over-expressing tumors. *J. Control. Release* **2018**, *288*, 84–95. [CrossRef]
37. Jonsson, A.; Dogan, J.; Herne, N.; Abrahmsén, L.; Nygren, P.Å. Engineering of a femtomolar affinity binding protein to human serum albumin. *Protein Eng. Des. Sel.* **2008**, *21*, 515–527. [CrossRef]

38. Salmiheimo, A.; Mustonen, H.; Stenman, U.-H.; Puolakkainen, P.; Kemppainen, E.; Seppänen, H.; Haglund, C. Systemic Inflammatory Response and Elevated Tumour Markers Predict Worse Survival in Resectable Pancreatic Ductal Adenocarcinoma. *PLoS ONE* **2016**, *11*, e0163064. [CrossRef]
39. Ding, H.; Altai, M.; Rinne, S.S.; Vorobyeva, A.; Tolmachev, V.; Gräslund, T.; Orlova, A. Incorporation of a Hydrophilic Spacer Reduces Hepatic Uptake of HER2-Targeting Affibody–DM1 Drug Conjugates. *Cancers* **2019**, *11*, 1168. [CrossRef]
40. Xu, T.; Ding, H.; Vorobyeva, A.; Oroujeni, M.; Orlova, A.; Tolmachev, V.; Gräslund, T. Drug conjugates based on a monovalent affibody targeting vector can efficiently eradicate HER2 positive human tumors in an experimental mouse model. *Cancers* **2021**, *13*, 85. [CrossRef]
41. Yin, W.; Xu, T.; Altai, M.; Oroujeni, M.; Zhang, J.; Vorobyeva, A.; Vorontsova, O.; Vtorushin, S.V.; Tolmachev, V.; Gräslund, T.; et al. The influence of domain permutations of an albumin-binding domain-fused her2-targeting affibody-based drug conjugate on tumor cell proliferation and therapy efficacy. *Pharmaceutics* **2021**, *13*, 1974. [CrossRef] [PubMed]
42. Vorobyeva, A.; Schulga, A.; Konovalova, E.; Güler, R.; Löfblom, J.; Sandström, M.; Garousi, J.; Chernov, V.; Bragina, O.; Orlova, A.; et al. Optimal composition and position of histidine-containing tags improves biodistribution of ^{99m}Tc-labeled DARPin G3. *Sci. Rep.* **2019**, *9*, 9405. [CrossRef] [PubMed]
43. Leitao, C.D.; Rinne, S.S.; Altai, M.; Vorontsova, O.; Dunås, F.; Jonasson, P.; Tolmachev, V.; Löfblom, J.; Ståhl, S.; Orlova, A. Evaluating the therapeutic efficacy of mono-and bivalent affibody-based fusion proteins targeting HER3 in a pancreatic cancer xenograft model. *Pharmaceutics* **2020**, *12*, 551. [CrossRef]
44. Thurber, G.M.; Schmidt, M.M.; Wittrup, K.D. Factors determining antibody distribution in tumors. *Trends Pharmacol. Sci.* **2008**, *29*, 57–61. [CrossRef]
45. Debie, P.; Lafont, C.; Defrise, M.; Hansen, I.; van Willigen, D.M.; van Leeuwen, F.W.B.; Gijsbers, R.; D'Huyvetter, M.; Devoogdt, N.; Lahoutte, T.; et al. Size and affinity kinetics of nanobodies influence targeting and penetration of solid tumours. *J. Control. Release* **2020**, *317*, 34–42. [CrossRef] [PubMed]
46. Bennett, G.; Brown, A.; Mudd, G.; Huxley, P.; Rietschoten, K.; Van Pavan, S.; Chen, L.; Watcham, S.; Lahdenranta, J.; Keen, N. MMAE delivery using the Bicycle toxin conjugate BT5528. *Mol. Cancer Ther.* **2020**, *19*, 1385–1394. [CrossRef] [PubMed]
47. Brandl, F.; Busslinger, S.; Zangemeister-Wittke, U.; Plückthun, A. Optimizing the anti-tumor efficacy of protein-drug conjugates by engineering the molecular size and half-life. *J. Control. Release* **2020**, *327*, 186–197. [CrossRef]
48. Schmidt, M.M.; Wittrup, K.D. A modeling analysis of the effects of molecular size and binding affinity on tumor targeting. *Mol. Cancer Ther.* **2009**, *8*, 2861–2871. [CrossRef]
49. Cardoso, V.M.; Campani, G.; Santos, M.P.; Silva, G.G.; Pires, M.C.; Gonçalves, V.M.; de Giordano, R.C.; Sargo, C.R.; Horta, A.C.L.; Zangirolami, T.C. Cost analysis based on bioreactor cultivation conditions: Production of a soluble recombinant protein using Escherichia coli BL21(DE3). *Biotechnol. Rep.* **2020**, *26*, e00441. [CrossRef]
50. Perols, A.; Karlström, A.E. Site-specific photoconjugation of antibodies using chemically synthesized IgG-binding domains. *Bioconjug. Chem.* **2014**, *25*, 481–488. [CrossRef]
51. Ding, H.; Xu, T.; Zhang, J.; Tolmachev, V.; Oroujeni, M.; Orlova, A.; Gräslund, T.; Vorobyeva, A. Affibody-derived drug conjugates targeting HER2: Effect of drug load on cytotoxicity and biodistribution. *Pharmaceutics* **2021**, *13*, 430. [CrossRef] [PubMed]

biomedicines

MDPI

Article

Potent Small-Molecule Inhibitors Targeting Acetylated Microtubules as Anticancer Agents against Triple-Negative Breast Cancer

Ahreum Kwon [1], Gwi Bin Lee [2], Taein Park [2], Jung Hoon Lee [3], Panseon Ko [4], Eunae You [4], Jin Hee Ahn [2], Soo Hyun Eom [2], Sangmyung Rhee [4,*] and Woo Keun Song [1,*]

1 Cell Logistics and Silver Health Research Center, School of Life Sciences, Gwangju Institute of Science and Technology, Gwangju 61005, Korea; kar3189@gist.ac.kr
2 Department of Chemistry, Gwangju Institute of Science and Technology, Gwangju 61005, Korea; hshmhshm@gist.ac.kr (G.B.L.); taeinpark@gist.ac.kr (T.P.); jhahn@gist.ac.kr (J.H.A.); eom@gist.ac.kr (S.H.E.)
3 Department of Biochemistry and Cell Biology, Geisel School of Medicine, Dartmouth College, Hanover, NH 03755, USA; Junghoon.Lee.Gr@dartmouth.edu
4 Department of Life Science, Chung-Ang University, Seoul 06974, Korea; kpskoh@hotmail.com (P.K.); yea108@naver.com (E.Y.)
* Correspondence: Sangmyung.rhee@cau.ac.kr (S.R.); wksong@gist.ac.kr (W.K.S.); Tel.: +82-62-715-2560 (S.R.); +82-2-820-5818 (W.K.S.)

Received: 29 July 2020; Accepted: 9 September 2020; Published: 9 September 2020

Abstract: Microtubules are one of the major targets for anticancer drugs because of their role in cell proliferation and migration. However, as anticancer drugs targeting microtubules have side effects, including the death of normal cells, it is necessary to develop anticancer agents that can target microtubules by specifically acting on cancer cells only. In this study, we identified chemicals that can act as anticancer agents by specifically binding to acetylated microtubules, which are predominant in triple-negative breast cancer (TNBC). The chemical compounds disrupted acetylated microtubule lattices by interfering with microtubule access to alpha-tubulin acetyltransferase 1 (αTAT1), a major acetyltransferase of microtubules, resulting in the increased apoptotic cell death of MDA-MB-231 cells (a TNBC cell line) compared with other cells, such as MCF-10A and MCF-7, which lack microtubule acetylation. Moreover, mouse xenograft experiments showed that treatment with the chemical compounds markedly reduced tumor growth progression. Taken together, the newly identified chemical compounds can be selective for acetylated microtubules and act as potential therapeutic agents against microtubule acetylation enrichment in TNBC.

Keywords: microtubule acetylation; triple-negative breast cancer; anti-cancer agent; apoptosis

1. Introduction

Breast cancer is the most frequently diagnosed cancer, and approximately 13% of women will be diagnosed with invasive breast cancer in their lifetime [1,2]. Breast cancer is a heterogeneous disease composed of several well-recognized subtypes with different clinical and prognostic characteristics [3,4]. To classify the molecular subtypes of breast cancer, the status of estrogen receptor (ER), progesterone receptor (PR), and human epidermal growth factor receptor 2 (HER2) has been considered [5], and the discrimination of these subtypes is crucial for determining treatment options and selecting anticancer agents.

Triple-negative breast cancer (TNBC), characterized by the absence of ER, PR, and HER2, represents approximately 15~20% of all breast cancers [6]; however, its metastasis and mortality are significantly worse than those of other subtypes [7]. Although various anticancer therapeutic agents that can target specific receptors in breast cancer have been developed [8], TNBC does not respond to hormonal therapies that target HER2 and ER [9]. In order to overcome the limitations of TNBC therapeutic options, platinum-based compounds, including cisplatin and carboplatin, have been used in therapeutic trials for TNBC, in which they can effectively interfere with DNA replication and related cellular processes [10–13]. In addition to the development and application of novel anticancer agents, knowledge of specific TNBC characteristics will provide more options in the selection of therapeutic trials. Therefore, it is crucial to identify therapeutic agents based on specific features for TNBC [14,15]. Microtubules are essential cytoskeletal components in eukaryotic cells, which are composed of α- and β-tubulin heterodimers arranged in the form of filamentous tubes [16], and their polymerization dynamics are firmly controlled spatially and temporally during cellular processes such as intracellular transport and cell division [17,18]. Microtubule disruption reagents such as paclitaxel and Vinca alkaloids have been successfully used as chemotherapeutic drugs against various cancers [19]; however, they can act on normal cells in other tissues, causing serious side effects such as nervous system adverse reactions [20] and abnormal changes in inflammatory cytokines in the plasma of cancer patients during chemotherapy [21,22]. Therefore, the development of anticancer agents that specifically target microtubules in cancer cells is required.

Microtubule dynamics and organization are mainly regulated through post-translational modifications (PTMs), such as detyrosination [23], glutamylation [24], and acetylation [15]. In contrast to other PTMs on the outer surface of polymerized microtubules, acetylation of the α-tubulin luminal residue lysine 40 (K40), a residue exposed to the microtubule lumen, is a well-known PTM [25,26]. In mammals, microtubule acetylation is an evolutionarily conserved modification regulated by alpha-tubulin acetyltransferase 1 (αTAT1) [27] and deacetylases including HDAC6, which is a NAD-independent deacetylase [28], and Sirt2, which is a NAD-dependent deacetylase [29]. Microtubule acetylation is considered as a hallmark of long-lived microtubules, which may be a consequence of microtubule stability rather than a contributing factor [30], and it has been reported to be involved in several biological processes related to microtubule flexibility [31], intracellular transport through kinesin-1 regulation [32], and cilia formation [33]. Furthermore, the relationship between microtubule acetylation and human diseases has revealed that the dysregulation of microtubule acetylation may be associated with Charcot-Marie-Tooth disease [34], Alzheimer's disease [35], and heart diseases [36]. Studies have been conducted on cell mobility according to the characteristics of microtubules [37], and some studies have focused on the relationship between microtubule acetylation and cancer progression. We reported recently that acetylated microtubules are required for TGF-β-induced cancer-associated fibroblast activation in breast cancer tissues [38], suggesting that microtubule acetylation seems to be a prerequisite cellular factor for breast cancer progression and that the downregulation of microtubule acetylation is important in breast cancer therapy. In this regard, we attempted to screen small chemical inhibitors for interfering with microtubule acetylation. The chemical inhibitors identified in this study showed the potential to inhibit αTAT1, a microtubule acetyltransferase, resulting in the downregulation of microtubule acetylation. We also demonstrated that the novel inhibitors successfully downregulated tumor growth in a mouse xenograft model. Therefore, the newly identified inhibitors could function as therapeutic agents against TNBC by directly inhibiting microtubule acetylation.

2. Materials and Methods

2.1. Sample and Data Collection

Paraffin-embedded human breast cancer tissue samples and patient pathological data were obtained from US Biomax, Inc., Derwood, MD, USA (BR10011). The Gwangju Institute of Science and Technology (GIST) Institutional Review Board approved this study (20191128-BR-49-05-02, approved on 28 November 2019).

2.2. Cell Culture

MDA-MB-231 cells were kindly provided by Dr. Jeong-Seok Nam (GIST, Gwangju, Korea) and maintained in Dulbecco's Modified Eagle's Medium (high glucose; Gibco, Billings, MT, USA) containing 10% fetal bovine serum (FBS), 50 μg/mL streptomycin, and 50 units/mL penicillin. ZR-75-1, T47D, HCC-1428, HCC-1419, and Hs578T were purchased from the Korean Cell Line Bank and maintained as specified in the datasheet.

2.3. Generation of αTAT1 Knockout (KO) Cells via CRISPR-Cas9

αTAT1 KO using the CRISPR-Cas9 system was performed as previously described [39]. Annealed sgRNA was cloned into the plasmid lentiCRISPR v2. For lentiviral production, the cloned αTAT1-targeting vector, psPAX2, and pMD2.G were co-transfected with polyethylenimine into HEK293T cells. After 48 h of incubation, the lentivirus produced from HEK293T cells was collected in conditioned media (CM). MDA-MB-231 cells were infected with the lentivirus in the presence of polybrene (8 μg/mL) for 48 h and selected with puromycin (1 μg/mL) for 2 weeks.

2.4. Chemical Library Screening

A ChemBridge chemical library (NM1117-1) containing 30,000 small synthetic compounds dissolved in dimethyl sulfoxide (DMSO) as a 10 mM stock solution was used. Spin90-KO mouse embryonic fibroblasts (MEFs) with consistently high expression of acetyl-α-tubulin [40] were cultured in 96-well plates, treated with each chemical compound at 10 μM, stained with antibodies against acetyl-α-tubulin (Sigma, St. Louis, MO, USA), and finally analyzed using an IX81 microscope. Selected candidates were analyzed quantitatively by western blotting.

2.5. Western Blotting

Collected cells were lysed in radioimmunoprecipitation assay buffer (RIPA buffer) containing 50 mM Tris-HCl (pH 7.4), 150 mM NaCl, 1% NP-40, 1% SDS, 0.25% sodium deoxycholate, 10 mM NaF, 1 mM Na_3VO_4, and protease inhibitor cocktail (Roche, Switzerland). Primary antibodies against acetyl-α-tubulin (Sigma), α-tubulin (Sigma), Cyclin Antibody Sampler Kit (Cell Signaling Technology, Danvers, MA, USA), and Acetyl-Histone H3 Sampler Kit (Cell Signaling Technology, USA) were used.

2.6. Immunofluorescence Staining

To detect acetyl-α-tubulin expression after treatment with the compounds, MDA-MB-231 cells treated with GM-90257 and GM-90631 were fixed with 4% paraformaldehyde in PBS and permeabilized with 0.1% triton-X 100 in PBS. After blocking with 0.1% bovine serum albumin, the cells were stained with primary antibodies against acetyl-α-tubulin (Sigma). To analyze the co-localization of αTAT1 and α-tubulin, MDA-MB-231 cells transfected with Dsred2-αTAT1 were prepared in the same manner and stained with primary antibodies against α-tubulin (Sigma). To stain the microtubule bundles shown, soluble tubulin proteins were extracted by dipping into 0.1% Triton X-100 for 15 s before fixation. Cells were fixed with ice-cold methanol at −20 °C for 5 min, after which the samples were treated with blocking solution (0.1% bovine serum albumin in PBST) for 1 h. The cells were stained with primary antibodies against α-tubulin (Sigma). All fluorescence microscopy images were obtained using a confocal microscope (FV1000; Olympus, Tokyo, Japan).

2.7. Soft Agar Colony Formation Assay

To confirm anchorage-independent cancer cell growth, a mixture of 1% agarose solution and 2× DMEM growth media (1:1 ratio) was dispensed into 6-well plates (1.5 mL per well) and allowed to solidify at room temperature for 30 min. MDA-MB-231 cells (5×10^3 cells/well) were resuspended in the growth media, mixed with 0.6% agar at 40 °C, and added to agar-coated 6-well plates. The plates were incubated at room temperature for 30 min, and 300 μL of complete growth media with chemical compounds was added to each well and incubated at 37 °C for 3 weeks.

2.8. Microtubule Polymerization Assay

Tubulin proteins purified from porcine brain (Cytoskeleton, Denver, CO, USA) were used for in vitro microtubule polymerization assay. The tubulin proteins were adjusted to 3 mg/mL and individually mixed with chemical compounds including paclitaxel, nocodazole, GM-90257, GM-90631, and DMSO as a control. Then, the OD value at 340 nm was determined every 30 s during the polymerization reaction at 37 °C for 1 h.

2.9. In Vitro Microtubule Acetylation Assay

Microtubules were purified from αTAT1 KO MDA-MB-231 cells. MTS buffer containing 80 mM PIPES, 1 mM EGTA, 1 mM $MgCl_2$, 40% glycerol, 1 mM PMSF, 1 mM Na_3VO_4, and 10 mM NaF 10 was added to the cells for 20 min at 37 °C in an incubator. After scraping the cells in MTS buffer, paclitaxel was added to a final concentration of 20 μM. Then, the solutions were homogenized three times with a 25G syringe and incubated for 10 min at room temperature. The solutions were separated into pellets and supernatants by centrifugation at 31,100 rpm using a Ti-70 rotor (Beckman Coulter, Brea, CA, USA) for 30 min at room temperature. Pellets were re-suspended with MTS buffer containing 20 μM paclitaxel. Then, 1 mg/mL His-αTAT1, 1 μM acetyl-CoA, and microtubule acetylation inhibitors (1 mM) were added and incubated for 30 min at 37 °C. SDS sample buffer (4×) was added to stop the enzymatic activity of αTAT1, and SDS-PAGE followed by western blotting was performed.

2.10. Immunohistochemistry

Tumors harvested from the xenograft breast cancer mouse model were fixed in 10% formalin and embedded in paraffin blocks. Paraffin sections were deparaffinized and rehydrated with ethanol. Antigen retrieval was performed by heating in retrieval solution (IHC World, Woodstock, MD, USA) for 30 min. The following primary antibodies were used for staining: acetyl-α-tubulin (Abcam, Cambridge, UK) and Ki67 (Sigma, St. Louis, MO, USA). Nuclear counterstaining was performed with Mayer's hematoxylin (Dako, Buffalo, NY, USA). Stained slide samples were scanned and analyzed using Aperio ImageScope (Leica Biosystems, Wetzlar, Germany).

2.11. Annexin V-PI Apoptosis Assay

Apoptotic and dead cells (MCF-10A, MCF-7, and MDA-MB-231) with chemical compounds were identified using an annexin V-PI apoptosis kit (BD Biosciences, San Jose, CA, USA). After 24 h of incubation with the chemical compounds, the cells were harvested and stained by FITC-annexin V and PI according to the manufacturer's protocol. Stained cells were analyzed by flow cytometry (BD Biosciences).

2.12. Xenograft Tumor Inoculation

Female NOD/SCID mice (8~12 weeks) were used for the breast cancer xenograft model. MDA-MB-231 cells (1×10^6 cells) mixed with Matrigel were injected into the left inguinal mammary fat pad, and developed tumors were quantified with the following equation: ($V = (L \times W^2)/2$, L; length, W; width). Each chemical compound was intraperitoneally injected for 15 days (once every 2 days) starting when the tumor reached 100 mm^3. All experimental procedures with animals were performed with the approval of the Animal Care and Ethics Committees of GIST (GIST-2017-003, approved on 7 March 2018).

2.13. Statistical Analysis

All experimental results were the average of three repetitions, and data were analyzed by two-tailed Student's *t*-test to confirm statistical significance between two groups (experimental and control group). The bars in graphs are shown as the mean ± standard deviation (SD), and all *p* values are two-sided. The data were regarded as statistically significant at *p* values less than 0.05: *, $p \leq 0.05$; **, $p \leq 0.01$; ***, $p \leq 0.001$.

3. Results

3.1. Acetylation of α-Tubulin on K40 for TNBC Progression

To determine the level of acetyl-α-tubulin in various types of breast cancer, we obtained tissue microarray slides derived from human cancer patients and performed staining for acetyl-α-tubulin (Figure 1A). Microarray samples were separated into two groups for analysis (a group with TNBC and a group with other subtypes including ER-, PR-, and HER2-positive cases). The proportion of cases with a high expression level of acetyl-α-tubulin was significantly higher in the group with TNBC than in the group with other subtypes. We also quantitatively analyzed the expression of acetyl-α-tubulin in various breast cancer cell lines. As shown in Figure 1B, the level of acetyl-α-tubulin was significantly higher in TNBC cell lines including Hs578T, BT549, and MDA-MB-231 compared with other cell lines.

To investigate the role of acetyl-α-tubulin in TNBC, we generated a MDA-MB-231 cell line with the KO of αTAT1, a major α-tubulin acetyltransferase [23], using the CRISPR-Cas9 system (Figure 1C) and carried out anchorage-independent growth assays with mock and αTAT1 KO MDA-MB-231 cell lines. In comparison with mock cells, αTAT1 KO MDA-MB-231 cells showed a significant reduction in colony formation on soft agar (Figure 1D). Subsequently, xenograft experiments also showed that αTAT1 deficiency in MDA-MB-231 cells markedly interrupted tumor growth (Figure 1E). Overall, these results indicated that microtubule acetylation in TNBC may be related to cancer progression.

3.2. Screening and Optimization of Microtubule Acetylation Inhibitors

As inhibitors that directly interfere with microtubule acetylation have not been developed, we attempted to identify chemical inhibitors capable of interfering with microtubule acetylation by cell-based screening using a chemical compound library (~30,000 compounds; ChemBridge). A total of 20 potential small chemical compounds, which were selected in the first screening, were analyzed quantitatively by western blotting, and GM-90257 was identified as a hit molecule (Figure 2A). GM-90257 was optimized by the introduction of various substituents at several positions (data not shown), which identified GM-90568 and GM-90631 (HCl salt of GM-90568) as potential compounds (Supplementary Materials Figure S1A). GM-90257 was synthesized as shown in Scheme 1. The commercially available compound **1** was converted to α-bromoketone **2**, which was coupled with thiourea **3** to give GM-90257. GM-90568 and GM-90631 were synthesized according to Scheme 2. 2-Aminothiazole **5** was reacted with α-bromoketone **6** to obtain compound **7**, followed by Friedel–Crafts acylation with chloroacetyl chloride **8** to give compound **9**. Compound **9** was cyclized with thiourea **3** to afford GM90631, which was basified by $NaHCO_3$ to finally obtain GM-90568.

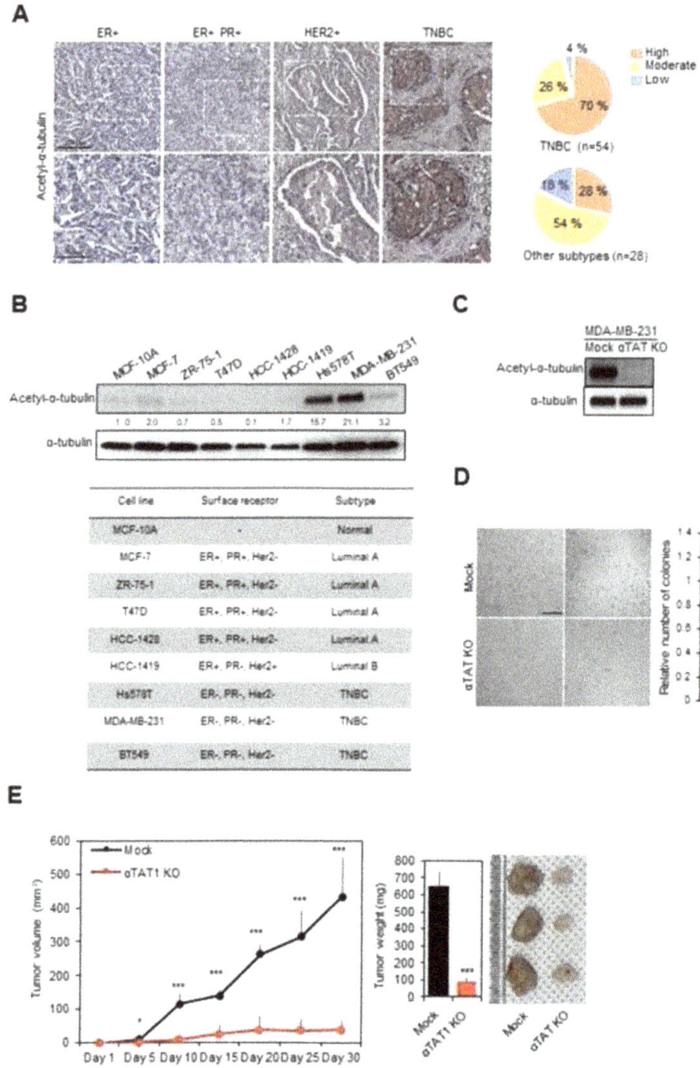

Figure 1. Inhibition of triple-negative breast cancer (TNBC) progression by blocking microtubule acetylation. (**A**) Representative images showing acetyl-α-tubulin expression in various subtypes of human breast cancer. Scale bar, 500 μm. The intensity was analyzed using Aperio Image Scope software. (**B**) Expression of microtubule acetylation in breast cancer cell lines (upper panel) categorized according to the subtype (lower panel). (**C**) Confirmation of acetyl-α-tubulin expression in alpha-tubulin acetyltransferase 1 (αTAT1) Knockout (KO) MDA-MB-231 cell lysates. (**D**) Anchorage-independent growth assays using αTAT1 KO MDA-MB-231 cells for 3 weeks. The number of colonies was quantified using ImageJ software. Scale bar, 1 mm. ***, $p \leq 0.001$. (**E**) Mock and αTAT1 KO MDA-MB-231 cells were injected into the mammary fat pad of non-obese diabetic/severe combined immunodeficiency (NOD/SCID) mice ($n = 5$, each). *, $p \leq 0.05$; ***, $p \leq 0.001$.

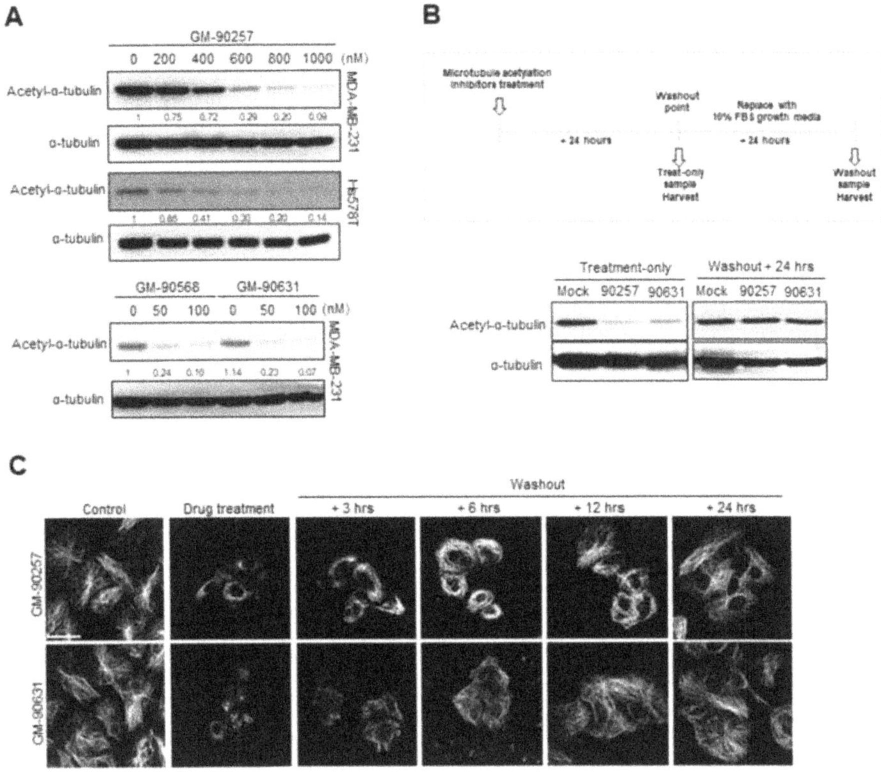

Figure 2. Reduction of microtubule acetylation in TNBC cell lines by selected microtubule acetylation inhibitors. (**A**) The expression of acetyl-α-tubulin in MDA-MB-231 and Hs578T cells treated with microtubule acetylation inhibitors was examined by western blotting. (**B**) GM-90257 (500 nM) and GM-90631 (50 nM) were added to MDA-MB-231 cells; one set of samples was harvested after 24 h of incubation, and the media of the other set of samples was replaced with new growth media followed by incubation for another 24 h. Western blotting for acetyl-α-tubulin was performed. (**C**) Confocal images showing the microtubules (white) in MDA-MB-231 cells treated with GM-90257 and GM-90631. Soluble tubulin in the cytoplasm was extracted before methanol fixation. After the washout of acetylation inhibitors, the cells were harvested in a time-dependent manner. Scale bar, 50 μm.

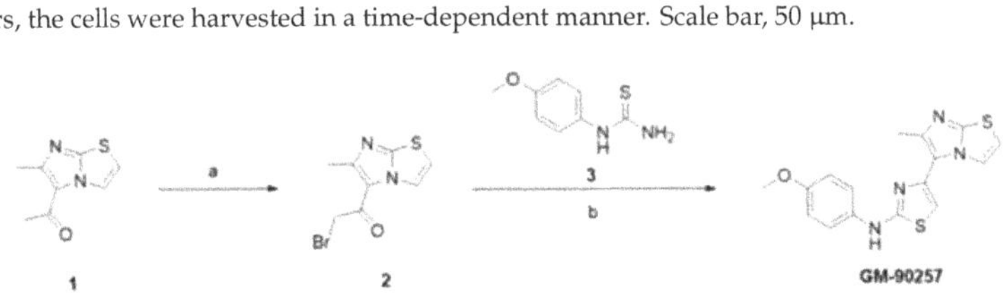

Scheme 1. Synthesis scheme of GM-90257. Reagents and conditions: (**a**) bromine, acetic acid, reflux to room temperature, overnight; (**b**) 1-(4-methoxyphenyl)thiourea, ethanol, reflux, 20 h.

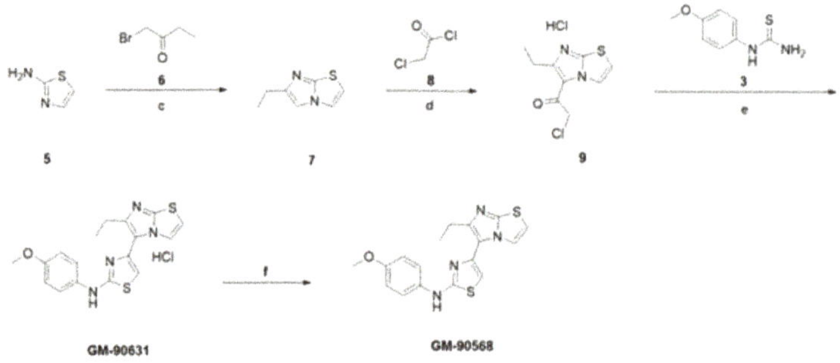

Scheme 2. Synthesis scheme of GM-90568 and GM-90631. Reagents and conditions: (**c**) 1-bromobutan-2-one, ethanol, reflux, 18 h; (**d**) chloroacetyl chloride, 1,4-dioxane, 120 °C, 10 min; (**e**) 1-(4-methoxyphenyl)thiourea, ethanol, reflux, 20 h; (**f**) aqueous $NaHCO_3$, room temperature, 30 min.

All three chemical compounds effectively reduced the level of acetyl-α-tubulin in MDA-MB-231 and Hs578T cells in a concentration-dependent manner. In particular, compared with GM-90257, GM-90568 and GM-90631 showed improved inhibitory effects (Figure 2A). The inhibition of acetyl-α-tubulin was abrogated within 24 h after the removal of the compounds from the cells (Figure 2B); however, there was no significant change in the level of histone H3 acetylation in the sample treated with GM-90631 (Supplementary Materials Figure S1B). Immunocytochemistry also showed that microtubule lattices were restored in a time-dependent manner after the washout of the inhibitors (Figure 2C), indicating that the inhibitors disrupted the microtubule structure. The expression of tubulin was not altered; however, both acetylated-α-tubulin and stable microtubule bundles were greatly reduced based on fluorescence microscopy (Supplementary Materials Figure S1C). Taken together, these results indicated that all three compounds could directly inhibit microtubule acetylation without any other side effects.

3.3. Inhibitory Effects of Potential Inhibitors on the Interaction between α-Tubulin and αTAT1

In order to determine which proteins may interact with our chemical compounds to inhibit microtubule acetylation, we used GalaxyWEB to predict ligand-protein interactions. First, we performed docking simulation with the active site of HDAC6, Sirt2, and αTAT1, which are involved in microtubule acetylation, to identify the target enzyme of potential inhibitors of microtubule acetylation [23]. However, there was no significant ligand-protein docking model. Then, molecular docking simulations were performed on α-tubulin to determine whether potential acetylation inhibitors can directly bind to α-tubulin and inhibit acetylation on the K40 residue. Both the GM-90257 and GM-90568 inhibitors were docked into an identical cleft and showed reasonable affinity. According to the docking results, GM-90257 was stably bound by a hydrogen bond with Gln85 and surrounded by a hydrophobic environment with various residues. GM-905680 formed a hydrogen bond with the backbone of Ser38 and the side chain of Asp39 and was predicted to be surrounded by a hydrophobic environment in the cleft, similar to GM-90257. When both inhibitors were bound to α-tubulin, it appeared to reduce the flexibility of the loop with the acetylation site, resulting in the inhibition of acetylation (Figure 3A). In particular, the binding affinity between GM-90568 and the K40 residue was predicted to be much higher than that for GM-90257, which is consistent with the result of the previous experiment as shown in Figure 2A.

As the predicted docking model between chemical compounds and α-tubulin prevented the recruitment of αTAT1 to the K40 residue in α-tubulin, we examined the localization of DsRed-tagged αTAT1 and α-tubulin with a fluorescence microscope in the presence of GM-90257 and GM-90631. Figure 3B showed that the DsRed-αTAT1 signal was overlapped with that of the microtubule structure in the mock-treated cells. However, in the presence of the inhibitors, DsRed-αTAT1 appeared as a puncta structure around the cytosol. Interestingly, the microtubules appeared broken in the presence of the inhibitors (Figure 3B). Both inhibitors also attenuated microtubule acetylation, which was increased by αTAT1 overexpression (Figure 3C). In vitro acetylation assay using the microtubules obtained from αTAT1 KO cells confirmed that the inhibitors directly reduced microtubule acetylation (Figure 3D).

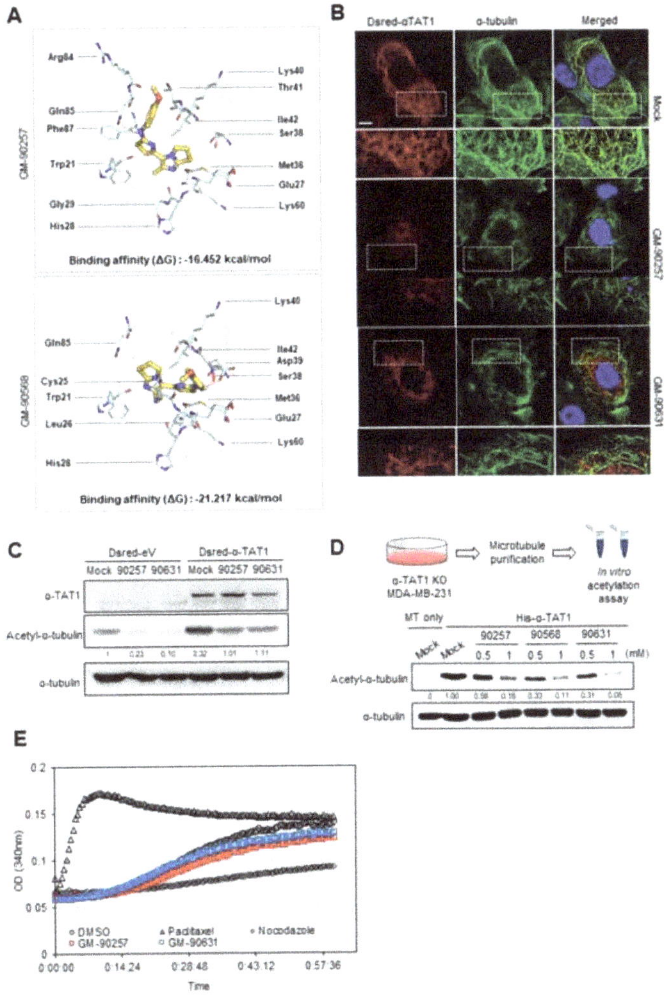

Figure 3. Prediction of the disruption of the interaction between microtubules and αTAT1 by microtubule acetylation inhibitors. (**A**) Predicted molecular docking models between acetylation inhibitors and the K40 residue in α-tubulin. Chemical-protein interactions and residues in the images were visualized using PyMOL program, and the binding affinity was obtained from GalaxyWEB. Hydrogen bonds are indicated by dashed red lines, and pale cyan sticks represent residues interacting with chemical compounds in α-tubulin. (**B**) Overexpressed αTAT1 (red), α-tubulin (green), and nucleus (blue) in MDA-MB-231 cells were detected by immunofluorescence staining. Scale bar, 50 μm. (**C**) DsRed-αTAT1-expressed MDA-MB-231 cell lysates were analyzed by western blotting. (**D**) Microtubules purified from αTAT1 KO MDA-MB-231 cells were incubated with His-αTAT1, acetyl-CoA, and one of the inhibitors (GM-90257, GM-90568, and GM-90631; 0.5 and 1 mM) or DMSO as a control for 30 min at 37 °C. (**E**) In vitro microtubule polymerization assay. DMSO was used as a vehicle control, and all reagents including paclitaxel, nocodazole, GM-90257, and GM-90631 were used at the same concentration of 1 μM.

To determine whether inhibitors can affect the rate of microtubule polymerization, we performed in vitro microtubule polymerization assay using the inhibitors with other microtubule-disrupting agents. Unlike paclitaxel or nocodazole, which are known to act directly on microtubule polymerization, the polymerization rate of microtubules was not different in the presence of GM-90257 and GM-90631 compared with the control (DMSO) (Figure 3E). Taken together, these results indicated that the inhibitors directly reduced microtubule acetylation by preventing the binding of αTAT1 to the microtubules.

3.4. Induction of the Apoptosis of TNBC Cells by GM-90257 and GM-90631

To clarify the functional effects of GM-90257 and GM-90631 in MDA-MB-231 cells, we performed anchorage-independent growth assays for 2 weeks with the chemical compounds. Both compounds significantly reduced the colony formation of MDA-MB-231 cells in a concentration-dependent manner (Figure 4A).

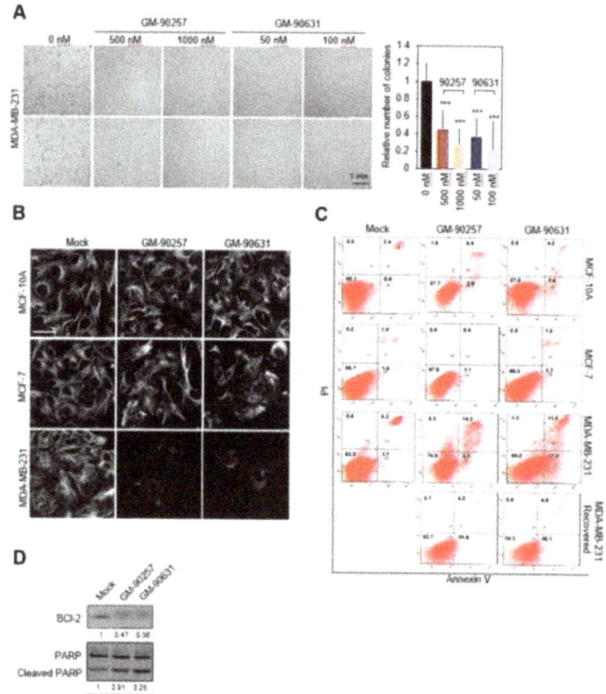

Figure 4. Induction of the apoptosis of TNBC cells by microtubule acetylation inhibitors. (**A**) Anchorage-independent growth assays using MDA-MB-231 cells treated with GM-90257 (500 and 1000 nM) and GM-90631 (50 and 100 nM) for 3 weeks. The number of colonies was quantified using ImageJ software. Scale bar, 1 mm. *** $p \leq 0.001$. (**B**) Confocal images showing microtubules (white) in MCF-10A, MCF-7, and MDA-MB-231 cells after GM-90257 (500 nM) and GM-90631 (50 nM) treatment. Scale bar, 50 μm. (**C**) MCF-10A, MCF-7, and MDA-MB-231 cells treated with GM-90257 (500 nM) and GM-90631 (50 nM) and drug-eliminated MDA-MB-231 cells were double-stained with annexin V and propidium iodide (PI) and analyzed by flow cytometry. The four populations of cells in the quadrants moving clockwise are as follows: non-apoptotic dead cells (Q1), late apoptotic cells (Q2), viable cells (Q3), and early apoptotic cells (Q4). (**D**) The expression levels of BCl-2, PARP, and cleaved PARP in MDA-MB-231 cells treated with GM-90257 (500 nM) and GM-90631 (50 nM) were examined by western blotting.

The previous result showed that the chemical compounds inhibited microtubule acetylation concomitant with the disruption of microtubule lattices (Figure 2C). Therefore, we examined whether the inhibitors can disrupt microtubule lattices by acting specifically on acetylated microtubules. Immunocytochemistry revealed that the microtubule structure in MDA-MB-231 cells was specifically disrupted by the inhibitors (Figure 4B). The result demonstrated that the inhibitors could act exclusively on acetylated microtubules considering the higher level of acetylated microtubules in untreated MDA-MB-231 cells compared with untreated MCF-10A and MCF-7 cells.

As the inhibitors specifically disrupted the microtubule structure of MDA-MB-231 cells, we investigated whether cell death caused by the inhibitors is also limited to MDA-MB-231 cells through FACS analysis. As shown in Figure 4C, treatment of the MCF-10A and MCF-7 cells having a low level of microtubule acetylation did not induce cell death; however, MDA-MB-231 cells showed increased apoptotic cell death following inhibitor treatment. Apoptotic cell death induced by the inhibitors was significantly reduced when the inhibitors were removed from MDA-MB-231 cells. In addition, GM-90257 and GM-90631 activated poly ADP ribose polymerase (PARP) cleavage in MDA-MB-231 cells but downregulated the anti-apoptotic protein, B-cell lymphoma (BCl-2) (Figure 4D). Taken together, these results demonstrated that the inhibitors could act specifically on cells with a high level of microtubule acetylation, leading to cell death through the disruption of microtubules.

3.5. Inhibition of TNBC Progression In Vivo by Blocking Microtubule Acetylation

To evaluate the anticancer effects of the inhibitors in vivo, we injected the inhibitors into tumors in the mammary fat pad of NOD/SCID mice, which were generated from MDA-MB-231 cells. Preliminary experiments to determine the effective concentration of GM-90257 and GM-90631 in vivo indicated biocompatibility in the range of 10~50 mg/kg. GM-90257 had no anticancer effects at 10 mg/kg in vivo, and mice administered with a dose of 50 mg/kg had a mortality rate of almost 50% when the drug injection was maintained for 2 weeks (data not shown). Based on these results, we decided to inject GM-90257 at 25 mg/kg. GM-90631, which was effective at a much lower concentration in vitro, inhibited tumor formation even at 10 mg/kg. After the tumor volume reached 100 mm^3 (average of 2 weeks after the injection of cancer cells), the mice were intraperitoneally (i.p.) administered GM-90257 and GM-90631 for 15 days; control mice were injected with DMSO once every 2 days (Supplementary Materials Figure S2A). The growth of tumors was significantly reduced in mice injected with GM-90257 and GM-90631 compared with mice injected with DMSO (Figure 5A), and there was no difference in the average weight of mice during the treatment period (Supplementary Materials Figure S2B). After treatment completion, tumors were separated from the mice, and the final tumor volume and weight were markedly reduced in a concentration-dependent manner (Figure 5B and Supplementary Materials Figure S2C). In addition, immunohistochemistry showed that the expression levels of acetyl-α-tubulin and Ki67, a representative proliferation marker, in sectioned tumor tissues were markedly decreased after treatment with GM-90257 and GM-90631 (Figure 5C).

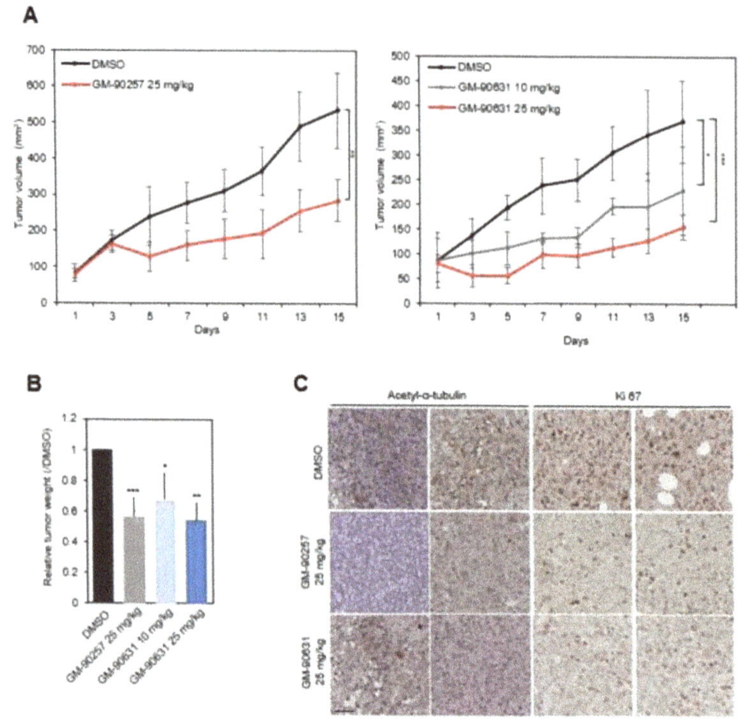

Figure 5. Inhibition of TNBC progression in vivo by GM-90257 and GM-90631. (**A**) MDA-MB-231 cells were injected into the mammary fat pads of NOD/SCID mice. The mice were randomly divided into two groups: control (DMSO) and 25 mg/kg GM-90257 ($n = 5$ per group, left graph). Each drug was intraperitoneally (i.p.) administered once every 2 days for 15 days. The right graph shows the experimental results for GM-90631, and the mice were randomly divided into three groups: control (DMSO), 10 mg/kg GM-90631, and 25 mg/kg GM-90631 ($n = 5$ per group). The graphs show the tumor volume growth for 15 days, and the bars in the graphs indicate the standard deviation. ** $p \leq 0.01$. (**B**) Harvested tumor weight. * $p \leq 0.05$; *** $p \leq 0.001$. (**C**) Immunohistochemical analysis of acetyl-α-tubulin and Ki67 in sectioned, paraffin-embedded tumor tissues. Scale bar, 100 μm.

4. Discussion

Although TNBC usually affects younger patients, the overall five-year survival rate for TNBC is around 77% versus 93% for other breast cancer types, and almost all women with metastatic TNBC will ultimately die from their disease [41]. For these reasons, the development of effective targeted therapeutic agents against TNBC has been actively conducted. In addition, studies have focused on specific features found in TNBC considering that anticancer agent efficacy is often diverse due to the molecular heterogeneity of TNBC with restricted treatment options [42]. For example, PARP inhibitors and platinum salt drugs are one of the most powerful anticancer agents but show different effects against TNBC depending on BRCA gene activation [43]. In order to address this problem, combined treatments with BET inhibitors regulating BRCA genes have been proposed [43]. Some studies of combined treatments using anticancer agents that can complement each other have been conducted. Here, we attempted to investigate significant factors that could be novel molecular features of TNBC and obtain a proof-of-principle for the development of novel agents against TNBC. In the study, we demonstrated that microtubule acetylation was specifically increased in TNBC patient tissues and a

TNBC cell line (MDA-MB-231), and the inhibition of microtubule acetylation in MDA-MB-231 cells by newly developed chemical inhibitors markedly reduced cancer progression in vitro and in vivo.

Microtubule acetylation is known to act as a double-edged sword in cancer research and treatment. The high expression HDAC6 has been suggested to promote the proliferation of lung adenocarcinoma cells [44], and HDAC6 inhibitors have been considered as effective anticancer agents against non-small cell lung cancer [45]. Moreover, another HDCA6 inhibitor, ACY-1215, could enhance anticancer activity against colorectal cancer [46]. On the other hand, the expression of microtubule acetylation is closely associated with several cancers. Acetyl-α-tubulin is considered as a prognostic marker of squamous cell carcinoma of the head and neck especially in metastatic cases [47]. As mentioned previously, microtubule acetylation is enhanced in metastatic breast cancer (especially in basal-like breast cancer types), promotes the formation of microtentacles (which are microtubule-based membrane protrusions), and contributes to cell migration [48]. As basal-like breast cancer frequently overlaps with the TNBC subtype, it provides strong supportive evidence for TNBC.

Although GM-90257, GM-90568, and GM-90631 are not direct inhibitors of the αTAT1 active site, they can be considered as competitive inhibitors as they inhibit αTAT1 from binding to the microtubules (Figure 3). Chemical compounds with a molecular weight of less than 900 Da are defined as small molecules, and over 100 small chemical compounds have been approved as oncology drugs by the FDA to date [49]. All of the three compounds identified in this study have a molecular weight of less than 900 Da; thus, they can be classified as small compounds (GM-90257: 342.4 Da; GM-90568: 356.4 Da; GM-90631: 392.9 Da). The biggest advantage of small compounds is that they can rapidly diffuse across cell membranes and localize in intracellular spaces easily. When TNBC cells were directly treated with our compounds, both microtubule acetylation and cancer cell proliferation were effectively inhibited. In addition, it was confirmed that the chemical backbone structure of GM-90568 and GM-90631 has a much higher binding affinity with microtubules (GM-90257: -16.452; GM-90568, GM-90631: -21.217); thus, they function effectively at lower concentrations. Furthermore, the results of in vivo experiments were statistically significant. Most small-molecule anticancer drugs currently in clinical use lack tumor selectivity and have various toxic side effects [50]. Based on the in vitro and in vivo results, the selected chemical compounds demonstrated potential as anticancer agents against TNBC.

Microtubules are cytoskeletal proteins, which play an essential role in cell survival, and they have attracted interest as a target in early anticancer drug development [22]. Microtubules consist of the α, β, γ, δ, and ε families of tubulins; tubulins γ, δ, and ε are typically located in the centrosome [51], and α- and β-tubulins are the main building blocks of microtubules. Established microtubule-targeting agents (MTAs) can be divided into two groups: stabilizers and destabilizers [19]. Paclitaxel and nocodazole, which were used in the in vitro microtubule polymerization assay (Figure 3E), are representative MTAs. Paclitaxel binds to the second globular domain of β-tubulins and stabilizes tubulin dimers without GTP binding, and nocodazole induces microtubule destabilization and disrupts mitotic spindles by binding to β-tubulins [52]. Similar to paclitaxel and nocodazole, acetylation inhibitors are thought to function by directly binding to microtubules. The microtubule bundles in MDA-MB-231 cells were reduced after treatment with inhibitors of α-tubulin acetylation, similar to the effect of destabilizing agents (Figure 4B). As microtubule acetylation plays a key role in maintaining long-lived microtubules [15], microtubule bundles would not be maintained when acetylation is inhibited in cells with constitutive acetyl-α-tubulin expression, such as MDA-MB-231 cells. The findings indicated that normal cells, or other carcinoma cells, which do not express acetyl-α-tubulin constitutively, would show a weaker response compared with that of acetyl-α-tubulin-expressing TNBC cells. Although it is not known why the microtubules in MCF-10A or MCF-7 cells were resistant to acetylation inhibitor treatment, other PTMs may participate in microtubule bundle formation in MCF-10A and MCF-7 cells. Therefore, the tested acetylation inhibitors might not significantly cause microtubule bundle disruption in MCF-10A and MCF-7 cells. Furthermore, there were significantly less apoptotic MCF-10A and MCF-7 cells than apoptotic MDA-MB-231 cells when the cells were treated with microtubule acetylation inhibitors at the same concentration (Figure 4C).

Thus far, both microtubule stabilizers and destabilizers are known to act as anticancer agents by regulating cell cycle events, such as mitosis. At a high concentration, stabilizers inhibit the depolymerization or destabilization of microtubules, thus, inducing gap 2 phase (G2)/mitosis(M) arrest, which would result in growth inhibition and cell death [53]. On the other hand, a high concentration of Vinca alkaloids (powerful destabilizers) promotes microtubule depolymerization, leaving the cancer cells blocked in mitosis with condensed chromosomes due to mitotic spindle degradation [54]. In our study, a high concentration of microtubule acetylation inhibitors was used (1 μM for GM-90257 and 100 nM for GM-90631), and after 12 h, it was observed that the cells were rounded for around 48 h (data not shown). However, in the case of αTAT1 KO MDA-MB-231 cells, where microtubule acetylation is not constitutive, the duration of proliferation was longer compared with that of mock MDA-MB-231 cells in a soft environment or in vivo (Figure 1). In addition, there was no critical defect in cell proliferation when normal breast cells (MCF-10A) and luminal-subtype breast cancer cells (MCF-7) were treated with microtubule acetylation inhibitors. Based on our experimental results, it is likely that the induction of apoptosis by microtubule acetylation inhibitors does not occur by merely blocking the cell cycle, and there might be unknown molecular mechanisms that contribute to apoptosis through the regulation of microtubule acetylation. In the case of MTAs, a different effect was observed at a low concentration compared with a high concentration; for example, when paclitaxel was added at a low concentration, it arrested cells in the G1 phase but not the M phase and inhibited cancer proliferation by increasing p53 and p21 expression levels [55]. A notable result of this study is that BCl-2 was significantly inhibited after treatment with microtubule acetylation inhibitors (Figure 4D). BCl-2 is one of the potent anti-apoptotic proteins in cells. BCl-2 family members are often amplified during carcinogenesis, and it may also cause MTA resistance; thus, BCl-2 regulation may be necessary to improve sensitivity to MTAs [56]. BCl-2 expression regulation by microtubule acetylation inhibitors should be explored in subsequent studies; nevertheless, it is expected that there would be no resistance problem with the effective inhibition of BCl-2 expression. Moreover, microtubule disruption is known to effectively reduce anticancer drug resistance caused by the long-term administration of Taxol [57]. In this study, treatment with microtubule acetylation inhibitors showed that microtubule bundles were disrupted in acetylation-enriched cells; thus, they could be used in combination therapy under the appropriate conditions.

5. Conclusions

It was confirmed that microtubule acetylation was specifically increased in TNBC compared with other subtypes of breast cancer through the analysis of tissue microarray slides derived from breast cancer patients and human breast cancer cell lines. In addition, the disruption of acetyl-α-tubulin by αTAT1 KO significantly reduced cancer progression in vitro and in vivo. We identified three chemical compounds (GM-90257, GM-90568, and GM-90631) that could effectively inhibit microtubule acetylation. Through molecular docking simulation analysis, all three small chemical compounds were found to bind directly to α-tubulin and prevent the recruitment of αTAT1 to the K40 residue in α-tubulin. In addition, treatment with these chemical inhibitors attenuated tumor growth in vivo and caused the apoptosis of MDA-MB-231 cells (TNBC cells) with weaker effects on MCF-10A or MCF-7 cells, which have a relatively low level of microtubule acetylation. Overall, we demonstrated the potential of the microtubule acetylation inhibitors GM-90257, GM-90568, and GM-90631 as targeted therapeutic agents against microtubule acetylation enrichment in TNBC.

Supplementary Materials: The following are available online at http://www.mdpi.com/2227-9059/8/9/338/s1, Figure S1: (A) Schematic flowchart for small chemical compound screening. (B) Acetyl-histone-H3 and total H3 expression in MDA-MB-231 cell lysates treated with GM-90631. (C) Fluorescence microscopy images showing microtubule bundles or β-tubulin and acetyl-α-tubulin in MDA-MB-231 cells after treatment with GM-90257 (500 nM) and GM-90631 (50 nM); Figure S2: (A) Photographs of mice after treatment with 25 mg/kg GM-90257 for 15 days. (B) The weight of mice showed no significant changes during treatment with GM-90257 (top) and GM-90631 (bottom). (C) Photographs of harvested tumors at the end of drug administration.

Author Contributions: A.K., S.R. and W.K.S. designed the study, analyzed the data, and prepared the manuscript. A.K. performed all experiments including in vitro and in vivo experiments and analyzed the data. G.B.L. and J.H.A. performed chemical synthesis and optimization of GM-90257, GM-90568, and GM-90631. J.H.L. performed the first chemical library screening using Spin90-KO MEFs. T.P. and S.H.E. performed protein docking simulation and analyzed the results. P.K., E.Y. and S.R. generated αTAT1 KO MDA-MB-231 cells using the CRISPR-Cas9 system. S.R. and W.K.S. supervised the work and provided conceptual development. All authors have read and agreed to the published version of the manuscript.

Funding: This research was supported by grants from the Cell Logistics Research Center (NRF-2016R1A5A1007318), the Basic Science Research Programs for Mid-career Researchers (NRF-2018R1A2B2005639), and the National Research Foundation of Korea (NRF) (NRF-2020R1A2C2007389, S. Rhee) funded by the Korean government (MSIT). This study was also supported by the "GIST Research Institute (GRI) ARI" grant funded by the GIST in 2020.

Acknowledgments: We would like to thank Jang-Soo Chun for providing the ChemBridge chemical library.

Conflicts of Interest: The authors declare no conflict of interest.

References

1. Emens, L.A. Breast cancer immunotherapy: Facts and hopes. *Clin. Cancer Res.* **2018**, *24*, 511–520. [CrossRef] [PubMed]
2. Sopik, V.; Sun, P.; Narod, S.A. Impact of microinvasion on breast cancer mortality in women with ductal carcinoma in situ. *Breast Cancer Res. Treat.* **2018**, *167*, 787–795. [CrossRef] [PubMed]
3. Enayatrad, M.; Amoori, N.; Salehiniya, H. Epidemiology and trends in breast cancer mortality in Iran. *Iran. J. Public Health* **2015**, *44*, 430–431. [PubMed]
4. Johansson, A.; Trewin, C.B.; Hjerkind, K.V.; Ellingjord-Dale, M.; Johannesen, T.B.; Ursin, G. Breast cancer-specific survival by clinical subtype after 7 years follow-up of young and elderly women in a nationwide cohort. *Int. J. Cancer* **2019**, *144*, 1251–1261. [CrossRef] [PubMed]
5. Garrido-Castro, A.C.; Lin, N.U.; Polyak, K. Insights into molecular classifications of triple-negative breast cancer: Improving patient selection for treatment. *Cancer Discov.* **2019**, *9*, 176–198. [CrossRef]
6. Lyons, T.G.; Traina, T.A. Emerging novel therapeutics in triple-negative breast cancer. *Adv. Exp. Med. Biol.* **2019**, *1152*, 377–399.
7. Dent, R.; Trudeau, M.; Pritchard, K.I.; Hanna, W.M.; Kahn, H.K.; Sawka, C.A.; Lickley, L.A.; Rawlinson, E.; Sun, P.; Narod, S.A. Triple-negative breast cancer: Clinical features and patterns of recurrence. *Clin. Cancer Res.* **2007**, *13*, 4429–4434. [CrossRef]
8. Draganescu, M.; Carmocan, C. Hormone therapy in breast cancer. *Chirurgia* **2017**, *112*, 413–417. [CrossRef]
9. Isakoff, S.J. Triple-negative breast cancer: Role of specific chemotherapy agents. *Cancer J.* **2010**, *16*, 53–61. [CrossRef]
10. Garutti, M.; Pelizzari, G.; Bartoletti, M.; Malfatti, M.C.; Gerratana, L.; Tell, G.; Puglisi, F. Platinum salts in patients with breast cancer: A focus on predictive factors. *Int. J. Mol. Sci.* **2019**, *20*, 3390. [CrossRef]
11. Malfatti, M.C.; Gerratana, L.; Dalla, E.; Isola, M.; Damante, G.; di Loreto, C.; Puglisi, F.; Tell, G. APE1 and NPM1 protect cancer cells from platinum compounds cytotoxicity and their expression pattern has a prognostic value in TNBC. *J. Exp. Clin. Cancer Res.* **2019**, *38*, 309. [CrossRef] [PubMed]
12. Gerratana, L.; Fanotto, V.; Pelizzari, G.; Agostinetto, E.; Puglisi, F. Do platinum salts fit all triple negative breast cancers? *Cancer Treat. Rev.* **2016**, *48*, 34–41. [CrossRef] [PubMed]
13. Corvaja, C.; Garutti, M.; Gerratana, L.; Pelizzari, G.; Puglisi, F. Hype or hope? The strange case of platinum salts' renaissance in breast cancer. *Expert Rev. Anticancer Ther.* **2019**, *19*, 1005–1008. [CrossRef] [PubMed]
14. Caviston, J.P.; Holzbaur, E.L. Microtubule motors at the intersection of trafficking and transport. *Trends Cell Biol.* **2006**, *16*, 530–537. [CrossRef]
15. Eshun-Wilson, L.; Zhang, R.; Portran, D.; Nachury, M.V.; Toso, D.B.; Löhr, T.; Vendruscolo, M.; Bonomi, M.; Fraser, J.S.; Nogales, E. Effects of α-tubulin acetylation on microtubule structure and stability. *Proc. Natl. Acad. Sci. USA* **2019**, *116*, 10366–10371. [CrossRef]
16. Bratman, S.V.; Chang, F. Mechanisms for maintaining microtubule bundles. *Trends Cell Biol.* **2008**, *18*, 580–586. [CrossRef]
17. Giodini, A.; Kallio, M.J.; Wall, N.R.; Gorbsky, G.J.; Tognin, S.; Marchisio, P.C.; Symons, M.; Altieri, D.C. Regulation of microtubule stability and mitotic progression by survivin. *Cancer Res.* **2002**, *62*, 2462–2467.

18. Schaap, I.A.; Carrasco, C.; de Pablo, P.J.; MacKintosh, F.C.; Schmidt, C.F. Elastic response, buckling, and instability of microtubules under radial indentation. *Biophys. J.* **2006**, *91*, 1521–1531. [CrossRef]
19. Long, H.J. Paclitaxel (Taxol): A novel anticancer chemotherapeutic drug. *Mayo Clin. Proc.* **1994**, *69*, 341–345. [CrossRef]
20. Toma, W.; Kyte, S.L.; Bagdas, D.; Alkhlaif, Y.; Alsharari, S.D.; Lichtman, A.H.; Chen, Z.J.; del Fabbro, E.; Bigbee, J.W.; Gewirtz, D.A.; et al. Effects of paclitaxel on the development of neuropathy and affective behaviors in the mouse. *Neuropharmacology* **2017**, *117*, 305–315. [CrossRef]
21. Pusztai, L.; Mendoza, T.R.; Reuben, J.M.; Martinez, M.M.; Willey, J.S.; Lara, J.; Syed, A.; Fritsche, H.A.; Bruera, E.; Booser, D.; et al. Changes in plasma levels of inflammatory cytokines in response to paclitaxel chemotherapy. *Cytokine* **2004**, *25*, 94–102. [CrossRef] [PubMed]
22. Rovini, A.; Savry, A.; Braguer, D.; Carré, M. Microtubule-targeted agents: When mitochondria become essential to chemotherapy. *Biochim. Biophys. Acta* **2011**, *1807*, 679–688. [CrossRef] [PubMed]
23. Nieuwenhuis, J.; Brummelkamp, T.R. The tubulin detyrosination cycle: Function and enzymes. *Trends Cell Biol.* **2019**, *29*, 80–92. [CrossRef]
24. Bompard, G.; van Dijk, J.; Cau, J.; Lannay, Y.; Marcellin, G.; Lawera, A.; van der Laan, S.; Rogowski, K. CSAP acts as a regulator of TTLL-mediated microtubule glutamylation. *Cell Rep.* **2018**, *25*, 2866–2877. [CrossRef] [PubMed]
25. Janke, C.; Montagnac, G. Causes and consequences of microtubule acetylation. *Curr. Biol.* **2017**, *27*, 1287–1292. [CrossRef] [PubMed]
26. Jenkins, B.V.; Saunders, H.; Record, H.L.; Johnson-Schlitz, D.M.; Wildonger, J. Effects of mutating α-tubulin lysine 40 on sensory dendrite development. *J. Cell Sci.* **2017**, *130*, 4120–4131. [CrossRef]
27. Kalebic, N.; Martinez, C.; Perlas, E.; Hublitz, P.; Bilbao-Cortes, D.; Fiedorczuk, K.; Andolfo, A.; Heppenstall, P.A. Tubulin acetyltransferase αTAT1 destabilizes microtubules independently of its acetylation activity. *Mol. Cell Biol.* **2013**, *33*, 1114–1123. [CrossRef]
28. Wang, B.; Rao, Y.H.; Inoue, M.; Hao, R.; Lai, C.H.; Chen, D.; McDonald, S.L.; Choi, M.C.; Wang, Q.; Shinohara, M.L.; et al. Microtubule acetylation amplifies p38 kinase signalling and anti-inflammatory IL-10 production. *Nat. Commun.* **2014**, *5*, 3479. [CrossRef]
29. Skoge, R.H.; Dölle, C.; Ziegler, M. Regulation of SIRT2-dependent α-tubulin deacetylation by cellular NAD levels. *DNA Repair* **2014**, *23*, 33–38. [CrossRef]
30. Li, L.; Yang, X.J. Tubulin acetylation: Responsible enzymes, biological functions and human diseases. *Cell Mol. Life Sci.* **2015**, *72*, 4237–4255. [CrossRef]
31. Xu, Z.; Schaedel, L.; Portran, D.; Aguilar, A.; Gaillard, J.; Marinkovich, M.P.; Théry, M.; Nachury, M.V. Microtubules acquire resistance from mechanical breakage through intralumenal acetylation. *Science* **2017**, *356*, 328–332. [CrossRef] [PubMed]
32. Reed, N.A.; Cai, D.; Blasius, T.L.; Jih, G.T.; Meyhofer, E.; Gaertig, J.; Verhey, K.J. Microtubule acetylation promotes kinesin-1 binding and transport. *Curr. Biol.* **2006**, *16*, 2166–2172. [CrossRef] [PubMed]
33. Piperno, G.; Fuller, M.T. Monoclonal antibodies specific for an acetylated form of alpha-tubulin recognize the antigen in cilia and flagella from a variety of organisms. *J. Cell Biol.* **1985**, *101*, 2085–2094. [CrossRef] [PubMed]
34. D'YDEWALLE, C.; Krishnan, J.; Chiheb, D.M.; van Damme, P.; Irobi, J.; Kozikowski, A.P.; Vanden-Berghe, P.; Timmerman, V.; Robberecht, W.; van den Bosch, L. HDAC6 inhibitors reverse axonal loss in a mouse model of mutant HSPB1-induced Charcot-Marie-Tooth disease. *Nat. Med.* **2011**, *17*, 968–974. [CrossRef]
35. Govindarajan, N.; Rao, P.; Burkhardt, S.; Sananbenesi, F.; Schlüter, O.M.; Bradke, F.; Lu, J.; Fischer, A. Reducing HDAC6 ameliorates cognitive deficits in a mouse model for Alzheimer's disease. *EMBO Mol. Med.* **2013**, *5*, 52–63. [CrossRef]
36. McLendon, P.M.; Ferguson, B.S.; Osinska, H.; Bhuiyan, M.S.; James, J.; McKinsey, T.A.; Robbins, J. Tubulin hyperacetylation is adaptive in cardiac proteotoxicity by promoting autophagy. *Proc. Natl. Acad. Sci. USA* **2014**, *111*, E5178–E5186. [CrossRef]
37. Bance, B.; Seetharaman, S.; Leduc, C.; Boëda, B.; Etienne-Manneville, S. Microtubule acetylation but not detyrosination promotes focal adhesion dynamics and astrocyte migration. *J. Cell Sci.* **2019**, *132*, 225805. [CrossRef]
38. You, E.; Ko, P.; Jeong, J.; Keum, S.; Kim, J.-W.; Seo, Y.-J.; Song, W.K.; Rhee, S. YAP nuclear translocation through dynein and acetylated microtubule controls fibroblast activation. *bioRxiv* **2019**, arXiv:693168. [CrossRef]

39. Oh, S.; You, E.; Ko, P.; Jeong, J.; Keum, S.; Rhee, S. Genetic disruption of tubulin acetyltransferase, αTAT1, inhibits proliferation and invasion of colon cancer cells through decreases in Wnt1/β-catenin signaling. *Biochem. Biophys. Res. Commun.* **2017**, *482*, 8–14. [CrossRef]
40. You, E.; Huh, Y.H.; Kwon, A.; Kim, S.H.; Chae, I.H.; Lee, O.J.; Ryu, J.H.; Park, M.H.; Kim, G.E.; Lee, J.S.; et al. SPIN90 depletion and microtubule acetylation mediate stromal fibroblast activation in breast cancer progression. *Cancer Res.* **2017**, *77*, 4710–4722. [CrossRef]
41. Abramson, V.G.; Mayer, I.A. Molecular heterogeneity of triple negative breast cancer. *Curr. Breast Cancer Rep.* **2014**, *6*, 154–158. [CrossRef]
42. Gerratana, L.; Basile, D.; Buono, G.; De Placido, S.; Giuliano, M.; Minichillo, S.; Coinu, A.; Martorana, F.; de Santo, I.; del Mastro, L.; et al. Androgen receptor in triple negative breast cancer: A potential target for the targetless subtype. *Cancer Treat. Rev.* **2018**, *68*, 102–110. [CrossRef] [PubMed]
43. Mio, C.; Gerratana, L.; Bolis, M.; Caponnetto, F.; Zanello, A.; Barbina, M.; Di Loreto, C.; Garattini, E.; Damante, G.; Puglisi, F. BET proteins regulate homologous recombination-mediated DNA repair: BRCAness and implications for cancer therapy. *Int. J. Cancer* **2019**, *144*, 755–766. [CrossRef] [PubMed]
44. Wang, Z.; Tang, F.; Hu, P.; Wang, Y.; Gong, J.; Sun, S.; Xie, C. HDAC6 promotes cell proliferation and confers resistance to gefitinib in lung adenocarcinoma. *Oncol. Rep.* **2016**, *36*, 589–597. [CrossRef] [PubMed]
45. Deskin, B.; Yin, Q.; Zhuang, Y.; Saito, S.; Shan, B.; Lasky, J.A. Inhibition of HDAC6 attenuates tumor growth of non-small cell lung cancer. *Transl. Oncol.* **2020**, *13*, 135–145. [CrossRef] [PubMed]
46. Lee, D.H.; Won, H.R.; Ryu, H.W.; Han, J.M.; Kwon, S.H. The HDAC6 inhibitor ACY1215 enhances the anticancer activity of oxaliplatin in colorectal cancer cells. *Int. J. Oncol.* **2018**, *53*, 844–854.
47. Saba, N.F.; Magliocca, K.R.; Kim, S.; Muller, S.; Chen, Z.; Owonikoko, T.K.; Sarlis, N.J.; Eggers, C.; Phelan, V.; Grist, W.J.; et al. Acetylated tubulin (AT) as a prognostic marker in squamous cell carcinoma of the head and neck. *Head Neck Pathol.* **2014**, *8*, 66–72. [CrossRef]
48. Boggs, A.E.; Vitolo, M.I.; Whipple, R.A.; Charpentier, M.S.; Goloubeva, O.G.; Ioffe, O.B.; Tuttle, K.C.; Slovic, J.; Lu, Y.; Mills, G.B.; et al. α-Tubulin acetylation elevated in metastatic and basal-like breast cancer cells promotes microtentacle formation, adhesion, and invasive migration. *Cancer Res.* **2015**, *75*, 203–215. [CrossRef]
49. Holbeck, S.L.; Camalier, R.; Crowell, J.A.; Govindharajulu, J.P.; Hollingshead, M.; Anderson, L.W.; Polley, E.; Rubinstein, L.; Srivastava, A.; Wilsker, D.; et al. The National Cancer Institute ALMANAC: A comprehensive screening resource for the detection of anticancer drug pairs with enhanced therapeutic activity. *Cancer Res.* **2017**, *77*, 3564–3576. [CrossRef]
50. Wang, S.; Mi, J.B.; Li, Y.Z.; Chang, W.B.; Ci, Y.X.; Zhao, M.Z.; Zhao, Y.K.; Zhu, L.Y.; Xu, G. Pharmacokinetics and tissue distribution of iv injection of polyphase liposome-encapsulated cisplatin (KM-1) in rats. *Acta Pharmacol. Sin.* **2003**, *24*, 589–592.
51. Chang, P.; Stearns, T. δ-tubulin and ε-tubulin: Two new human centrosomal tubulins reveal new aspects of centrosome structure and function. *Nat. Cell Biol.* **2000**, *2*, 30–35. [CrossRef] [PubMed]
52. Wloga, D.; Joachimiak, E.; Fabczak, H. Tubulin post-translational modifications and microtubule dynamics. *Int. J. Mol. Sci.* **2017**, *18*, 2207. [CrossRef] [PubMed]
53. Jordan, M.A. Mechanism of action of antitumor drugs that interact with microtubules and tubulin. *Curr. Med. Chem. Anticancer Agents* **2002**, *2*, 1–17. [CrossRef] [PubMed]
54. Jordan, M.A.; Wilson, L. Microtubules as a target for anticancer drugs. *Nat. Rev. Cancer* **2004**, *4*, 253–265. [CrossRef]
55. Chen, J.G.; Yang, C.P.; Cammer, M.; Horwitz, S.B. Gene expression and mitotic exit induced by microtubule-stabilizing drugs. *Cancer Res.* **2003**, *63*, 7891–7899.
56. Whitaker, R.H.; Placzek, W.J. Regulating the BCL2 family to improve sensitivity to microtubule targeting agents. *Cells* **2019**, *8*, 346. [CrossRef] [PubMed]
57. Wang, Y.; Zhou, Y.; Zheng, Z.; Li, J.; Yan, Y.; Wu, W. Sulforaphane metabolites reduce resistance to paclitaxel via microtubule disruption. *Cell Death Dis.* **2018**, *9*, 1134. [CrossRef]

© 2020 by the authors. Licensee MDPI, Basel, Switzerland. This article is an open access article distributed under the terms and conditions of the Creative Commons Attribution (CC BY) license (http://creativecommons.org/licenses/by/4.0/).

biomedicines

MDPI

Article

Drug Resistance in Glioma Cells Induced by a Mesenchymal–Amoeboid Migratory Switch

Sophie E. Ketchen [1], Filomena O. Gamboa-Esteves [2], Sean E. Lawler [3], Michal O. Nowicki [4], Arndt Rohwedder [1], Sabine Knipp [5], Sally Prior [5], Susan C. Short [2], John E. Ladbury [1] and Anke Brüning-Richardson [5,*,†]

1 Light Laboratories, School of Molecular and Cellular Biology, University of Leeds, Leeds LS2 9JT, UK; s.e.ketchen@leeds.ac.uk (S.E.K.); A.Rohwedder@leeds.ac.uk (A.R.); J.E.Ladbury@leeds.ac.uk (J.E.L.)

2 Leeds Institute of Medical Research at St. James's, University of Leeds, Leeds LS9 7TF, UK; F.E.O.Esteves@leeds.ac.uk (F.O.G.-E.); s.c.short@leeds.ac.uk (S.C.S.)

3 Brown University Cancer Center, Pathology & Laboratory Medicine, Brown University, Providence, RI 02903, USA; sean_lawler@brown.edu

4 Harvey Cushing Neuro-Oncology Research Laboratories, Department of Neurosurgery, Brigham and Women's Hospital, Harvard Medical School, Boston, MA 02115, USA; mnowicki@bwh.harvard.edu

5 School of Applied Sciences, University of Huddersfield, Huddersfield HD1 3DH, UK; S.Knipp@hud.ac.uk (S.K.); u1973968@unimail.hud.ac.uk (S.P.)

* Correspondence: a.bruning-richardson@hud.ac.uk; Tel.: +44-01484-472518

† Current address: Department of Biological and Geographical Sciences, University of Huddersfield, Queensgate, Huddersfield HD1 3DH, UK.

Abstract: Cancer cell invasion is a precondition for tumour metastasis and represents one of the most devastating characteristics of cancer. The development of drugs targeting cell migration, known as migrastatics, may improve the treatment of highly invasive tumours such as glioblastoma (GBM). In this study, investigations into the role of the cell adhesion protein Cellular communication network factor 1 (CCN1, also known as CYR61) in GBM cell migration uncovered a drug resistance mechanism adopted by cells when treated with the small molecule inhibitor CCG-1423. This inhibitor binds to importin α/β inhibiting the nuclear translocation of the transcriptional co-activator MKL1, thus preventing downstream effects including migration. Despite this reported role as an inhibitor of cell migration, we found that CCG-1423 treatment did not inhibit GBM cell migration. However, we could observe cells now migrating by mesenchymal–amoeboid transition (MAT). Furthermore, we present evidence that CCN1 plays a critical role in the progression of GBM with increased expression in higher-grade tumours and matched blood samples. These findings support a potential role for CCN1 as a biomarker for the monitoring and potentially early prediction of GBM recurrence, therefore as such could help to improve treatment of and increase survival rates of this devastating disease.

Keywords: migration; invasion; glioblastoma; CCN1; mesenchymal–amoeboid transition; biomarker

Citation: Ketchen, S.E.; Gamboa-Esteves, F.O.; Lawler, S.E.; Nowicki, M.O.; Rohwedder, A.; Knipp, S.; Prior, S.; Short, S.C.; Ladbury, J.E.; Brüning-Richardson, A. Drug Resistance in Glioma Cells Induced by a Mesenchymal–Amoeboid Migratory Switch. *Biomedicines* **2022**, *10*, 9. https://doi.org/10.3390/biomedicines10010009

Academic Editor: Simon J Allison

Received: 17 September 2021
Accepted: 19 December 2021
Published: 22 December 2021

Publisher's Note: MDPI stays neutral with regard to jurisdictional claims in published maps and institutional affiliations.

Copyright: © 2021 by the authors. Licensee MDPI, Basel, Switzerland. This article is an open access article distributed under the terms and conditions of the Creative Commons Attribution (CC BY) license (https://creativecommons.org/licenses/by/4.0/).

1. Introduction

Localisation and the invasive behaviour of the most aggressive form of gliomas, glioblastoma (GBM), contributes to the challenge in treating brain tumour patients, in fact, there is currently no curative approach available. Recent trends in GBM research aim to provide a better understanding of the molecular pathology and genomic landscape of GBM, which will enable tailored therapeutic approaches. Though much progress has been made in cancer treatments in general, substantial challenges remain for GBM treatment including tumour location, relapse, tumour heterogeneity and drug resistance. The current standard of care consists of surgical resection followed by radiotherapy and adjuvant chemotherapy with temozolomide [1]. Unfortunately, despite these therapeutical approaches, nearly all patients will develop a recurrent tumour and treatment options become limited to palliative care [2]. It is known, that the cell adhesion protein CCN1 (also known as cysteine-rich angiogenic inducer 61 or CYR61) stimulates mesenchymal migration in fibroblasts, endothelial cells, smooth muscle cells and some types of cancer cells [3]. Studies have also shown that CCN1 is associated with cancer progression and invasion [4,5]. The tight association between CCN1, the extracellular matrix (ECM), and the cell surface stimulates adhesive signalling and supports cell adhesion [3]. This results in the formation of focal adhesions and activation of focal adhesion kinase (FAK); paxillin and Rac are induced leading to reorganisation of the actin cytoskeleton and lamellipodia and filopodia formation [6–8].

In tumour progression, CCN1 has been shown to promote invasion, angiogenesis, cell proliferation, cell survival and metastasis [3]. Silencing CCN1 expression in vitro leads to loss of invasion [9]. In addition, in immunodeficient mice implanted with human gastric, naturally CCN1 deficient, adenocarcinoma cells induced expression of the CCN1 cDNA is an effective promoter of angiogenesis, concomitant with tumour vascularisation and growth [10,11]. CCN1 is highly expressed in primary gliomas, as well as in high-grade glioma cell lines [12]. High expression levels were also observed in the astrocytoma cell lines U87, U373 and T98G and only low expression levels in the less invasive U343 cell line. Forced expression of CCN1 in U343 cells increased proliferation, angiogenesis and tumorigenicity in vivo. U343 cells overexpressing CCN1 migrated more readily and produced larger, more vascularised tumours in nude mice [12].

Here, we used the small molecule inhibitor CCG-1423 to investigate the effects of CCN1 on cell migration in GBM cell lines. CCG-1423 blocks the production of CCN1. It binds to MKL1, preventing its nuclear import and therefore activation of MKL1/SRF-dependent transcription and CCN1 production [13]. Contrary to other studies, our in vitro results did not show CCG-1423 to have an anti-migratory effect as previously suggested [13,14]. In a recent publication, using iSIM microscopy to analyse cell morphology during migration we could show that GBM cells can respond to CCG-1423 treatment by switching their migratory mode [15]. Therefore, since our previous study indicated that CCN1 is required for mesenchymal migration, we hypothesise that glioma cells can overcome the anti-migratory drug treatment by inducing a mesenchymal–amoeboid transition (MAT) for continued migration and invasion. We also present clear evidence that CCN1 has a pivotal role in the progression of GBM, as we present evidence that its increased expression in higher grade tumours is correlated with higher concentrations in patient blood. This could provide a valuable molecular marker for the characterisation, progression, and possibly early detection of GBM. Furthermore, we present evidence that the cellular phenomenon of MAT is in fact being utilised by cancer cells as a drug resistance mechanism. This enables the cells to 'side-step' the action of drugs that only target one type of migration, allowing continued spread and invasion.

2. Materials and Methods

2.1. Cell Culture

The glioma cell lines U87 (U-87MG, RRID:CVCL_0022) and U251 (U-251MG, RRID: CVCL_0021) were obtained from ATCC and ECACC. U87-NT cell lines were created to stably express a GFP control construct, a kind gift from Dr Chiara Galloni (University of Leeds). The primary human oligodendroglioma cell line HOG (RRID:CVCL_D276) (gifted by Diane Jaworski, University of Vermont) and adult glioma cell line G44 [16], the murine glioma cell line GL261 (RRID:CVCL_Y003) and the rat glioma cell line CNS1 (RRID:CVCL_5276) were all maintained and grown in the Neurosurgery department at Harvard Medical School, Brigham and Women's Hospital. U251, U87, GL261 and CNS1 cell lines were grown in Dulbecco's Modified Eagle's Medium (Sigma, Poole, Dorset, UK) with 10% heat-inactivated foetal calf serum (Sigma, Poole, Dorset, UK) and 0.5% penicillin-streptomycin (Sigma, Poole, Dorset, UK) (complete medium) and cultured in a Sanyo CO_2 incubator at 37 °C in a humidified atmosphere of 5% CO_2 (in air).

HOG and G44 were grown as neurospheres suspended in neurobasal medium (ThermoFisher, Altrincham, UK) containing B27 supplement, 0.5% penicillin-streptomycin (ThermoFisher, Altrincham, UK), glutamine (ThermoFisher, Altrincham, UK), EGF and FGF (ThermoFisher, Altrincham, UK).

All human cell lines were recently authenticated using short tandem repeat profiling by the University of Leeds and Harvard Medical School (July 2021), and all experiments were performed with mycoplasma-free cells.

2.2. *Tissue Microarray (TMA) Staining*

A commercially available brain cancer tissue microarray (TMA) (Biomax BS17015a, Derwood, MD 20855, USA) including 63 cases/cores was stained for CCN1. The 63 cores included 38 astrocytoma grades 1–3, 14 GBM, 6 oligodendroglioma, 1 ependymoma, 1 medulloblastoma, 3 cases of cancer adjacent brain tissue (negative control) and 1 pheochromocytoma (adrenal gland tumour, positive control). Because of the small sample size of the ependymoma and medulloblastoma, these were not included in the data analysis. The TMA slide was dewaxed and underwent antigen retrieval and staining with the primary antibody anti-CYR61/CCN1 Rb (Abcam, Cambridge, UK) at 1:250 following the protocol described by Cheng et al. [17]. The slide was analysed by calculating and combining two sub-scores for the CCN1 staining: staining intensity and percentage of core stained. Staining intensity was scored out of 3 with 0 being no staining and 3 being strong staining. Percentage of staining was scored out of 4 with 0 being 0% stained, 1 being 1–25% stained, 2 being 26–50% stained, 3 being 51–75% stained and 4 being 75%+ stained. Thus, the final maximum combined score is 7.

2.3. *Immunohistochemistry (IHC) of Patient Samples*

Patient GBM samples were collected, processed and stored by the Leeds Multidisciplinary Routine Tissue Banking (RTB) service, obtained from GBM patients undergoing surgery at the Leeds General Infirmary (ethical approval no. RTB 15/YH/0080). Tumour tissues were embedded in paraffin wax, sectioned on a microtome (Leica Biosystems, Newcastle Upon Tyne, UK) at 8 μm and dried on glass microslides for 24 h. The slides were deparaffinized and rehydrated following standard procedures, then dewaxed by 2× 1-min washes in xylene (Sigma, Poole, Dorset, UK), followed by 2× 1 min washes in ethanol (Sigma, Poole, Dorset, UK) and 3× 1 min washes in water. Antigen retrieval was performed using Tris EDTA, pH 9 (Abcam, Cambridge, UK) in a pressure cooker for 2 min. The primary antibody anti-CYR61/CCN1 rabbit polyclonal (Abcam, Cambridge, UK) was used at a concentration of 1:250. A secondary antibody (anti-rabbit IgG HRP polymer (ready to use, Vector, 2BScientific, Upper Heyford, UK)) was used for signal amplification and DAB (Abcam, Cambridge, UK) reaction (10 min) was used for signal detection. Slide analysis was carried out by calculating and combining two scores for the CCN1 staining as described above for TMA analysis. The samples were randomised prior to scoring to prevent bias.

2.4. *Enzyme-Linked Immunosorbent Assay (ELISA)*

GBM patient blood samples matched to tumour samples were collected, processed and stored by the Leeds Multidisciplinary Routine Tissue Banking (RTB) service from GBM patients undergoing surgery at the Leeds General Infirmary (ethical approval no. RTB 15/YH/0080). A commercially available Human CYR61/CCN1 Quantikine ELISA kit (R&D Systems, Abingdon, UK) was used for the solid-phase sandwich ELISA for CCN1. The ELISA was carried out following the manufacturer's instructions, using 100 μL of assay diluent per well, followed by 50 μL of patient blood serum in duplicate, control blood serum (from healthy donors), or standard supplied in the kit. After the final incubation step the plate was immediately read out on a plate reader at 540 nm (ThermoFisher, Altrincham, UK).

For determination of CCN1 secretion from established cell lines, the same ELISA kit was used. U251 and U87 cells were split to a density of 2×10^3 into 6-well plates (Costar, Corning Lifesciences, New York, NY, USA) and incubated for 24 h at 37 °C. Following incubation, the media was removed and replaced with 2 mL of complete medium (control + DMSO only) or medium with 500 nM of CCG-1423 (originally resuspended in DMSO) (Tocris, Bristol, UK). The cell supernatants were collected at 24, 48 and 72-h time points. For the ELISA, three wells per time point per cell line and 8 for the standard curve were set up and 100 µL of assay diluent was added to each well, followed by 50 µL of cell supernatant, control or standard. The same protocol as for the blood samples was followed.

2.5. Live Cell Imaging

U251 and U87 cells were adjusted to a density of 1.5 × 10^3/mL in complete medium. 100 µL of either cell suspension was pipetted into each well of a flat bottomed 96 well plate (half a plate per cell line). Cells were allowed to settle for 24 h at 37 °C (humidified atmosphere with 5% CO_2). For the experiments, 100 µL of complete medium containing either DMSO (control) or inhibitor diluted in DMSO (20 mM lithium chloride (LiCl), 5 µM 6-bromo-indirubin-3′-oxime (BIO), Selleckchem, Cambridge, UK), 500 nM CCG-1423, all diluted in DMSO) was added. Live cell time-lapse imaging using the IncuCyte Zoom System (Essen BioScience, Royston, UK) started immediately after drug application for 72 h at 37 °C (humidified atmosphere with 5% CO_2). Movies were generated using the IncuCyte Zoom System software. For analysis and quantification, cell motility was tracked and measured with the MTrackJ plugin for ImageJ (Rasband, W.S., ImageJ, U.S. National Institutes of Health, Bethesda, MA, USA, https://imagej.nih.gov/ij/, 1997–2018, accessed on 14 December 2021). Migration distance was determined as the accumulated total migration distance over the given time. Displacement as the length of the vector from start to end-point of measured migration was calculated to give an indication of directionality of movement.

2.6. Immunofluorescence (IF) and Analysis of MKL1 Intracellular Distribution

U251 cells were cultured from a density of 2 × 10^3 on sterile 25 × 25 coverslips in 6 well plates (Costar, Corning Lifesciences, New York, NY, USA) allowed to settle for 24 h at 37 °C. Cell culture media was replaced with 2 ml of complete medium with DMSO (control) or medium with 500 nM of CCG-1423 (originally resuspended in DMSO, Tocris, Bristol, UK). Cells were incubated for a further 48 h with control/inhibitor medium at 37 °C. The cells were then fixed using 4% PFA (Thermo Scientific, Altrincham, UK) and blocked for 1 h using 5% normal goat serum (Abcam, Cambridge, UK). The cells were then labelled for MKL1 (anti-MKL1, made in Rb, Abcam, 1:100 in blocking solution, Cambridge, UK) and incubated for 24 h at 4 °C. After 3 PBS washes and incubation for 1 h with the secondary antibody (anti-rabbit AlexaFluor 488 (Abcam, Cambridge, UK), 1:500) the cells were washed again 3 times in PBS and mounted on glass microslides using Mowiol containing DAPI nucleus stain (Sigma, Poole, Dorset, UK). Samples were imaged on a Zeiss LSM880 inverted confocal microscope (Zeiss, Birmingham, UK) and resulting images were analysed in ImageJ, as follows: Threshold images were created using standard ImageJ settings for the DAPI and 488 channel separately. These were used to create ROIs outlining the nucleus (DAPI channel) or the respective (single) cell (488 channel). These ROIs were then used to calculate the mean grey values of MKL1 fluorescence in the cytoplasm (ROI cell minus nucleus) and nucleus (ROI nucleus).

2.7. Western Blots

Cell lysates were obtained from cells grown in 25 cm^2 plastic tissue culture flasks (Corning) treated with 500 nM of CCG-1423 (Tocris, Bristol, UK) for 24, 48 and 72 h, respectively. Cell culture vessels were placed on ice and cell cultures were washed twice with ice-cold PBS. Cells were collected and pelleted at 400× g for 5 min. PBS was removed and the cell pellet was resuspended in 0.5 mL of Tris-HCl lysis buffer, containing 25 µL/mL protease inhibitor (Sigma, Poole, Dorset, UK) and incubated on ice for 5 min. For cytoplasmic separation, the solution was centrifuged at 400× g for 10 min. Supernatants containing the cytoplasmic extract were collected and transferred into clean tubes. The protein concentration of each cell lysate was determined by Bradford protein assay (BioRad, Watford, UK). For Western blotting, proteins were separated by SDS-PAGE and transferred to PVDF membranes (BioRad, Watford, UK). Antibodies used for immuno labelling of Western blots were rabbit anti-MKL1 (1:300, Abcam, Cambridge, UK) and HRP-conjugated donkey anti-rabbit IgG (1:1000, GE Healthcare, Chalfont Saint Giles, UK). Proteins were detected by SuperSignal West Femto maximum sensitivity substrate (ThermoFisher, Altrincham, UK), according to the provided protocol and visualised using a ChemiDoc MP imaging system (BioRad, Watford, UK) and associated Image Lab software (BioRad, Watford, UK). Protein band intensity was analysed using ImageJ software and normalised to the corresponding β-actin control.

2.8. 3D Spheroid Generation and Invasion Assay

U87-NT cells were seeded at 3×10^3/well in low adherence 96 well plates (ThermoFisher, Altrincham, UK) as previously described [14]. Three days after seeding, spheroids contained within the wells were embedded in rat tail collagen V (Corning Life Science, Glendale, AZ 85301, USA), polymerisation was achieved with 1 M NaOH. The inhibitors CCG-1423 and Rhosin-HCl (Tocris, Bristol, UK) were resuspended in DMSO and added at a predetermined anti-migratory concentration (CCG-1423: 500 nM, Rhosin: 1 μM). Control spheroids were mock-treated with DMSO-supplemented medium only. Spheroids were allowed to grow for 72 h in collagen, fixed with 4% PFA and labelled with Phalloidin-TRITC (ThermoFisher, Altrincham, UK) in their collagen plugs. Image z-stacks of spheroids were acquired on a Zeiss 880 LSM confocal microscope (Zeiss, Birmingham, UK) with an EC plan-neofluar 10× objective. For analysis maximum projections of the z-stacks were used. The resulting grey-scale images were inverted and using the free-hand selection tool in ImageJ, the outline of the spheroid as well as of the gaps between migration cells were drawn and the area measured (for illustration see Supplement Figure S2).

2.9. Data Analysis

For all experiments, quantified data were statistically analysed using RStudio 1.4.1106 (RStudio Inc., Boston, MA, USA) using the R packages ggplot2, tidyverse, ggbeeswarm, rstatix, and ggpubr. Data were tested for normality using the Shapiro Wilk test. Accordingly, ANOVA with Tukey post hoc test or *t*-test for normally distributed data, otherwise Kruskal–Wallis with post hoc Dunn test, or Wilcoxon–Mann–Whitney-test were performed. Used scripts are available on request. p-values < 0.05 were considered statistically significant.

3. Results

3.1. CCN1 Expression Correlates with Tumour Grades in Patient Tissues

To obtain an overview of CCN1 expression in different types and grades of brain tumours, we employed a commercially available TMA assay. In general, cytoplasmic, and membranous CCN1 staining was visible in tumour cells, with strong staining in elongated cells and cells in close proximity to blood vessels (data not shown), suggesting a correlation between CCN1 and invasive cell migration. The clinical characteristics associated with the samples, i.e., age, sex, and tumour grade, are summarised in a table in Figure 1A. Samples were dichotomised into low and high CCN1 expression with low scoring samples having a score of 3 or less and high scoring samples having a score of 4 or more. High CCN1 expression correlated with high astrocytoma grade (e.g., 50% astrocytoma grade 4 samples were scored 4 or higher), while low CCN1 expression was observed within low astrocytoma grades. Quantification and comparison of the different brain tumour types revealed a clear correlation of high CCN1 expression with GBM (tumour grade 4) that was significantly different from the negative control and low-grade astrocytoma (Figure 1B).

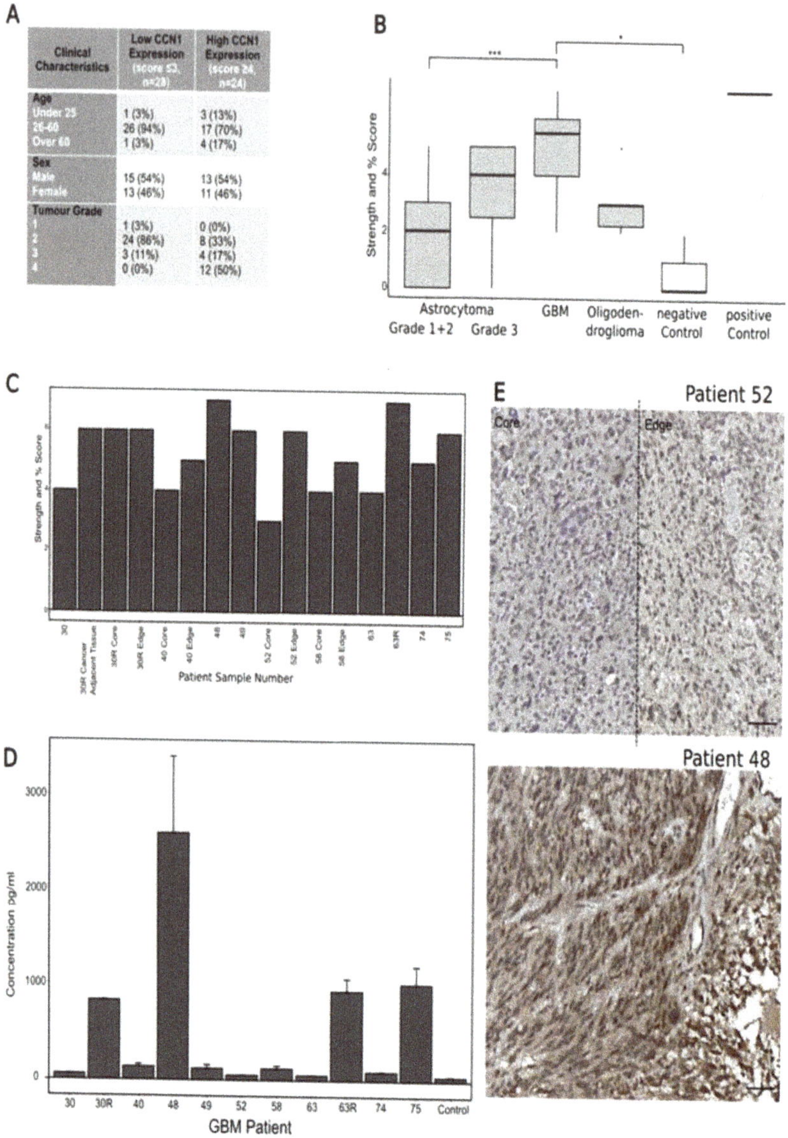

Figure 1. CCN1 expression in patient samples. (**A**) Combined scoring of CCN1 staining strength and percentage of the core covered on a TMA containing astrocytomas grade 1–2, GBM, oligodendroglioma, cancer adjacent brain tissue as a control and an adrenal core as a positive control. (**B**) The clinical characteristics of the TMA astrocytoma samples. The results reveal associations of CCN1 expression levels with tumour grade. 86 % of the samples with low CCN1 expression are grade 2 astrocytomas, whilst 50% of the samples with high CCN1 expression are grade 4 astrocytomas (GBM). There were no associations of CCN1 levels with age or sex. (**C**) Combined scores of CCN1 staining intensity and percentage of staining covering the samples collected from 9 GBM patients during surgery. Recurrent tumour samples were also collected for patients 30 and 63 (30R and 63R). Samples of the core and edge of the tumour were collected for patients 30, 40, 52 and 58. Patient 30 also had a sample of cancer adjacent tissue collected. (**D**) CCN1 concentration in 9 patient blood serum samples.

Blood samples from patient no. 30 and 63 were also collected at their recurrent GBM tumour surgery (30R and 63R). The percentage increase of each patient compared to the control was also calculated. (**E**) CCN1 IHC staining correlates with the CCN1 concentration determined for the matched blood samples. The GBM tissue sample from patient no. 52 presented with the lowest concentration of CCN1 in the blood (51.51 pg/mL) which correlates with the low intensity CCN1 staining in the tissue samples. The GBM tissue sample from patient no. 48 presented with the highest concentration of CCN1 in the blood (2588.041 pg/mL), which correlates with the high intensity CCN1 staining in the tissue sample. Scale bar = 50µm, Magnification 20×. Asterisks show statistical significance. * $p \leq 0.05$, *** $p \leq 0.001$, Kruskal–Wallis: $p = 0.0000344$. Significant differences were marked in the graphs when available.

We further analysed CCN1 expression judged from IHC staining on a total of 16 GBM samples collected from 9 patients undergoing surgery. We followed the procedure and scoring for the TMA samples. Tumour edge samples, defined as 200 µm from the tumour boundary, and tumour core samples, defined as the tumour region outside the edge margins [18], were collected from 4 of the 9 patients, 1 of which also had a sample of cancer adjacent tissue collected. Two of the patients had an additional later surgery for recurrent tumours, from which samples were also collected. Figure 1C summarises the resulting scoring for these samples. In brief, all samples but one were scored as highly expressing CCN1 (score 4 or higher). Interestingly, where core and edge samples were available, the edge sample scored equal (patient 30) or higher than the core (patient 40, 52, 58, Figure 1C). Moreover, samples from recurrent samples always scored higher, than the primary sample. This trend also emerged in the matched blood samples, collected from each patient at the time of surgery (Figure 1D). These matched samples allowed us to investigate a possible correlation between CCN1 levels in the blood of cancer patients and tumour aggressiveness/severity. CCN1 is known to be associated with angiogenesis and we did observe distinct CCN1 labelling around vessels in TMA samples. Thus, we assumed CCN1 blood levels may be detectable if CCN1 was able to cross the blood–brain barrier, as the blood–brain barrier is known to be compromised during GBM formation and progression. Indeed, differences in CCN1 concentration in serum samples, compared to the control, were seen in four patients. As mentioned previously, the CCN1 concentration of two patients became especially elevated at the time of recurrent tumours (patients 30 and 63, Figure 1D). Comparing the IHC staining in tissue samples from patient 48, with the highest measured CCN1 blood concentration, to patient 52 with the lowest concentration, showed a low intensity of CCN1 immunostaining in the latter, but strong staining in patient 48 tissue.

These findings highlight a correlation between CCN1 staining intensity and concentration of CCN1 in the serum of patients. Moreover, the results suggest a correlation between tumour grade and CCN1 concentration in tissue and blood of patients. This opens the interesting opportunity that CCN1 may be a useful liquid biopsy marker for GBM diagnosis.

3.2. *Effects of CCG-1423 on Cell Migration in Live Cell Imaging*

To better understand the relevance of our findings on the correlation between CCN1 expression and tumour aggressiveness/grade and concentration in matched patient serum, we next investigated the effects of blocking CCN1 in the established glioma cell lines U251 and U87. CCG-1423 is known to inhibit CCN1 production [13] and was therefore used to observe effects on cell migratory behaviour by live-cell imaging. The commercially available inhibitors BIO and Lithium chloride (LiCl) are well known for their inhibitory effects on migration in glioma cells [19,20] and were used as positive controls for comparison with CCG-1423 effects. Migratory distance (i.e., accumulated travelled distance over time), and cell displacement (i.e., vector length between cell positions at start and end of experiment) of treated cells are shown in Figure 2A,B and Supplementary Figure S2.

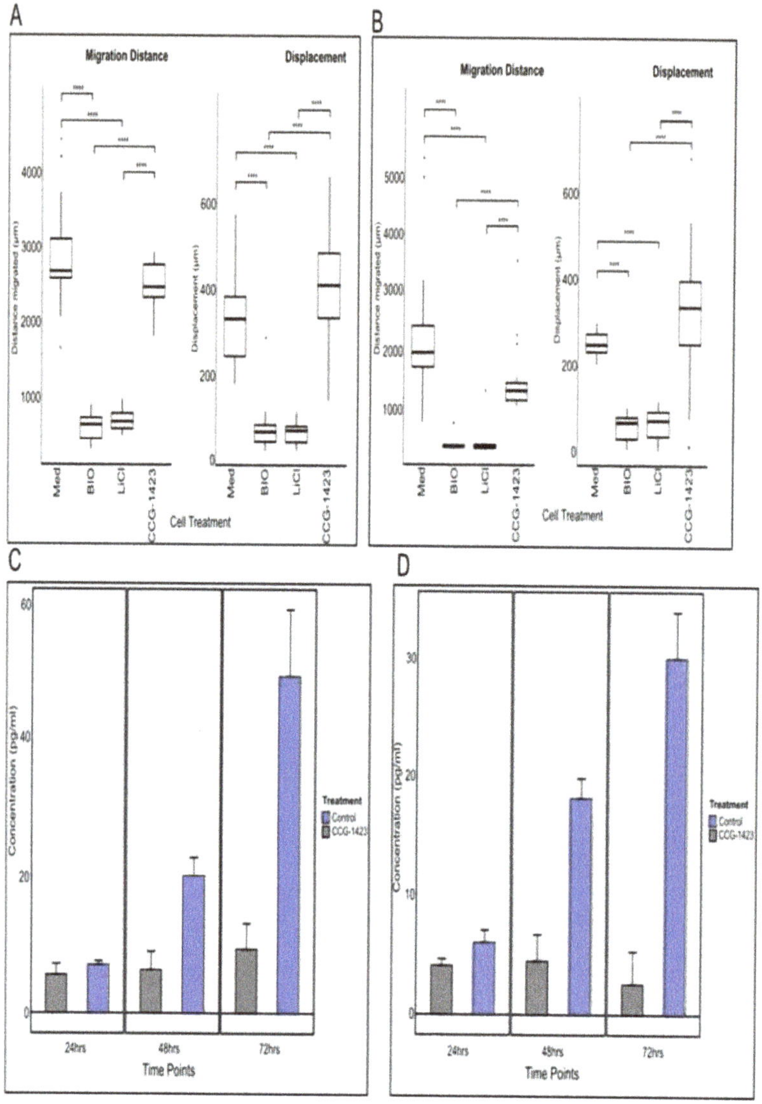

Figure 2. Effect of CCG-1423 on cell migration and CCNI secretion. The effects of inhibitors on cell migration as measured by distance travelled and cell displacement in (**A**) U251 and (**B**) U87 cell lines. Experiments were repeated in triplicate. For sample specific tracking examples see Supplementary Figure S2B. Asterisks indicate statistically significant results compared to the control (Med). **** $p \leq 0.0001$, Kruskal–Wallis for A, Distance: $p = 4.18 \times 10^{-13}$, Displacement: $p = 1.52 \times 10^{-12}$; for B, Distance: $p = 6.62 \times 10^{-13}$, Displacement: $p = 2.17 \times 10^{-10}$. An ELISA was carried out to determine CCN1 levels in the supernatants of (**C**) U251 and (**D**) U87 cell lines over a 72-h period. Cells were incubated for 24, 48 and 72 h +/− CCG-1423. The supernatants were collected for each time point and tested for CCN1 by ELISA. The results suggest reduced CCN1 concentration levels in the supernatants of treated U251 and U87 cells, at the 48 and 72-h time points, when compared to the untreated controls as indicated by decreased concentration. Significant differences were marked in the graphs when available. **** $p \leq 0.0001$.

After treatment with BIO or LiCl, migration (both in terms of distance and displacement) was significantly reduced in both cell lines in comparison to the control, as one would expect. The effects of BIO were observed almost immediately, within 10 min of treatment, in both cell lines and were maintained throughout the 72 h of imaging. Although the effect of LiCl U87 appeared to wane around the 48 h mark, the cells started to recover and adopt their normal cell migratory behaviour. In comparison, U251 cells did not recover for the duration of the imaging. Unlike the other inhibitors, the CCG-1423 treatment did neither result in changes in migration distance nor displacement, compared to the control cells. These results were surprising in the light of the previously reported data suggesting a significant decrease in PC3 cell invasion when treated with CCG-1423 [13].

To ascertain further the impact of treatment of CCG-1423 in blocking CCN1 production, we investigated the levels of secretion of CCN1 pre- and post-application of CCG-1423. As CCN1 is secreted by cells, a sandwich ELISA was used to detect the presence of CCN1 in the supernatant cell culture medium of CCG-1423-treated and untreated U251 and U87 cell cultures over a time span of 72 h. We found, indeed that the concentration of CCN1 in the medium did not increase in treated cells, whereas CCN1 levels clearly increased in supernatants from untreated cells over time (Figure 2C,D). From these results, it was evident that CCG-1423 had an effect on the secretion and/or production of CCN1; despite cells continuing to migrate, as illustrated by the live-cell imaging results. This suggested a previously unreported relationship between CCG-1423 exposure and CCN1 expression.

3.3. *Loss of Nuclear Import of MKL-1 Is Facilitated by CCG-1423 Activity*

To investigate this further, we decided to investigate the regulation of CCN1 production. CCG-1423 binds to the transcriptional co-activator MKL1, preventing its nuclear import and therefore activation of MKL1/SRF-dependent transcription and CCN1 production [15]. Immunofluorescence labelling was used to visualise the accumulation of MKL1 in the cytoplasm of U251 cells treated with CCG-1423 in comparison to untreated control cells. As U87 and U251 cell lines had shown so far equal reactions to CCG-1423 treatment and U251 is the more representative cell line in terms of adopting mesenchymal migration with a pronounced lamellipodia front as observed by live-cell imaging, only results for U251 is shown here. There appeared to be a marked increase in cytoplasmic MKL1 in cells treated for 72 h with CCG-1423 with a concomitant loss of nuclear MKL1 in comparison to the control (Figure 3A,B).

To confirm the results from the IF imaging of MKL1, Western blots were carried out to determine cytoplasmic protein levels of MKL1 in untreated control and CCG-1423 treated U251 cells over a 72 h period. During sample preparation of cell lysates, nuclear-associated components were separated and discarded (as described in materials and methods), therefore the results from the Western blot analysis represent solely cytoplasmic MKL1. This biochemical analysis of U251 cells showed an increase in MKL1 accumulation in the cytoplasm when treated with CCG-1423 (Figure 3C,D), supporting the IF results. The signal intensity of MKL1 was clearly stronger (after normalisation with actin) in CCG-1423 treated U251 cells at 24 and 48 h time points. After 72 h treatment, signal strengths converged, which is consistent with the live-cell imaging results, showing that cells returned to normal behaviour after 72 h CCG-1423 treatment. Western blot analysis for cytoplasmic MKL1 in untreated control versus CCG-1423-treated cells was further carried out on the cell lines HOG and G44 grown as neurospheres, the murine glioma cell line GL261 and rat glioma cell line CNS1 (Supplementary Figure S1). As for the U251 cell line, all additional cell lines showed clearly elevated signals for the intensity of the MKL1 band in CCG-1423 treated cells compared to the control at all analysed time points (Figure S1). These results, along with the MKL1 immunofluorescence analysis of cellular localisation and CCN1 ELISA, illustrate that CCG-1423 can prevent CCN1 production through inhibition of the nuclear import of MKL1.

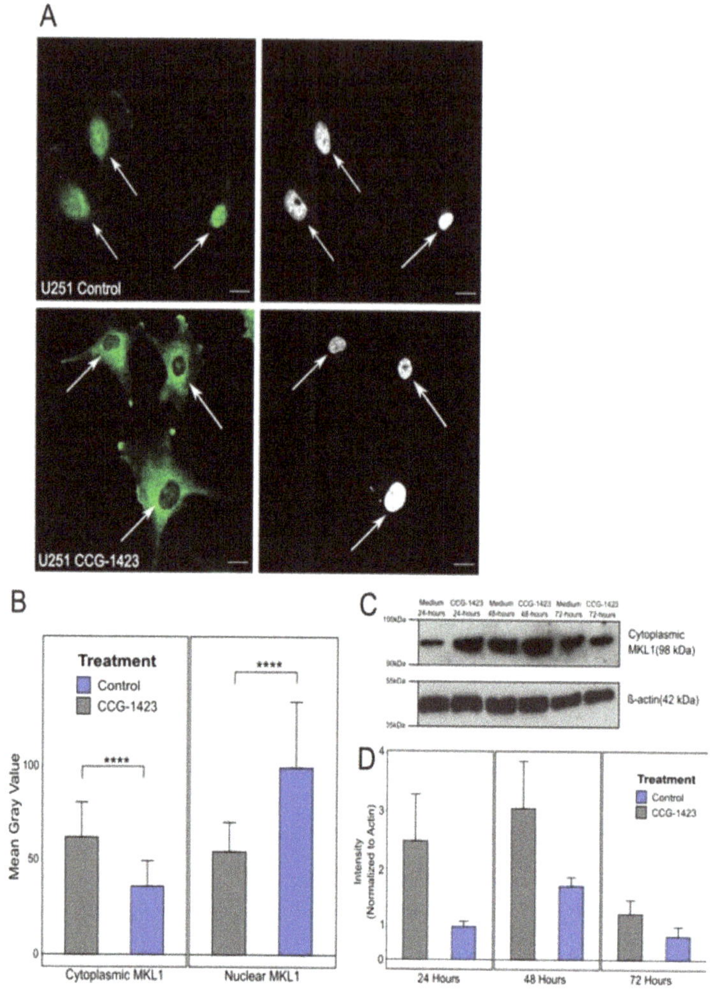

Figure 3. Immunofluorescence staining and analysis of MKL1 in untreated and CCG-1423 treated U251 cells. (**A**) Arrows show a build-up of MKL1 in the cytoplasm of treated cells compared to the control cells which show pronounced nuclear staining. Imaged on a confocal microscope at a magnification of x63. Green fluorescent stain = MKL1; right panel: nuclear DAPI stain; Scale bar = 20 µm. (**B**) Quantification of immunofluorescence staining of MKL1 to determine differences in nuclear and cytoplasmic MKL1 localisation in untreated (control) U251 cells and U251 cells treated with CCG-1423. The images were analysed using ImageJ where a mean gray value was calculated for the nucleus and the cytoplasm of each cell. Kruskal–Wallis: $p = 1.96 \times 10^{-31}$ (**C**) The U251 cell line was treated with 500 nM of CCG-1423 and incubated for 24, 48 and 72 h before cell lysates were obtained. Protein levels of cytoplasmic MKL1 were then determined by western blot. A beta-actin control was included as a protein loading and transfer control. (**D**) Western blot analysis of the effects of CCG-1423 on U251 cells. MKL1 band intensity was quantified and normalised to the β-actin control using Image J software. Graphs show mean ± SD of 3 individual experiments. Asterisks show statistical significance where **** $p \leq 0.0001$. Significant differences were marked in the graphs when available. **** $p \leq 0.0001$.

3.4. Targeting RhoA Signalling and Rac Activation Reduces Cell Migration by a Switch to Collective Cell Migration

In our previous publication [15], we reported on a mesenchymal–amoeboid switch in the U251 cell line characterised by loss of major protrusions/lamellipodia in cells treated with CCG-1423 and a striking increase in the number of filopodia, shown here at high resolution to highlight the phenotypic differences (Figure 4A). Following on these original findings, we reasoned that targeting Rac activation and RhoA-driven cell migration by a combination treatment with CCG-1423 and a RhoA inhibitor, such as Rhosin HCL, will reduce cell migration in the cells. We used a combination treatment and then analysed U87-generated spheroids to assess the effect of the combination treatment by confocal microscopy. We utilised U87 as a proof of principle here as U87 rapidly migrates in a 3D, collagen-based environment characterised by producing extensive protrusions emanating from the original spheroid. Our preliminary findings indicate that treatment of the cells with a combination of CCG-1423 and Rhosin HCl induced a third shift potentially towards a collective cell migration as indicated by our findings that gap areas within the protrusions were significantly reduced (Figure 4B,C).

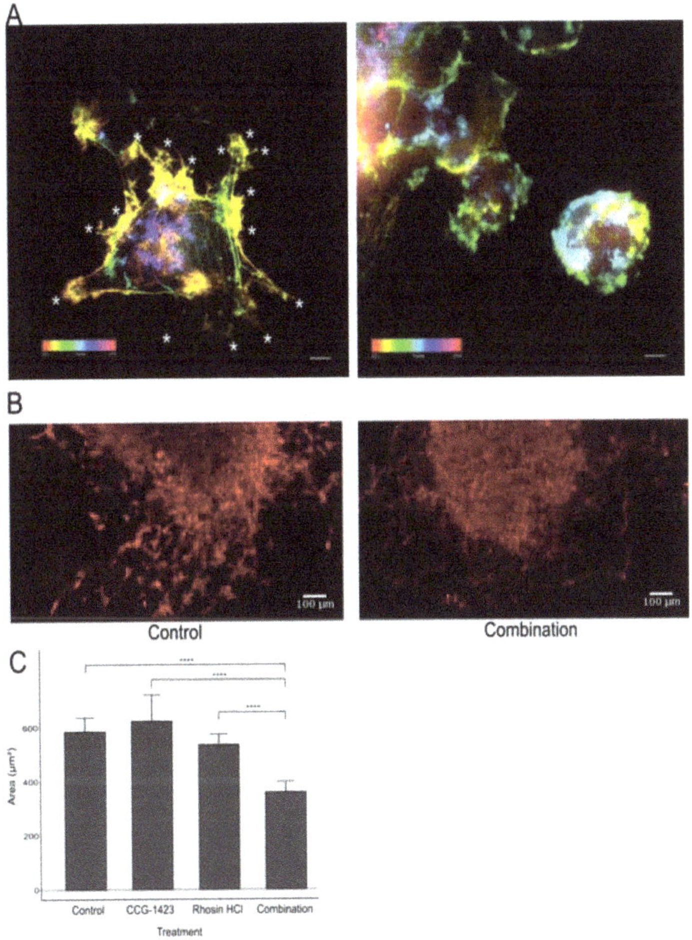

Figure 4. High resolution iSIM imaging and analysis of fixed spheroids and migrating cells within collagen plugs. (**A**) Top image: a single cell in untreated spheroids emanating from the

original spheroid core. Asterisks show filopodia. Right hand image: the elongated shape disappears after treatment with inhibitor and is replaced by a rounded phenotype. Fluorescent label = Alexa Fluor 488 phalloidin. Colour bar: sample depth. Scale bar = 10 µm. (**B**) Confocal microscopy visualizes the effect of drug treatment on U87 spheroids. Images of fixed U87 treated spheroids, in their collagen plugs, taken on a Zeiss LSM 880 with an EC Plan-Neofluar 10× objective lens. Cells express GFP (488 nm) and are labelled with TRITC conjugated phalloidin (594 nm), to highlight the actin cytoskeleton. Migration appears more sheet-like in response to the combination treatment, in comparison to the mesenchymal and amoeboid extensions seen migrating from the control spheroid. Scale bar represents 100 µm. (**C**) Average U87 collagen gap area. U87 cells were treated with CCG-1423 (500 nM), Rhosin HCl (1 µM) or a combination of both. The combination of both inhibitors show a significant ($p < 0.0001$) difference to untreated (Control) and all single inhibitor treatments (Kruskal–Wallis $p = 3.08 \times 10^{-10}$). See Supplementary Figure S2 for quantification illustration. Significant differences were marked in the graphs when available. **** $p \leq 0.0001$.

4. Discussion

CCN1 is known to be overexpressed in a number of different cancer types and to drive invasion [21–24]. Elevated levels of CCN1 in laryngeal squamous cell cancer promoted the induction of epithelial–mesenchymal transition (EMT), thus leading to invasion and metastasis and resulting in a poor prognosis [25]. In addition, CCN1 (as well as other CCN family members) has been shown to be dysregulated in colorectal cancer and to be involved in the initiation and development as well as the promotion of this disease [26]. Furthermore, a study in osteosarcoma tumours determined that silencing CCN1 reduces tumour vascularisation and slows the growth of osteosarcoma cells resulting in a reduction in subsequent lung metastases [27]. These few examples indicate towards diverse actions of CCN1 on cancer progression.

In the current study, CCN1 levels were explored in patient-derived glioma tissue and GBM cell lines. CCN1 levels were elevated both in its secreted form in blood samples and in tumour tissues from glioma patients. In tissues, CCN1 was expressed in the cytoplasm of tumour cells, with strongly stained cells associated with blood vessels. Moreover, CCN1 expression levels were concomitant with high-grade tumours, especially in comparison to expression levels in lower-grade gliomas. This data suggests that CCN1 expression is correlated with tumour aggressiveness, which confirms a previous study that performed a semi-quantitative IHC analysis of gliomas and normal brain samples correlating CCN1 with tumour grade: In 88% of the WHO grade IV samples studied, CCN1 was overexpressed, and expression was positively correlated with the expression of c-Met, a receptor tyrosine kinase involved in proliferation, migration and invasion. When investigated in vivo using a U87 xenograft in mouse models, CCN1 siRNAs significantly inhibited proliferation by 57% compared to the control [11].

Interestingly, in our study levels of secreted CCN1 in the GBM tissue, samples were mirrored in the matched patient blood in comparison to blood samples from healthy volunteers. In addition, samples from patients with recurrent tumours showed a significant increase in CCN1 levels both in tumour tissues as well as in blood samples in comparison to the primary tumour and blood sample. As the observed increases in CCN1 levels are considerable, the use of larger sample sizes and further investigations could prove CCN1 to be a potential blood biomarker for the monitoring of brain tumour patients to predict disease recurrence and help adapt further treatment.

Moreover, improved sensitivity to enhance the detection levels of an ELISA-based screening system may lead to the development of a cost-effective and non-invasive diagnostic tool for advanced early detection of brain tumours, an unmet need in glioma. The molecular subtyping of brain tumours themselves has advanced diagnosis and treatment options; however, a conclusive diagnosis is still largely based on the interpretation of the histopathology of tumour samples, which may be open to human error and bias. The growing identification and use of biomarkers and genomics data will enhance brain tumour diagnosis by providing more detailed information when paired with the traditional histopathology method [28].

The distribution of CCN1 identified in our clinical sample derived data provided the basis for our functional studies. Previous investigations revealed that the small molecule inhibitor CCG-1423 inhibits the production of CCN1, followed by a reduction in cell migration, [11,12]. Unexpectedly, our findings showed seemingly unaffected migration despite CCG-1423 treatment. However, indeed, we observed a striking relocalisation of the co-transcriptional activator MKL1 from a nuclear to a cytoplasmic localisation as a result of CCG-1423 treatment, e.g., in U251 cells. Western blot analysis of U251 cell lysates also uncovered a build-up of MKL1 in the cytoplasm of cells treated with CCG-1423 at 24 and 48 h time points with cells appearing to return to pre-treatment balance after 72 h. This is consistent with the original CCG-1423 inhibition studies conducted by Evelyn et al. (2007) on PC-3 cells, whereby cells recovered following the withdrawal of CCG-1423 [13]. Furthermore, we identified a significant decrease in CCN1 secretion in the supernatants of U251 and U87 cells treated with CCG-1423, confirming the action of the inhibitor. However, there was no apparent overall effect on cell migration in cells treated with CCG-1423, judged from our migration studies. In our previous study, using iSIM imaging and quantification of cell extensions, we showed a direct effect on the frequency of filopodia and major cell protrusions of U251 cells due to incubation with CCG-1423 [15]. We included additional representative iSIM images from this data set, a quantification of the data was reported in this previous publication. U251 cells reacted to the inhibitor with a significant reduction in major protrusions and adopted a more rounded shape, which indicates a transition from mesenchymal to amoeboid migration [29]. These results led us to hypothesise a potential induction of MAT due to CCN1 inhibition as a drug resistance mechanism adopted by glioma cells.

Previous studies suggest that CCN1 secretion results in inhibition of Rac activation leading to reduced mesenchymal migration and loss of focal adhesions with a concomitant increase in rounded cell movement [30]. The extensions of the plasma membrane in lamellipodia of cells are primarily driven by Rac-activated actin polymerisation. An essential part of migration involving lamellipodia is integrin-mediated adhesion, as this maintains activated Rac in a positive feedback loop whereby integrins at the leading edge of a cell stimulate Rac activation [31]. Rac recruits activated integrins to the leading edge to promote cell migration [32]. Among these is the integrin $\alpha v\beta 3$, whose expression is in turn stimulated by CCN1 [33]. By inhibiting CCN1, integrin $\alpha v\beta 3$ will not be activated and therefore, will not be recruited by Rac to stimulate the positive feedback loop and promote elongated cell migration. Because of this reduced Rac activation, rounded motility of cells will be elevated via a bleb-driven cell migration, such as amoeboid migration. This motility type correlates with a high level of active RhoA/ROCK signalling and is driven by cortical actomyosin contractility [34]. This migratory activity is often seen in vivo and in low-adhesion in vitro systems. Cells have the ability to readily transition between bleb-based migration (amoeboid) and lamellipodia-based migration (mesenchymal) in vivo in order to adapt to their surroundings as they migrate [35].

CCG-1423 activity targets events downstream of the RhoA signalling pathway but upstream of CCN1 production and therefore Rac activation. Not only does this explain the reduction in CCN1 cell secretion but also the changes in phenotype from elongated to rounded. This correlates with the concept that CCG-1423 is inducing MAT in these cells in vitro, allowing them to continue to migrate after treatment with CCG-1423 by amoeboid migration. We were able to observe a distinct loss of major protrusions/lamellipodia in treated cells compared to the control cells, which was concurrent with an increase in the number of filopodia in treated cells compared to control cells, a striking feature of MAT (Figure 4A). These results support a direct effect on actin dynamics and cell morphology indicative of MAT when cells are treated with CCG-1423. Anti-migratory combination treatments to prevent tumour cells from migrating to healthy parts of the brain will need to be developed to circumvent a potential migratory switch from one phenotype to another.

Finally, based on findings by Butler [36], that a combination treatment of U87 with CCG-1423 and an inhibitor targeting RhoA-driven migration reduced cell migration in 2D and 3D, we investigated this effect by confocal microscopy. In a preliminary experiment we treated U87 tumour spheroids with either CCG-1423, or the RhoA inhibitor Rhosin HCl, or a combination of both inhibitors, which revealed a strong effect of both inhibitors in combination on cell invasion into collagen. In addition, by confocal microscopy, for the first time, we were able to uncover that the combination of both inhibitors leads to a more sheet-like collective migration compared to the loose single-cell migration in control or single inhibitor groups. Cancer cells have been reported to adopt this type of cell migration highlighting great plasticity in migratory patterns [37]. This apparent ability of glioma cells to respond to extracellular challenges and adopt to continuous cell migration and invasion requires further detailed investigation including additional in vitro methods to characterise RhoA activation and MMP activity as reported by Chikina and Alexandrova (2018) [38]. In addition, the downstream events regulated by MKL1 activity should also be further investigated.

5. Conclusions

The studies presented here provide novel and compelling evidence that glioma cells can undergo MAT in response to pharmacological intervention intended to prevent migration, resulting in continued migration and invasion, which supports the notion of combination treatments with migrastatic inhibitors. This also highlighted the importance of moving cell migration research into a 3D setting to obtain more accurate drug efficacy results in a more tumour-relevant environment. In addition, our study has uncovered an important role for CCN1 in disease progression supporting CCN1 as a potential biomarker for GBM, which could help disease monitoring and improve survival rates of this devastating disease.

Supplementary Materials: The following are available online at https://www.mdpi.com/article/10.3390/biomedicines10010009/s1, Figure S1: Western blot analysis of the effects of CCG-1423 on glioma cell lines; Figure S2: Quantification illustration of cell gap area measurement.

Author Contributions: Conceptualization, S.E.K., A.B.-R., S.E.L. and J.E.L.; methodology, S.E.K., F.O.G.-E. and M.O.N.; software, A.R.; validation, S.E.K.; formal analysis, S.E.K., A.R. and S.P.; investigation, S.E.K.; resources, S.E.L.; data curation, S.E.K. and A.R.; writing—original draft preparation, S.E.K.; writing—review and editing, S.E.K., A.B.-R., S.E.L., S.C.S., J.E.L. and S.K.; visualization, S.E.K., A.R. and S.K.; supervision, A.B.-R., S.C.S. and J.E.L.; project administration, S.E.K.; funding acquisition, S.C.S. and A.B.-R. All authors have read and agreed to the published version of the manuscript.

Funding: This research was funded by The PPR Foundation, The Musella Foundation, and the North American Foundation for the University of Leeds.

Institutional Review Board Statement: The study was conducted according to the guidelines of the Declaration of Helsinki and approved by the Institutional Review Board (or Ethics Committee) of the University of Leeds. Patient samples were collected, processed, and stored by the Leeds Multidisciplinary Routine Tissue Banking service from GBM patients undergoing surgery at the Leeds General Infirmary (ethical approval no. RTB 15/YH/0080).

Informed Consent Statement: All patients gave written informed consent.

Data Availability Statement: The data presented in this study are available on request from the corresponding author.

Acknowledgments: We would like to acknowledge and thank FRAME (https://frame.org.uk/, accessed on 15 December 2021) for granting Sally Prior a summer scholarship (summer 2021) to allow her to gain expertise in the skills needed to contribute to this publication.

Conflicts of Interest: The authors declare no conflict of interest. The funders had no role in the design of the study; in the collection, analyses, or interpretation of data; in the writing of the manuscript, or in the decision to publish the results.

References

1. Glioblastoma Multiforme Treatment & Management. Medical Care, Surgical Care, Consultations. 1 June 2016. Available online: http://emedicine.medscape.com/article/283252-treatment (accessed on 6 July 2016).
2. Preusser, M.; de Ribaupierre, S.; Wöhrer, A.; Erridge, S.C.; Hegi, M.; Weller, M. Current concepts and management of glioblastoma. *Ann. Neurol.* **2011**, *70*, 9–21. [CrossRef] [PubMed]
3. Lin, B.-R.; Chang, C.-C.; Chen, L.-R.; Wu, M.-H.; Wang, M.-Y.; Kuo, I.-H. Cysteine-Rich 61 (CCN1) Enhances Chemotactic Migration, Transendothelial Cell Migration, and Intravasation by Concomitantly Up-Regulating Chemokine Receptor 1 and 2. *Mol. Cancer Res.* **2007**, *5*, 1111–1123. [CrossRef]
4. Xie, L.; Song, X.; Lin, H.; Chen, Z.; Li, Q.; Guo, T.; Xu, T.; Su, T.; Xu, M.; Chang, X.; et al. Aberrant activation of CYR61 enhancers in colorectal cancer development. *J. Exp. Clin. Cancer Res.* **2019**, *38*, 213. [CrossRef]
5. Li, Z.Q.; Wu, W.R.; Zhao, C.; Zhao, C.; Zhang, X.L.; Yang, Z.; Pan, J.; Si, W.K. CCN1/Cyr61 enhances the function of hepatic stellate cells in promoting the progression of hepatocellular carcinoma. *Int. J. Mol. Med.* **2018**, *41*, 1518–1528. [CrossRef] [PubMed]
6. Lau, L.F. CCN1/CYR61: The Very Model of a Modern Matricellular Protein. *Cell. Mol. Life Sci.* **2011**, *68*, 3149–3163. [CrossRef]
7. Vial, E.; Sahai, E.; Marshall, C.J. ERK-MAPK signaling coordinately regulates activity of Rac1 and RhoA for tumor cell motility. *Cancer Cell.* **2003**, *4*, 67–79. [CrossRef]
8. Frantz, C.; Stewart, K.M.; Weaver, V.M. The extracellular matrix at a glance. *J. Cell Sci.* **2010**, *123*, 4195–4200. [CrossRef] [PubMed]
9. Kassis, J.N.; Virador, V.M.; Guancial, E.A.; Kimm, D.; Ho, A.S.; Mishra, M. Genomic and phenotypic analysis reveals a key role for CCN1 (CYR61) in BAG3—modulated adhesion and invasion. *J. Pathol.* **2009**, *218*, 495–504. [CrossRef]
10. Babic, A.M.; Kireeva, M.L.; Kolesnikova, T.V.; Lau, L.F. CYR61, a product of a growth factor-inducible immediate early gene, promotes angiogenesis and tumor growth. *Proc. Natl. Acad. Sci. USA* **1998**, *95*, 6355–6360. [CrossRef] [PubMed]
11. Goodwin, C.R.; Lal, B.; Zhou, X.; Ho, S.; Xia, S.; Taeger, A. Cyr61 mediates HGF-dependent tumor cell growth, migration and Akt activation. *Cancer Res.* **2010**, *70*, 2932–2941. [CrossRef]
12. Xie, D.; Yin, D.; Tong, X.; O'Kelly, J.; Mori, A.; Miller, C. Cyr61 Is Overexpressed in Gliomas and Involved in Integrin-Linked Kinase-Mediated Akt and β-Catenin-TCF/Lef Signaling Pathways. *Cancer Res.* **2004**, *64*, 1987–1996. [CrossRef]
13. Evelyn, C.R.; Wade, S.M.; Wang, Q.; Wu, M.; Iñiguez-Lluhí, J.A.; Merajver, S.D. CCG-1423: A small-molecule inhibitor of RhoA transcriptional signaling. *Mol. Cancer Ther.* **2007**, *6*, 2249–2260. [CrossRef] [PubMed]
14. Hayashi, K.; Watanabe, B.; Nakagawa, Y.; Minami, S.; Morita, T. RPEL Proteins Are the Molecular Targets for CCG-1423, an Inhibitor of Rho Signaling. *PLoS ONE* **2014**, *9*, e89016. [CrossRef] [PubMed]
15. Ketchen, S.; Rohwedder, A.; Knipp, S.; Esteves, F.; Struve, N.; Peckham, M. A novel workflow for three-dimensional analysis of tumour cell migration. *Interface Focus.* **2020**, *10*, 20190070. [CrossRef] [PubMed]
16. Ricklefs, F.L.; Alayo, Q.; Krenzlin, H. Immune evasion mediated by PD-L1 on glioblastoma-derived extracellular vesicles. *Sci. Adv.* **2018**, *4*, eaar2766. [CrossRef]
17. Cheng, V.; Esteves, F.; Chakrabarty, A.; Cockle, J.; Short, S.; Brüning-Richardson, A. High-content analysis of tumour cell invasion in three-dimensional spheroid assays. *Oncoscience* **2015**, *2*, 596–606. [CrossRef]
18. Lin, C.-M.; Yu, C.-F.; Huang, H.-Y.; Chen, F.-H.; Hong, J.-H.; Chiang, C.-S. Distinct Tumor Microenvironment at Tumor Edge as a Result of Astrocyte Activation Is Associated with Therapeutic Resistance for Brain Tumor. *Front Oncol.* **2019**, *9*, 307. Available online: https://www.frontiersin.org/articles/10.3389/fonc.2019.00307/full (accessed on 15 July 2019). [CrossRef]
19. Williams, S.P.; Nowicki, M.O.; Liu, F.; Press, R.; Godlewski, J.; Abdel-Rasoul, M. Indirubins Decrease Glioma Invasion by Blocking Migratory Phenotypes in Both the Tumor and Stromal Endothelial Cell Compartments. *Cancer Res.* **2011**, *71*, 5374–5380. [CrossRef]
20. Nowicki, M.O.; Dmitrieva, N.; Stein, A.M.; Cutter, J.L.; Godlewski, J.; Saeki, Y. Lithium inhibits invasion of glioma cells; possible involvement of glycogen synthase kinase-3. *Neuro Oncol.* **2008**, *10*, 690–699. [CrossRef]
21. Sun, Z.-J.; Wang, Y.; Cai, Z.; Chen, P.-P.; Tong, X.-J.; Xie, D. Involvement of Cyr61 in growth, migration, and metastasis of prostate cancer cells. *Br. J Cancer.* **2008**, *99*, 1656–1667. [CrossRef]
22. Gery, S.; Xie, D.; Yin, D.; Gabra, H.; Miller, C.; Wang, H. Ovarian Carcinomas: CCN Genes Are Aberrantly Expressed and CCN1 Promotes Proliferation of these Cells. *Clin. Cancer Res.* **2005**, *11*, 7243–7254. [CrossRef] [PubMed]
23. Xie, D.; Miller, C.W.; O'Kelly, J.; Nakachi, K.; Sakashita, A.; Said, J.W. Breast cancer cyr61 is overexpressed, estrogen-inducible, and associated with more advanced disease. *J. Biol. Chem.* **2001**, *276*, 14187–14194. [CrossRef]
24. Kok, S.-H.; Chang, H.-H.; Tsai, J.-Y.; Hung, H.-C.; Lin, C.-Y.; Chiang, C.-P. Expression of Cyr61 (CCN1) in human oral squamous cell carcinoma: An independent marker for poor prognosis. *Head Neck* **2010**, *32*, 1665–1673. [CrossRef] [PubMed]
25. Liu, Y.; Zhou, Y.-D.; Xiao, Y.-L.; Li, M.-H.; Wang, Y.; Kan, X. Cyr61/CCN1 overexpression induces epithelial-mesenchymal transition leading to laryngeal tumor invasion and metastasis and poor prognosis. *Asian Pac. J. Cancer Prev.* **2015**, *16*, 2659–2664. [CrossRef]
26. Chang, C.-C.; Lin, B.-R.; Wu, T.-S.; Jeng, Y.-M.; Kuo, M.-L. Input of microenvironmental regulation on colorectal cancer: Role of the CCN family. *World J. Gastroenterol.* **2014**, *20*, 6826–6831. [CrossRef] [PubMed]
27. Habel, N.; Vilalta, M.; Bawa, O.; Opolon, P.; Blanco, J.; Fromigué, O. Cyr61 silencing reduces vascularization and dissemination of osteosarcoma tumors. *Oncogene* **2015**, *34*, 3207–3213. [CrossRef]
28. Alcantara Llaguno, S.R.; Parada, L.F. Cell of origin of glioma: Biological and clinical implications. *BJC* **2016**, *115*, 1445–1450. [CrossRef]

29. Bergert, M.; Chandradoss, S.D.; Desai, R.A.; Paluch, E. Cell mechanics control rapid transitions between blebs and lamellipodia during migration. *Proc. Natl. Acad. Sci. USA* **2012**, *109*, 14434–14439. [CrossRef]
30. Wang, X.; Deng, Y.; Mao, Z.; Ma, X.; Fan, X.; Cui, L. CCN1 promotes tumorigenicity through Rac1/Akt/NF-κB signaling pathway in pancreatic cancer. *Tumor Biol.* **2012**, *33*, 1745–1758. [CrossRef]
31. Lawson, C.D.; Burridge, K. The on-off relationship of Rho and Rac during integrin-mediated adhesion and cell migration. *Small GTPases* **2014**, *5*, e27958. [CrossRef]
32. Kiosses, W.B.; Shattil, S.J.; Pampori, N.; Schwartz, M.A. Rac recruits high-affinity integrin αvβ3 to lamellipodia in endothelial cell migration. *Nat. Cell Biol.* **2001**, *3*, 316–320. [CrossRef]
33. Vellon, L.; Menendez, J.A.; Lupu, R. αVβ3 integrin regulates heregulin (HRG)-induced cell proliferation and survival in breast cancer. *Oncogene* **2005**, *24*, 3759–3773. [CrossRef]
34. Ruprecht, V.; Wieser, S.; Callan-Jones, A.; Smutny, M.; Morita, H.; Sako, K. Cortical Contractility Triggers a Stochastic Switch to Fast Amoeboid Cell Motility. *Cell* **2015**, *160*, 673–685. [CrossRef]
35. Charras, G.; Paluch, E. Blebs lead the way: How to migrate without lamellipodia. *Nat. Rev. Mol. Cell Biol.* **2008**, *9*, 730. [CrossRef] [PubMed]
36. Butler, H. Combination treatment with migrastatic inhibitors to target brain tumour spread. *Fields J. Hudders. Stud. Res.* **2001**, *7*. [CrossRef]
37. Jia-Shun, W.; Jiang, J.; Chen, B.J.; Wang, K.; Tang Yl Liang, X.H. Plasticity of cancer cell invasion: Patterns and mechanisms. *Transl. Oncol.* **2021**, *14*. [CrossRef]
38. Chikina, A.S.; Alexandrova, A.Y. An In Vitro System to Study the Mesenchymal-to-Amoeboid Transition. In *Methods in Molecular Biology*; Cell Migration; Gautreau, A., Ed.; Humana Press: New York, NY, USA, 2018; Volume 1749. [CrossRef]

biomedicines

MDPI

Article

Anticancer Effects with Molecular Docking Confirmation of Newly Synthesized Isatin Sulfonamide Molecular Hybrid Derivatives against Hepatic Cancer Cell Lines

Mahmoud Eldeeb [1,†], Eman F. Sanad [2,†], Ahmed Ragab [3], Yousry A. Ammar [3], Khaled Mahmoud [4], Mamdouh M. Ali [1] and Nadia M. Hamdy [2,*]

1 Department of Biochemistry, Biotechnology Research Institute, National Research Centre, 12622 Giza, Egypt; me.eldeeb@nrc.sci.eg (M.E.); dr-mamdouh-moawad@nrc.sci.eg (M.M.A.)
2 Biochemistry and Molecular Biology Department, Faculty of Pharmacy, Ain Shams University, 11566 Cairo, Egypt; dr.emansanad@pharma.asu.edu.eg
3 Department of Chemistry, Faculty of Science (for Boys, Cairo Branch), Al-Azhar University, 11884 Cairo, Egypt; ahmed_ragab@azhar.edu.eg (A.R.); yossry@azhar.edu.eg (Y.A.A.)
4 Department of Pharmacognosy, Pharmaceutical and Drug Industries Research Institute, National Research Centre, 12622 Giza, Egypt; km.hanafi@nrc.sci.eg
* Correspondence: nadia_hamdy@pharma.asu.edu.eg
† These authors contributed equally to this work.

Abstract: The current study investigated the cytotoxic effect of ten sulfonamide-derived isatins, following molecular hybridization, based on the association principles, on hepatocellular carcinoma (HCC) HepG2 and Huh7 cell lines, compared for safety using human normal retina pigmented epithelial (RPE-1) cells. The ten compounds showed variable in vitro cytotoxicity on HepG2 and Huh7 cells, using the MTT assay. Four compounds (4/10) were highly cytotoxic to both HepG2 and HuH7. However, only 3 of these 4 were of the highest safety margin on RPE-1 cells in vitro and in the *in vivo* acute (14-day) oral toxicity study. These later, superior three compounds' structures are 3-hydroxy-3-(2-oxo-2-(p-tolyl)ethyl)-5-(piperidin-1-ylsulfonyl)indolin-2-one (**3a**), N-(4-(2-(2-oxo-5-(piperidin-1-ylsulfonyl)indolin-3-ylidene)acetyl)phenyl)acetamide (**4b**), and N-(3-(2-(2-oxo-5-(piperidin-1-ylsulfonyl)indolin-3-ylidene)acetyl)phenyl)acetamide (**4c**). The half-maximal inhibitory concentration (IC50) of the tested compounds (**3a**, **4b**, and **4c**) on HepG2 cells were approximately 16.8, 44.7, and 39.7 μM, respectively. The **3a**, **4b**, and **4c** compounds significantly decreased the angiogenic marker epithelial growth factor receptor (EGFR) level and that was further confirmed via molecular docking inside the EFGR active site (PDB: 1M17). The binding free energies ranged between −19.21 and −21.74 Kcal/mol compared to Erlotinib (−25.65 Kcal/mol). The most promising compounds, **3a**, **4b**, and **4c**, showed variable anticancer potential on "hallmarks of cancer", significant cytotoxicity, and apoptotic anti-angiogenic and anti-invasive effects, manifested as suppression of Bcl-2, urokinase plasminogen activation, and heparanase expression in HepG2-treated cells' lysate, compared to non-treated HepG2 cells. In conclusion, compound "**3a**" is highly comparable to doxorubicin regarding cell cycle arrest at G2/M, the pre-G0 phases and early and late apoptosis induction and is comparable to Erlotinib regarding binding to EGFR active site. Therefore, the current study could suggest that compound "**3a**" is, hopefully, the most safe and active synthesized isatin sulfonamide derivative for HCC management.

Keywords: apoptosis; HepG2; Huh7; isatin sulfonamides; angiogenesis; invasion; cancer hallmarks; molecular docking; EGFR tyrosine kinase inhibitor

Citation: Eldeeb, M.; Sanad, E.F.; Ragab, A.; Ammar, Y.A.; Mahmoud, K.; Ali, M.M.; Hamdy, N.M. Anticancer Effects with Molecular Docking Confirmation of Newly Synthesized Isatin Sulfonamide Molecular Hybrid Derivatives against Hepatic Cancer Cell Lines. *Biomedicines* **2022**, *10*, 722. https://doi.org/10.3390/biomedicines10030722

Academic Editor: Simon J Allison

Received: 5 February 2022
Accepted: 18 March 2022
Published: 20 March 2022

Publisher's Note: MDPI stays neutral with regard to jurisdictional claims in published maps and institutional affiliations.

Copyright: © 2022 by the authors. Licensee MDPI, Basel, Switzerland. This article is an open access article distributed under the terms and conditions of the Creative Commons Attribution (CC BY) license (https://creativecommons.org/licenses/by/4.0/).

1. Introduction

"Hallmarks of cancer" include the programmed molecular cell death mechanism "Apoptosis" [1]. Apoptosis is related to other cancer hallmarks such as progression and metastasis [1–3]. Therefore, it is important to design more drug moieties targeting apoptosis and/or tumor invasiveness.

Isatin (1H-indole-2,3-dione) is a well-known natural product found in many plants which has also been a common scaffold in various anticancer drugs [4,5]. Many isatin derivatives display diverse pharmacological activities, including anti-viral, anti-convulsant, anti-bacterial, and anti-fungal activities [6–8]. Because of its unique size and privileged electronic properties, there is considerable interest in pharmaceutical drug development and the biochemical pharmacology of isatin anticancer drug derivatives [9]. The isatin scaffold is incorporated in many synthetic anti-cancer drugs such as Sunitinib V maleate and Toceranib phosphate [10]. These inhibitors with isatin moiety have shown outstanding anti-cancer effects in clinical trials, including multi-receptor tyrosine kinase inhibitor activity [11]. Therefore, anti-tumor drugs containing isatin scaffolds could possess a broader spectrum of cytotoxicity against cancer cells, namely, apoptotic, anti-proliferative, and anti-migratory effects, collectively, anti-tumor invasiveness.

The α, β-unsaturated ketone (Michael Acceptor) pharmacophore is presented as a common skeleton in many naturally occurring agents including isatin [10,12]. Therefore, incorporating a "Michael Acceptor" into designed molecules results in products with improved cytotoxicity [10,13]. Hence, "α, β-unsaturated ketone" can be considered as a cytotoxic drug design "a functionality structure".

Molecular hybrids are one of the most popular strategies to develop new anticancer agents, based on combining structural features of two different active fragments. This reduces the risk of drug–drug interactions and improves the biochemical/molecular outcome(s) [14,15]. During the last decade, many isatin-based hybrids have been developed as promising anti-cancer agents, including Erlotinib, Vandetanib, Olaparib, and others [16]. Several published datasets have reported that isatin derivatives have potential toward different biological targets such as tubulin, tyrosine kinase, and histone deacetylase, leading to apoptosis and, moreover, influencing apoptosis-related gene expression [17–20]. Interestingly, incorporation of sulfonamide moiety to a benzene ring of isatin showed increased anti-tumor activities with a potential inhibitory effect against epidermal growth factor receptors (EGFRs) [21,22].

This study aimed to characterize ten novel synthesized isatin sulfonamide-molecular-hybrid derivatives to target EGFRs and evaluate their in vitro antiproliferative activities. Based on the previous promising anti-tumor activities of some morpholinosulfonyl isatin derivatives on HepG2 [22,23], we tested their anti-cancer activities against two hepatocellular carcinoma (HCC) cell lines: the HepG2 and Huh7 cell lines, which are different in their drug-metabolizing activities as well as their p53 expression. To ensure the safety of the tested compounds, the most active derivatives were tested in vitro against normal human cell line, and an in vivo acute oral toxicity study was performed. The most active and safest synthesized compounds have been evaluated for their possible apoptotic effect, cellular cytotoxicity, and whether they cause angiogenesis arrest or invasion inhibition. The cell death mechanism underlying this activity was investigated via cell cycle analysis and apoptotic studies. Doxorubicin (Dox) was used as an internationally accepted cytotoxic agent to be the positive control to compare the cytotoxic potential of the tested drugs. In addition, inhibitory mechanism confirmation using molecular docking (MD) was carried out to study the interaction of promising compounds with EGFRs.

2. Materials and Methods

2.1. Biochemical Reagents, Chemicals, Solvents

Unless otherwise specified, chemicals and reagents were purchased from Sigma-Aldrich Chemical Co. (St. Louis, MO, USA) and were of analytical grade. Dimethyl sulfoxide (DMSO) was used as the vehicle for isatin sulfonamide derivatives per their molecular wt. Dox was used as a small anti-cancer cytotoxic reference drug (positive control), Erlotinib as reference EGFR inhibitor. Other chemicals and solvents such as xylene, paraffin, trypsin, and alcohol were of the highest grade commercially available.

2.2. Biochemical/Molecular Assay Kits

MTT assay kit and propidium iodide (PI) staining FC assay kit was purchased from Abcam (Boston, MA, USA). AST, ALT, and bilirubin (total and direct) kits and creatinine, urea kits, and GSH kit were all purchased from Biodiagnostics Company (Cairo, Egypt).

Heparanase ELISA kit (Abcam, Cambridge, UK), Annexin V/PI double staining kit (eBiosciencesDx), EGFR 96-well plate kit (Abcam, Cambridge, UK), uPA ELISA kit (Creative Diagnostics, New York, NY, USA), and BcL-2 ELISA kit (Zymed Laboratories, Carlsbad, CA, USA) were used.

2.3. Cell Lines

The cell lines used were obtained from ATCC including Hep G2 [HEPG2] (ATCC HB-8065), Huh7 cell line-615, and the normal retina pigmented epithelial cells immortalized with hTERT; hTERT RPE-1 (ATCC CRL-4000). Handling procedure for flask cultures and subculturing, media (DMEM, bovine serum albumin), and incubations were all performed according to www.ATCC.org (accessed on April 2021).

2.4. Animals

5–7-week-old mice (weighing 15–20 g, females) were obtained from Nile Co. for Pharmaceutical and Chemical Industries (Egypt). Mice were acclimatized for 1 week under standard laboratory conditions in cages in a room with 12 h light/dark cycles, at 25 °C ± 2 °C and 55% ± 5% relative humidity. Mice were fed on standard diet pellets (El Nasr Company for Intermediate Chemicals, Giza, Egypt) containing no less than 20% protein, 3.5% fat, 6.5% ash, 5% fiber, and a vitamin mixture. Animals were allowed free access to drinking water ad libitum.

2.5. Isatin Sulfonamide Derivative Synthesis

In this study, motivated by the structural features of isatin and α, β-unsaturated ketone, we proposed combining these two bioactive scaffolds into a single chemical entity. We centered on molecular hybridization strategy [23] to design and synthesize a novel series of isatin sulfonamide derivatives carrying the α, β-unsaturated ketone scaffold.

Quantitative Structure–Activity Relationships (QSAR) model was utilized to get a more profound understanding of the molecular description of compounds' activities. Few QSAR models that clarify the anti-cancer activity of isatin analogues were reported [24].

2.6. Chemical Synthetic Pathway

The routes adopted to synthesize isatin sulfonamide derivative compounds were as described in Figure 1. The parent molecule is 5-(piperidin-1-ylsulfonyl) indoline-2, 3-dione. This compound was derivatized into three compounds: **2a-b**, **3a-d**, and **4a-d**. These newly designed compounds were structurally recognized and affirmed by using different spectroscopic analysis as IR, ^{1}H NMR, and ^{13}C NMR spectra as described in organic synthesis literature article [25]. Additionally, all the spectra investigations (in the Supplementary File) and were in great conformity with the proposed structures. These compounds were prepared by the Organic Chemistry authors in the Organic Chemistry Lab, Faculty of Science (Boys-Branch), Al Azhar University, Cairo, Egypt. Then, these three compounds were derivatized to produce ten final isatin sulfonamide-derivative compounds to be tested further by various biochemical/molecular assays.

2.7. Cancer Cell Proliferation Inhibition

MTT-based cell viability assay was carried out to evaluate the antiproliferation potential of the target compounds on HCC cell lines (HepG2 and Huh7). The half maximal inhibitory concentration (IC50) values were calculated against DMSO non-treated HepG2 cells (negative control) and Dox-treated HepG2 cells (positive control). The assay was conducted using 3-(4, 5-di-MethylThiazol-2-yl)-2,5-diphenyl Tetrazolium bromide (MTT) to test cell viability, according to the method described by Mosmann [26] with minor changes [27]. The yellow MTT is reduced in the viable cells to purple formazan. The insoluble formazan is solubilized by DMSO to colored solution measured spectrophotometrically. Briefly, cells were seeded in a 96-well plate at density of 1×10^4 cells/well for 24 h in 5% CO_2 incubator at 37 °C to allow their adherence. Then, cells were incubated for 48 h with different concentrations of test compounds ranging from 100 to 1.56 μM in 5% CO_2 incubator at 37 °C. After incubation, 20 μL of a 5 mg/mL MTT solution were mixed with the contents of each well, and the plate was incubated for another 4 h. Then, the medium was suctioned, and the wells were washed with PBS. After 2 h of drying, 200 μL of DMSO was added to each well. The plate was set on a shaker to dissolve the formazan crystals. Then, the absorbance was measured spectrophotometrically at 570 nm with a reference wavelength of 630 nm using ELX800 UV universal microplate reader (BioTek Instruments Inc., VT, Santa Clara, CA, USA).

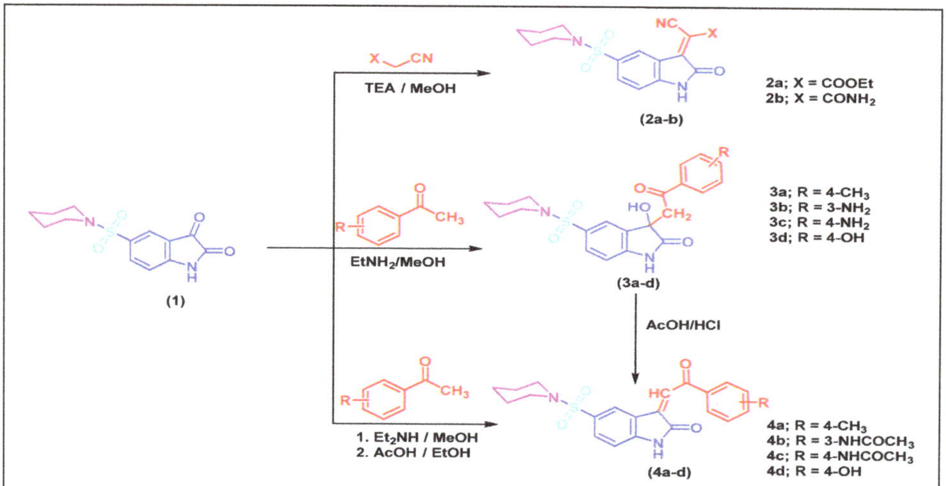

[isatin derivatives (1), triethyl amine (TEA)]

Chemical Names	**Compound no.**
Ethyl I-2-cyano-2-(2-oxo-5-(piperidin-1-ylsulfonyl) indolin3-ylidene) acetate	**(2a)**
2-Cyano-2-(2-oxo-5-(piperidin-1-ylsulfonyl) indolin-3-ylidene) acetamide	**(2b)**
3-Hydroxy-3-(2-oxo-2-(p-tolyl) ethyl)-5-(piperidin-1-ylsulfonyl) indolin-2-one	**(3a)**
3-(2-(3-Aminophenyl)-2-oxoethyl)-3-hydroxy-5-(piperidin-1-ylsulfonyl) indolin-2-one	**(3b)**
3-(2-(4-Aminophenyl)-2-oxoethyl)-3-hydroxy-5-(piperidin-1-ylsulfonyl) indolin-2-one	**(3c)**
3-hydroxy-3-(2-(4-hydroxyphenyl)-2-oxoethyl)-5-(piperidin-1-ylsulfonyl) indolin-2-one	**(3d)**
3-(2-Oxo-2-(p-tolyl) ethylidene)-5-(piperidin-1-ylsulfonyl) indolin-2-one	**(4a)**
N-(4-(2-(2-Oxo-5-(piperidin-1-ylsulfonyl) indolin-3-ylidene) acetyl) phenyl) acetamide	**(4b)**
N-(3-(2-(2-Oxo-5-(piperidin-1-ylsulfonyl) indolin-3-ylidene) acetyl) phenyl) acetamide	**(4c)**
3-(2-(4-Hydroxyphenyl)-2-oxoethylidene)-5-(piperidin-1-ylsulfonyl) indolin-2-one	**(4d)**

Figure 1. Synthetic routes scheme for arylidine and 3-hydroxy-3-substituted isatin sulfonamide and 2-oxindole as well as 3-phenacylidene-2-indolinone derivatives with their chemical names.

All experiment conditions were tested in three replicates. The % cell viability was calculated from the following equation: (mean absorbance of treated sample/mean absorbance of negative control sample) × 100. The % death rate = 100 -% cell viability, and the inhibitory concentration of 50% (IC50) was measured from the exponential curve of viability against concentration (dose–response curve) [28].

2.8. *In Vitro Safety/Cell Viability Assay towards Human Healthy RPE1 Cell Line*

Toxicity of the most active six isatin sulfonamide derivatives (**2a**, **3a**, **4a**, **4b**, **4c**, **4d**) were evaluated using noncancerous cells (RPE1) to examine the safety of these newly synthesized isatin sulfonamide derivatives towards normal human cell line, again using the MTT test.

2.9. *In Vivo Acute Oral Toxicity Study*

GPower 3.0.10 software was used to estimate the sample size needed for the experiment and the number of mice per group but, according to the Organization for Economic Cooperation and Development (OECD) Test Guidelines No. 407 (OECD, 2008 and updated 2017) [29] and under the International Academy of Science's Guide for the Care and Use of Laboratory Animals, as well as the ARRIVE guidelines, to use the least number of mice per group for the acute single dose oral toxicity test.

2.10. *Ethics Statement*

Animal care and all experimental protocols were approved and conducted in accordance with the ethical guidelines approved by the Institutional Review Board of the Faculty of Pharmacy, Ain Shams University, Cairo, Egypt (Ethics Committee Approval, 10/2019/1).

2.11. *Experimental Design*

Animals were divided randomly into nine groups (5 animals per group). Before the experiment, the body weight (BW) of all mice was recorded. All mice were made to fast for 24 h and treated once (day zero) then carefully monitored for 14 days.

Normal control group: mice received saline orally once;

Negative control group: mice received vehicle (DMSO) orally once;

Positive control group: mice were injected with Dox (15 mg/kg BW, i.p.) once [30,31].

Isatin-sulfonamide-derivative-compound-treated groups (**2a**, **3a**, **3d** and **4b**, **4c** and **4d**) of mice received 250 mg/kg BW of each orally once [29,32].

After single oral dose, animals were observed for the first 30 min and 4–5 times at intervals of 48 h to record any signs of abnormality. The animals' BWs were recorded at the end of the 14 days of observation. Blood samples were collected from the retro-orbital plexus and allowed to clot. Sera were prepared by centrifugation at 4000 rpm for 15 min and then kept frozen at − 80 °C for liver and kidney function tests (LFTs and KFTs). Thereafter, mice were deprived of food overnight, euthanized, and sacrificed by cervical dislocation. Liver and kidney tissues were collected, washed with ice-cold saline, and weighed. Excised weighed livers and kidneys were fixed in a suitable buffer for histological examination.

N.B. the objective of the acute toxicity study was not the determination of LD50 values.

2.12. *Histopathological Examination*

Liver and kidney tissues were fixed in 10% neutral buffered formaldehyde overnight and then embedded in paraffin, deparaffinized in serial grades of alcohols, cleared in xylene, and were subjected to ultramicrotomy, where 4 µm thick tissue sections were cut by rotatory microtome. Tissue sections were stained with hematoxylin and eosin (H and E) for histological examination using full HD light microscopic imaging system (Leica Microsystems GmbH, Wetzlar, Germany). Standard procedures for sample fixation and staining were done as previously described [33,34].

2.13. *Flow Cytometric Analysis and Apoptotic Studies*

Targeting cell cycle and induction of apoptosis are promising strategies to develop potential anticancer agents. Therefore, the 3 promising isatin sulfonamide derivatives were investigated for their apoptotic activities.

2.13.1. Cell Cycle Analysis

In 6-well plate, HepG2 (1×10^5 cells per well) were treated with either (0.1% DMSO) vehicle or 5 µM of **3a**, **4b**, and **4c** compounds. After 48 h incubation, cells were harvested and fixed for 12 h using ice-cold 70% ethanol at 4 °C. Ethanol was removed and cells were washed by PBS and incubated for 30 min at 37 °C in 0.5 mL of PBS containing 1 mg/mL RNase. Then, cells were stained for 30 min with PI in the dark and the DNA content was determined using flow cytometer [35].

2.13.2. Annexin V/FITC Apoptosis Assay

To apprehend the mechanism of the chosen newly synthesized isatin sulfonamide-derivative compounds and to recommend further in vivo testing as new chemotherapeutics, HepG2 cells were incubated with IC50 dose of compounds **3a**, **3b**, and **3c**, against the positive chemotherapeutic control, Dox. Annexin V/PI double staining kit (eBiosciencesDx) was used to detect apoptotic activity of the selected isatin sulfonamide derivatives. The Annexin V corresponding signal provides a very sensitive marker for cellular apoptosis, while PI was used to detect necrotic or late apoptotic cells, identified by the loss of plasma integrity and loss of nuclear membranes.

Briefly, 1×10^5 cells/mL of HepG2 cells were seeded into a 6-well plate and incubated at 37 °C, 5% CO_2, overnight. Cells were treated with either 0.1% DMSO or IC50 of the three selected compounds for 48 h (50 µL from each compound solution in DMSO; IC50 dose) in triplicate. After time elapsed, trypsinization of cells was carried out. Washing was done by PBS once, followed by fixation using 70% ethanol. Fixed cells were stored at 4 °C for 2 h, then centrifuged at 500× *g* for 5 min. Again, cells were washed with PBS prior to staining. An amount of 200 µL of the staining solution (annexin V- fluorescein 5-isothiocyanate (FITC) and PI in binding buffer (10 mM HEPES, 140 mM NaCl, and 2.5 mM $CaCl_2$ at pH 7.4)) were added to the cell suspension and incubated at 37 °C for 20 to 30 min in dark and analyzed with a fluorescence-activated cell sorter (FACS) (BD FACSCalibur, BD Biosciences, California, USA). Data were analyzed using ModFit LT v2.0 (Verity Software House, Topsham, Maine, USA).

2.14. Biochemical Evaluation

As previously described [36], with little modification, HepG2 cells were cultured in T25 flask until they reached 30% confluency. Then, cells were incubated separately with selected isatin derivatives (**3a**, **4b**, and **4c**) at their IC50 for 48 h at 37 °C under 5% CO_2. After treatment, cells were washed by PBS, trypsinized and centrifuged at 10,000 rpm. The formed pellets were washed twice with PBS and lysed in 1 mL ice-cold radio immunoprecipitation assay (RIPA) lysis buffer (50 mM Tris-HCl, pH 8.0, 150 mM NaCl, 1% Triton X-100, 0.5% sodium deoxycholate, 0.1% sodium dodecyl sulfate) and kept on ice for 20 min. Then, the mixture was centrifuged at 15,000 rpm for 15 min to remove any cell debris. The cell lysate was aliquoted and stored at −80 °C for determination of protein using Pierce BCA Protein Assay Kit (Thermo Fisher Scientific, Waltham/Boston, USA) according to manufacturer's recommendations and determination of other biochemical parameters including EGFR, urokinase Plasminogen Activator (uPA), B-Cell Lymphoma-2 (Bcl-2), heparanase, and oxidative stress markers (glutathione; GSH and malonaldehyde; MDA).

Levels of EGFR [37] and heparanase [38] in cell lysate were investigated using commercially available enzyme-linked immunosorbent assay (ELISA) kits supplied by Abcam (Cat. No. ab269558, ab256401, respectively). Levels of uPA [39] were measured using ELISA kit supplied by Creative Diagnostics (New York, USA, Cat. No. DEIA1630). Bcl-2 levels [36,40] were determined using ELISA kit supplied by Zymed Laboratories, Carlsbad, California (Cat. No. 99-0042). Briefly, standards and samples were added to corresponding 96-well plate coated with suitable antibody. After the washing step, biotinylated anti-human antibody was added to each well. After washing away unbound biotinylated antibody, horseradish peroxidase (HRP)-conjugated streptavidin was pipetted into each well. A peroxidase substrate solution was added to all wells, after washing, and the color developed was proportionate to the amount of EGFR bound. Finally, the stop solution was added and the color intensity was measured at 450 nm.

Glutathione levels were determined using GSH Biodiagnostics kit (Cairo, Egypt). The method is based on the reduction of 5,5′dithiobis (2-nitrobenzoic acid) (DTNB) with glutathione (GSH) to produce a yellow compound. The reduced chromogen is directly proportional to GSH concentration and its absorbance was measured at 405 nm [41,42].

Levels of MDA, the marker of lipid peroxidation, were measured by thiobarbituric acid (TBA) method. TBA reacts with MDA in an acidic medium at temperature of 95 °C for 30 min to form colored thiobarbituric acid reactive (TBAR) product, the absorbance of which was measured at 534 nm. The MDA kit was purchased from Biodiagnostics, Cairo, Egypt [43,44].

2.15. Molecular Docking Simulation

MD was performed using Molecular Operating Environmental (MOE) software (Version10.2008, Chemical Computing Inc., Montreal, Quebec, Canada) targeting EGFRs. The macromolecule structure for the EGFR tyrosine kinase domain inhibitor Erlotinib (PDB: 1M17) was obtained from the protein data bank, February 2021 (https://www.rcsb.org/structure/1M17) (accessed on 5 February 2022). The docking process and the generation of the active site were performed according to the previously reported method [22,40,45]. The structure of the newly designed and most promising compound was drawn using ChemDraw14.0 then exported to MOE. The new structure was potentate 3D and minimized energy using MMFF94x force field. The co-crystalized ligand was exposed to validation process with RSMD 1.99 Å using Triangle Matcher placement method and London dG as docking score energy. Validation process of the co-crystalized ligand Erlotinib showed binding energy S = −25.65 Kcal/mol with RMSD = 0.89 °A Å, through one hydrogen bond backbone donor, between Met769 and nitrogen of quinazoline, with bond length 3.04 Å and strength 27%.

2.16. Statistical Analysis

Data are presented as the mean ± SEM. All experiments were performed in triplicate and repeated twice. Testing of data normality was carried out using Kolmogorov–Smirnov test. Multiple comparisons were done using a one-way ANOVA test followed by Tukey–Kramer as a post hoc test. The statistical significance criterion used was the 0.05 level of probability *p*. Statistical analyses were carried out using GraphPad Prism v 5 software (ISI software, San Diego, California, USA).

3. Results

3.1. Biological Evaluation

3.1.1. Antiproliferative and Anticancer Activities of Isatin Sulfonamide Derivatives on HepG2 and Huh7 Cell Lines

The cytotoxic effect of the ten compounds was assessed by the MTT assay using two human hepatocellular carcinoma cell lines, HepG2 and Huh7. The assay was performed in triplicate to calculate the IC50 (µM), the concentration required for 50% cell cytotoxicity and a % death rate of 100 p.p.m (µg/mL) after the specified time (48 h) against DMSO (vehicle, negative control) and Dox (positive control drug) (Table 1). The compounds showed more selectivity to HepG2. Tumor cells showed normal growth in the culture system, and DMSO did not seem to have any noticeable effect on cellular growth. Only six compounds (**2a**, **3a**, **4a**, **4b**, **4c**, and **4d**) had cytotoxic activity compared to the DMSO-treated cells (HepG2 or Huh7).

Before proceeding further to test these latter six compounds as potential selective anticancer agents, their safety was examined in vitro and in vivo.

Table 1. The in vitro cytotoxic activity of isatin sulfonamide derivatives on HepG2 and Huh7 cell lines (10 compounds) and in vitro safety assay towards normal human retina pigmented epithelial (RPE1) cell line (6 compounds) using MTT assay.

Cells	HepG2		Huh7		RPE1	
Compound no.	IC50 (μM)	% Death Rate at 100 p.p.m (μg/mL)	IC50 (μM)	% Death Rate at 100 p.p.m (μg/mL)	CC50 (μM)	% Cell Viability at 100 p.p.m (μg/mL)
2a	54.60 ± 2.00	81.50 *	40.00 ± 3.80	100 *	40.30 ± 1.61	4.40
2b	>100	10.20	>100	N.A	-	-
3a	16.80 ± 1.44	70.10 *	40.00 ± 2.20	100 *	>100	74 #
3b	>100	1.20	>100	N.A	-	-
3c	>100	11.50	>100	N.A	-	-
3d	>100	19.50	>100	N.A	-	-
4a	12.00 ± 0.40	100 *	>100	N.A	21.90 ± 1.15	0
4b	44.70 ± 1.55	94.20 *	53.00 ± 3.00	76.00 *	>100	66.40 #
4c	39.70 ± 1.90	95.60 *	35.00 ± 1.90	75.60 *	>100	80.70 #
4d	13.30 ± 0.75	100 *	18.76 ± 0.8	0	11.90 ± 0.70	0
DMSO	>100	1	>100	5	>100	95
Doxorubicin	21.60 ± 0.81	100	11.60 ± 0.90	100	-	-

Results are expressed as the mean ± SEM of three separate experiments. IC50, half-maximal inhibitory concentration (used for HepG2 and Huh7 cell lines); CC50, half-maximal cytotoxic dose (used for healthy RPB1 cell line); N.A, no activity; ppm, parts per million (equivalent to μg/mL). * highly active compounds in case of HepG2 and Huh7 cells, and # highly safe compounds in case of RPE1.

3.1.2. In Vitro Safety Assay towards Human Healthy Retina Pigmented Epithelial (RPE1) Cell Line

As shown in Table 1, three superior isatin sulfonamide derivatives (**3a**, **4b**, and **4c**) showed a non-cytotoxic effect towards the noncancerous cell line (RPE1), proving a high safety profile as ***potential selective anticancer agents*** to be tested later. In addition, compound **4c** was better than **4b** and **3a** regarding both the safety and death rate on HepG2. However, **4b** and **4c** showed lower IC50 and higher death rates than **3a** on Huh7 cells (Table 1).

The results were expressed as MTT assay IC50 (μM) and % cell viability at 100 ppm values (Table 1). The three superior isatin sulfonamide derivatives (**3a**, **4b**, and **4c**) showed non-cytotoxic effects towards the noncancerous cell line (RPE1), thereby proving a high safety profile as potential selective anti-cancer agents to be tested further.

3.1.3. In Vivo Acute Oral Toxicity Assay

After giving isatin sulfonamide-derivative compounds **2a**, **3a**, **4a**, and **4b**, **4d**, and **4c** as a single acute oral dose of 250 mg/kg [29,32], all the mice groups survived the treatment period (14 days). No physical or abnormal changes were observed in the body weight (BW), skin, fur, eyes, mucus membranes, tremors, salivation, behavior patterns, sleep patterns, or the animals' overall appearance. Kidney and liver biochemical analyses (liver and kidney function tests) were reported as normal (Table 2). Histopathology analyses (H and E staining) of the liver and kidney tissues are presented in Figures 2 and 3, respectively. There were no differences in the hepatic or renal tissue structures between the treated (**3a**, **4b**, and **4c** groups), the untreated normal mice control group, and the negative control group which received DMSO, and these were superior to the positive control Dox-treated animals, where hepatic and renal pathology were obvious.

Table 2. Effects of the acute oral toxicity study of isatin derivatives (**2a, 3a, 4a, and 4b, 4d,** and **4c**) on liver and kidney function tests of mice.

Tests		**liver function**				**kidney function**	
Parameters		**S.ALT**	**S.AST**	**S.T.Bilirubin**	**S.D.Bilirubin**	**S.Creatinine**	**S.Urea**
Groups	**/Units**	**(U/L)**	**(U/L)**	**(mg/dL)**	**(mg/dL)**	**(mg/dL)**	**(mg/dL)**
Control	Normal	28.92 ± 0.51	36.24 ± 0.94	0.69 ± 0.016	0.18 ± 0.005	0.67 ± 0.03	53.50 ± 1.01
	Negative	30.06 ± 0.72	34.40 ± 0.89	0.74 ± 0.016	0.19 ± 0.002	0.79 ± 0.03	52.38 ± 0.69
	Positive	52.97 ± 0.92 *,#	46.93 ± 0.30 *,#	1.45 ± 0.072 *,#	0.36 ± 0.026 *,#	1.57 ± 0.02 *,#	83.27 ± 0.95 *,#
Treated	**2a**	28.42 ± 0.27 ^	37.11 ± 0.35 ^	0.65 ± 0.018 ^	0.16 ± 0.003 ^	0.59 ± 0.04 ^	52.24 ± 0.65 ^
	3a	27.05 ± 0.81 ^	36.78 ± 0.33 ^	0.65 ± 0.026 ^	0.16 ± 0.002 ^	0.82 ± 0.03 ^	51.25 ± 0.36 ^
	4a	28.14 ± 0.56 ^	36.51 ± 0.15 ^	0.72 ± 0.019 ^	0.17 ± 0.004 ^	0.70 ± 0.02 ^	55.10 ± 2.96 ^
	4b	27.95 ± 0.34 ^	35.18 ± 0.47 ^	0.70 ± 0.013 ^	0.17 ± 0.001 ^	0.88 ± 0.06 ^	62.60 ± 3.64 ^
	4c	28.41 ± 0.28 ^	37.35 ± 1.47 ^	0.73 ± 0.014 ^	0.16 ± 0.007 ^	0.79 ± 0.04 ^	55.76 ± 5.64 ^
	4d	30.14 ± 0.93 ^	38.72 ± 1.37 #,^	0.69 ± 0.016 ^	0.18 ± 0.004 ^	0.85 ± 0.03 ^	48.18 ± 0.53 ^

Data are expressed as mean ± SEM. Statistical analyses were carried out using ANOVA followed by the Tukey–Kramer post hoc test. Acute toxic doses of selected isatin derivatives (**2a**, **3a**, **4a**, and **4b**, **4d**, and **4c**, 250 mg/kg) were given to mice as a single oral dose, and the impact on their liver and kidney function tests was evaluated after 14 days of dose intake in comparison to normal healthy non-treated mice, DMSO-treated mice as negative control, and doxorubicin-treated mice as positive control (15 mg/kg, I.P, once). * $p < 0.05$ compared with the normal control; # $p < 0.05$ compared with the negative control group (DMSO-treated); ^ $p < 0.05$ compared with the positive control group (doxorubicin-treated).

3.1.4. Isatin Sulfonamide Molecular Hybrids Effect Cell Cycle Analysis

Flow cytometry was done to characterize whether the synthetic isatin sulfonamide-derivative compounds exerted an anticancer effect via disrupting the cell cycle compared to untreated controls. Moreover, flow cytometry would prove whether isatin sulfonamide-derivative hybrids activate the programmed cell death pathway, inducing apoptosis, with an overall gain of "cell death", or not. The results are presented in Table 3, where the % cell cycle arrest in the different cell cycle phases showed that treatment of HepG2 cells with compounds **3a** and **4c** arrested the cells at the G2-M phase by 44.58 and 37.62 %, respectively, which is 6.32- and 5.30-fold higher compared to the negative control, HepG2 (7.1%). It was also noticed that the effect of compound **3a** was superior to the effect of Dox (39%) on arresting the cells at the G2-M phase. In addition, compound **4b** caused the accumulation of cells at the G0-G1 and S phases by 41.25 and 53.26%, respectively. Interestingly, the three compounds, **3a**, **4b**, and **4c**, induced apoptosis at the pre-G1 phase, and the percentages of cell death were 46.29, 28.14, and 32.02, respectively, in comparison to the negative control (1.47%). Again, induction of apoptosis at the pre-G1 phase by compound **3a** was more than the induction induced by doxorubicin (46.29% versus 42.38%). These results indicate that compounds **3a**, **4b**, and **3c** could target the cancer cell cycle and induce apoptosis at different phases.

The cell cycle arrest in HepG2 is expressed as % cells in each phase after treatment with 5 μM isatin sulfonamide tested compounds, **3a**, **4b**, and **4c**, against positive and negative controls. The apoptotic activity of compounds **3a**, **4b**, and **4c** is expressed as % cell death induction after treatment of HepG2 with IC50 of the selected compounds for 24 h. The positive control was Dox-treated HepG2 cells; the negative control was DMSO-treated HepG2 cells.

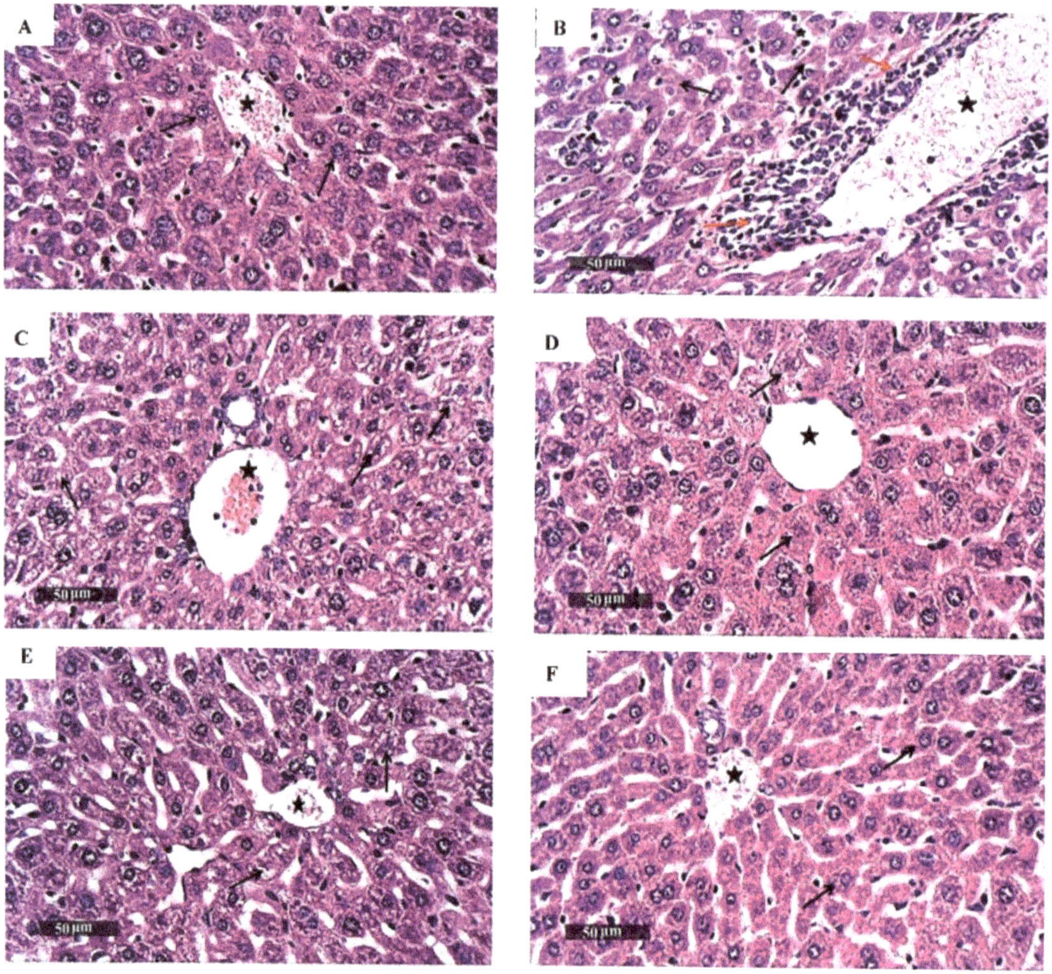

Figure 2. Photomicrographs of mice liver sections stained by H and E from (**A**) the normal control group, which demonstrated normal histological features of the liver parenchyma with apparently intact well-organized hepatocytes with intact subcellular details (black arrow), normal hepatic sinusoids, vasculatures (star), and portal tracts; (**B**) Dox-treated mice as the positive control group, which showed hepatocellular necrotic changes (arrow) accompanied with moderate dilatation of the hepatic blood vessels and hepatic sinusoids (star) associated with focal perivascular mixed inflammatory cell infiltrates (red arrow); (**C**) mice treated with compound **3a**, which showed intact vasculatures (star) as well as hepatic sinusoids with no inflammatory cell infiltrates; (**D**) mice group treated with compound **4b**, which demonstrated apparently intact hepatocytes and allover hepatic parenchyma with intact vasculatures (star), and almost no record of inflammatory cell infiltrates; (**E**) mice treated with compound **4c**, and (**F**) negative control group that received DMSO. Both (**E**,**F**) groups showed almost intact histological features of hepatic parenchyma (as normal control; (**A**)). [scale bars 50 µm].

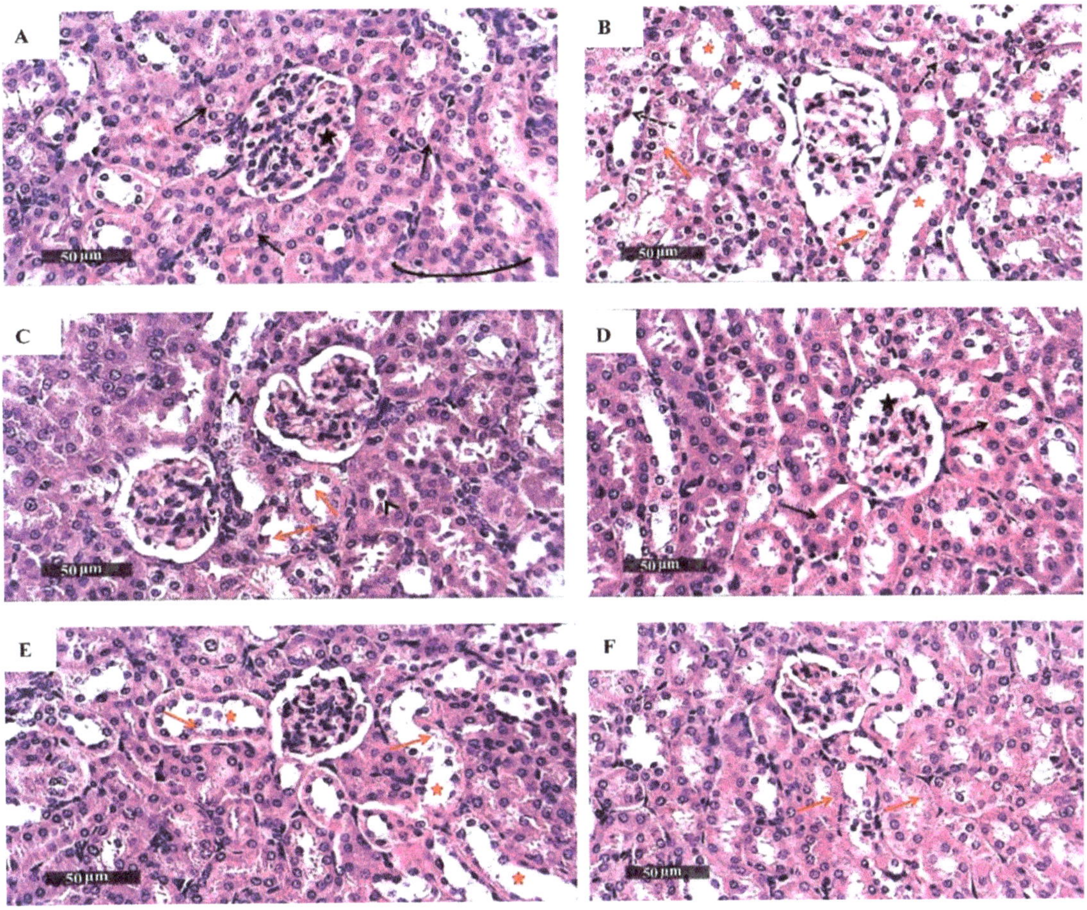

Figure 3. Photomicrographs of mice kidney sections, where renal tissue stained by H and E from (**A**), the normal control group, demonstrated normal histological features of the renal parenchyma with many intact renal corpuscles (star) and nephron segments with intact tubular epithelium (black arrow) and intact interstitial tissues and vasculatures. (**B**) Positive control treated with Dox, showing dilatation of the cortical tubular segments (red star) with tubular vacuolar degenerative changes (red arrow), loss of luminal border integrity, and occasional tubular epithelial necrotic cells records (dashed black arrow); (**C**) group treated with compound **3a**, showing intact well-organized histological features of renal parenchyma with minimal records of sporadic tubular epithelial changes (red arrow); (**D**) group treated with compound **4b**, with almost intact histological features of renal parenchyma; (**E**) group treated with compound **4c**, where most of the tubular segments showed almost intact well-organized morphological features (as in (**A**)), and (**F**) negative control group that received DMSO, which showed almost intact histological features of renal parenchyma with minimal records of tubular degeneration (red arrow). [scale bars 50 µm].

Table 3. Cell cycle analysis using flow cytometry in HepG2 expressed as % of cells in each phase after treatment with isatin sulfonamide tested compounds, **3a**, **4b**, and **4c**, and controls, and Annexin V/FITC apoptotic activity of selected isatin compounds is expressed as % of cell death induction.

Compound no.	%Cell Cycle Arrest/Phase				%Cell Death			
	%G0-G1	%S	%G2-M	%pre-G1	Total	Early Apoptosis	Late Apoptosis	Necrosis
3a	31.57	23.58	44.85	46.29	46.29	3.56	29.58	13.15
4b	41.25	53.26	5.49	28.14	28.14	6.12	12.61	9.41
4c	36.44	25.94	37.62	32.02	32.02	2.88	18.08	11.06
Positive control	29.74	31.26	39	42.38	42.38	1.85	26.27	14.26
Negative control	55.29	37.61	7.1	1.47	1.47	0.36	0.12	0.99

Data are expressed as mean percentage of three separate experiments.

3.1.5. Effect of Isatin Sulfonamide Derivatives on Apoptotic Assay

The mechanism of cell death was analyzed by flow cytometry using Annexin V/PI double staining. The data generated were plotted in two-dimensional dot plots, in which PI is represented versus Annexin V-FITC (Figure 4).

These plots were divided into four regions corresponding to:

(1) Viable cells, which are negative to both probes (PI/FITC −/−),
(2) Apoptotic cells, which are PI negative and Annexin positive (PI/FITC −/+),
(3) Late apoptotic cells, which are both PI and Annexin positive (PI/FITC +/+),
(4) Necrotic cells, which are PI positive and Annexin negative (PI/FITC+/−).

As shown in Table 4 and Figure 4, the % of total apoptosis induced by the tested compounds (**3a**, **4b**, and **4c**) were 46.3%, 28.1%, and 32.0%, respectively, compared to the negative control (1.47%). Compound 3a exhibited the highest induction of total apoptosis, 31.49-fold greater than that of negative control HepG2 cells and higher than the total apoptosis induced by Dox (42.38%). Moreover, compound **3a** was the most potent compound for inducing total apoptosis or necrosis (46.29 % and 13.15%, respectively) in comparison to **4b** and **4c**. In addition, the percentage of late apoptosis varies from 29.58% to 12.61% and 18.08% for compounds **3a**, **4b**, and **3c**, respectively, when compared to both negative control HepG2 cells (0.12%) and Dox-treated cells (26.27%).

Table 4. Effects of treatment with the superior isatin sulfonamide derivatives (**3a**, **4b**, and **4c** compounds) on the levels of EGFR, uPA, BcL-2, heparanase, GSH, and MDA in cell lysate from HepG2 cells treated with the selected compounds at their IC50.

HepG2 Cell Line	Control		Treated		
Parameter/Groups	**Positive**	**Negative**	**3a**	**4b**	**4c**
EGFR (pmol/mg protein)	27.5 ± 1.7 ^{a*}	306 ± 20	42 ± 2.3 ^{a*,b*}	87.49 ± 3.4 ^{a*,b*}	54 ± 2.4 ^{a*,b*}
uPA (nmol/mg protein)	1244 ± 18 ^{a*}	3149 ± 111	1258 ± 15 ^{a*}	1916 ± 40 ^{a*,b*}	1729 ± 38 ^{a*,b*}
Bcl-2 (nmol/mg protein)	1.7 ± 0.10 ^{a*}	7.8 ± 0.07	4.2 ± 0.10 ^{a*,b*}	2.5 ± 0.04 ^{a,b*}	3.6 ± 0.11 ^{a*,b*}
Heparanase (pmol/mg protein)	902.1 ± 21 ^{a*}	3097 ± 160	1262 ± 35 ^{a*,b*}	1827 ± 30 ^{a*,b*}	1449 ± 12.5 ^{a*,b*}
GSH (μmol/mg protein)	0.8 ± 0.1 ^{a*}	1.7 ± 0.2	1.6 ± 0.7	1.1 ± 0.7	2.1 ± 0.3 ^{b*}
MDA level (μmol/mg protein)	18.5 ± 1.5 ^{a*}	13.8 ± 1.0	2.0 ± 0.5 ^{a*,b*}	3.7 ± 0.4 ^{a*,b*}	14.5 ± 1.4

Data are expressed as mean ± SEM of three separate experiments. Results were computed by a one-way ANOVA test followed by Tukey–Kramer as a post-hoc test. * p values < 0.05 were considered significant. EGFR, epidermal growth factor receptor; uPA, urokinase plasminogen activator; Bcl-2, B-cell lymphoma; GSH, reduced glutathione; MDA, malondialdehyde. [a] difference from negative control (non-treated HepG2 cells); [b] difference from positive control (doxorubicin-treated HepG2 cells).

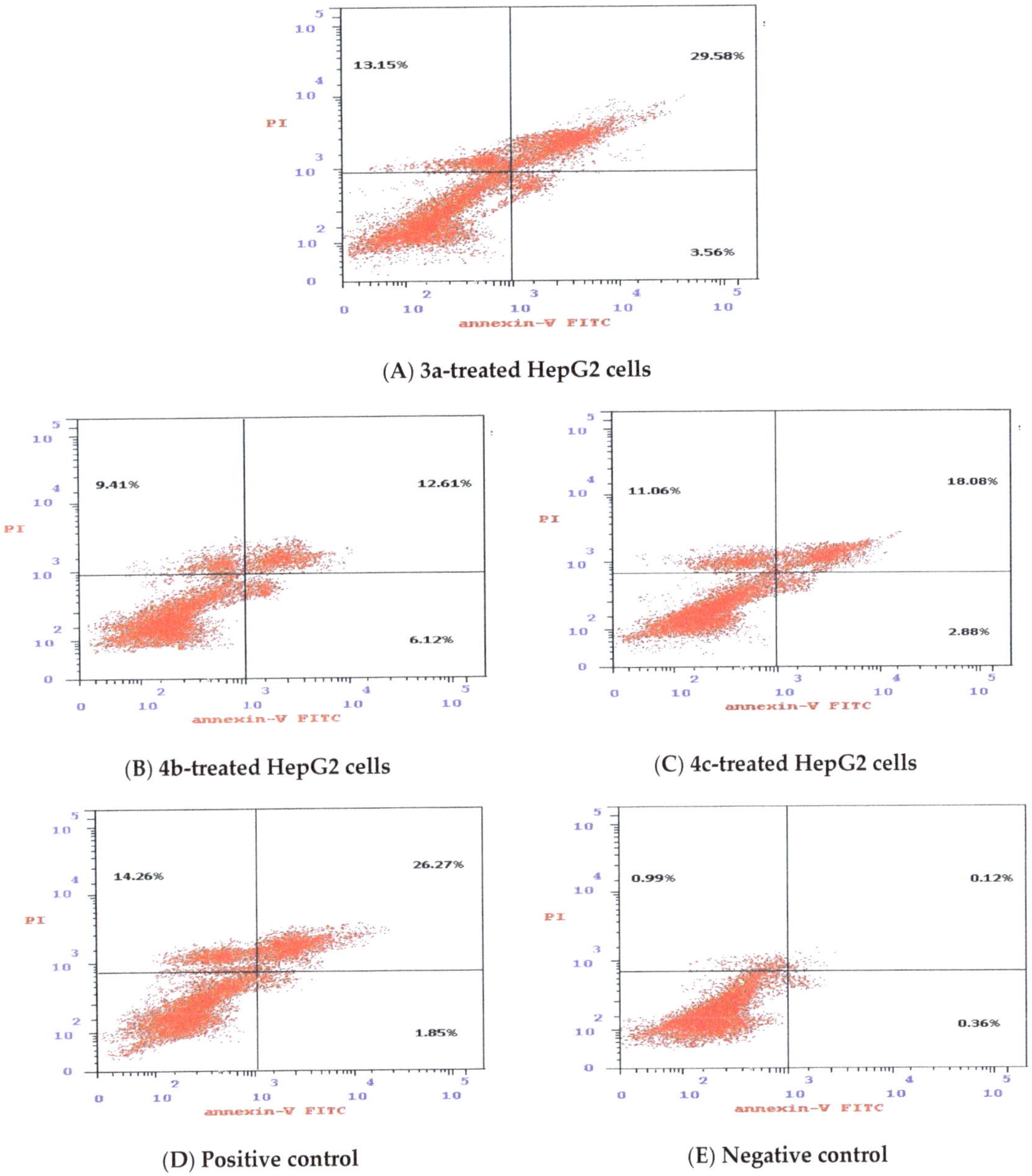

Figure 4. Apoptosis analysis by flow cytometry using the Annexin V/PI double staining method ((**A**) compound **3a**-treated HepG2 cells, (**B**) compound **4b**-treated HepG2 cells, (**C**) compound **4c**-treated HepG2 cells, (**D**) positive control; Dox-treated HepG2 cells, and (**E**) negative control; non-treated HepG2 cells). Apoptosis analysis depends on the quantitation of DNA content after PI staining, being stoichiometric, i.e., cells in the S phase will have more DNA than cells in the G1 phase and will take up proportionally more dye and will fluoresce more brightly. Cells in the G2 phase will be approximately twice as bright as cells in the G1 phase.

3.1.6. Effect of Isatin Sulfonamide Molecular Hybrids on Other Cancer Hallmark Markers Assay

As illustrated in Table 4, compounds **3a-**, **4b-**, and **4c-**treated HepG2 cells showed significant reductions in the EGFR levels, where **3a-**treated cells reported the most reduction to 42 ± 2.3 pmol/mg protein in comparison to untreated HepG2 cells (306 ± 20 pmol/mg protein). This signifies the potent anti-angiogenic effect of the selected promising isatin sulfonamide derivatives.

As clarified in Table 4, cells treated with compounds **3a**, **4b**, and **4c** showed significantly lower levels of uPA (1258 ± 15, 1916 ± 40, and 1729 ± 38 nmol/mg protein, respectively) than the non-treated negative control HepG2 cells (3149 ± 111 nmol/mg protein). These levels, except those for 3a-treated cells, were significantly higher than the Dox-treated cells uPA levels (1244 ± 18 nmol/mg protein).

Anti-apoptotic repression via Bcl-2 antigen level restoration was also evaluated. Compounds **3a**, **4b**, and **4c** showed significant differences in levels of Bcl-2 (4.2 ± 0.1, 2.5 ± 0.04, and 3.6 ± 0.11 nmol/mg protein, respectively) when compared to the negative control (7.8 ± 0.11 nmol/mg protein). This points to the fact that one role of these three selected isatin derived compounds as anti-cancer promising drugs is via apoptosis induction.

Heparanase expression levels were assessed as markers of metastasis. Compound 3a showed the lowest levels of heparanase (1262 ± 35 pmol/mg protein) in comparison to the negative HepG2 cell control (3097 ± 160 pmol/mg protein). Compounds **4b** and **4c** lowered heparanase expression levels in HepG2 cell lysates as well, to 1827 ± 30 and 1449 ± 12.5 pmol/mg protein, respectively.

With respect to markers of oxidative stress, the three investigated compounds had no significant effect on GSH levels compared to non-treated HepG2 cells. On the other hand, only compounds **3a** and **4b** had remarkable reducing effects on MDA levels in HepG2 cell lysates, to 2.0 ± 0.5 and 3.7 ± 0.4 μmol/mg protein, respectively, in comparison to both non-treated HepG2 cells (13.8 ± 1.0 μmol/mg protein) and Dox-treated HepG2 cells (18.5 ± 1.5 μmol/mg protein).

3.2. Molecular Docking Studies and Binding to EGFR

Next, molecular docking studies were carried out to better understand the binding energies (Kcal/mol) and amino acid residue interactions of the most active isatin sulfonamide derivatives in the current study (**3a**, **4b**, and **4c**). The molecular docking results were compared to Dox and the co-crystalized ligand (Erlotinib) entering into the ATP binding site inside the active site of an EGFR (PDB: 1M17) retrieved from the protein data bank (https://www.rcsb.org/structure/1M17) (accessed on 5 February 2022) to explore the binding mode.

As shown in Table 3, **3a**, **4b**, and **4c**, when docked to the same binding site as Erlotinib (Table 5) revealed binding free energy with a minus score, showing quick fitting into the EGFR binding site, with free energies of −21.74, −19.21, and −20.80 kcal/mol, respectively. The results confirm the promiscuity of these three isatin sulfonamide derivatives to form good binding within the EGFR binding site [22,37]. Dox displayed lower binding energy of S = −22.82 Kcal/mol, through five hydrogen bonds, as did Erlotinib (S= −25.65 Kcal/mol). Compound **3a** displayed a hydrophobic interaction with piperidin-1-ylsulfonyl, phenyl derivatives of isatin's bioactive core, as well as the carbonyl and tolyl groups (Figure 5). The hydrophobic interactions of both compounds **4b** and **4c** appear on the acetyl, piperidin-1-ylsulfonyl, and phenyl derivatives of the isatin scaffold. The synthesized isatin sulfonamide derivatives showed a hydrogen bond backbone acceptor between Met769 and a carbonyl of isatin with a low bond length of less than 3.04 Å, as did Erlotinib, that formed one hydrogen bond between the nitrogen of quinazoline and Met769, providing more evidence of good activity against EGFRs as promising new tyrosine kinase inhibitors.

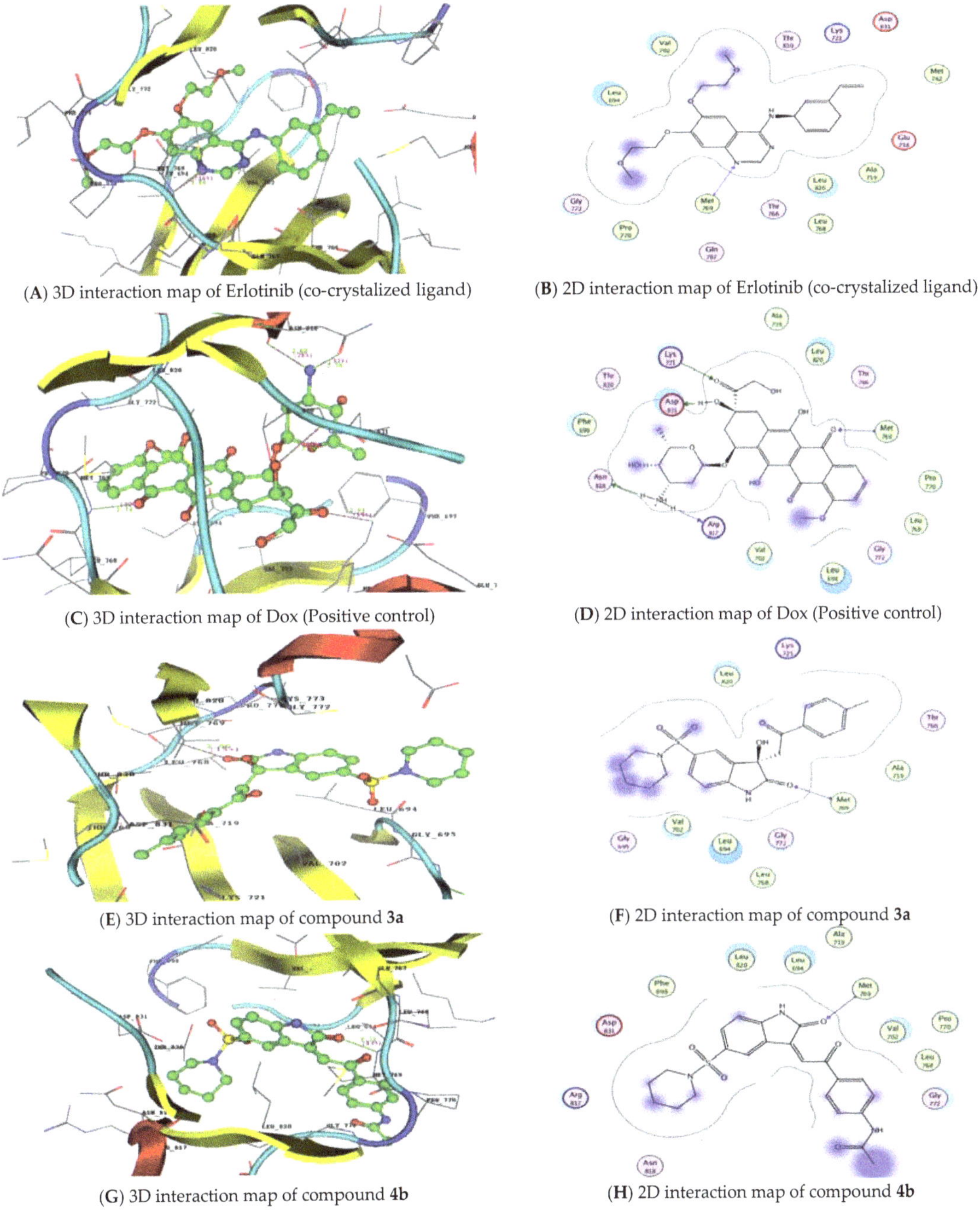

(**A**) 3D interaction map of Erlotinib (co-crystalized ligand)

(**B**) 2D interaction map of Erlotinib (co-crystalized ligand)

(**C**) 3D interaction map of Dox (Positive control)

(**D**) 2D interaction map of Dox (Positive control)

(**E**) 3D interaction map of compound **3a**

(**F**) 2D interaction map of compound **3a**

(**G**) 3D interaction map of compound **4b**

(**H**) 2D interaction map of compound **4b**

Figure 5. *Cont.*

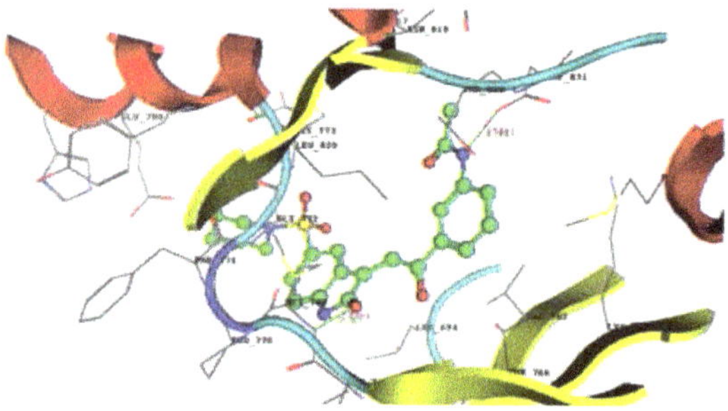

(**I**) 3D interaction map of compound **4c**

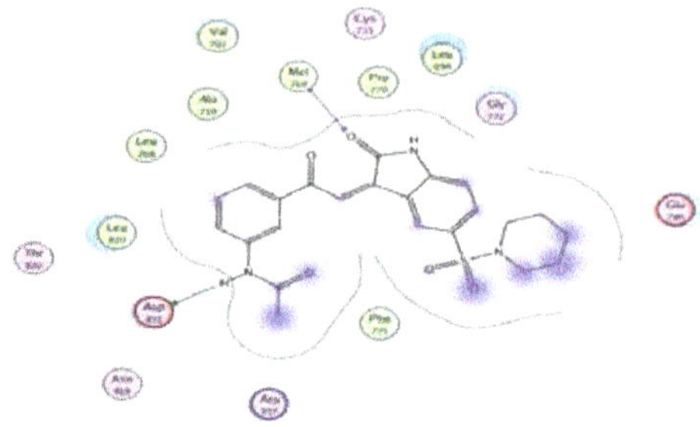

(**J**) 2D interaction map of compound **4c**

Figure 5. Molecular docking study: 3D and 2D interaction maps of the co-crystalized ligand (Erlotinib) (**A**,**B**), the positive control (doxorubicin) (**C**,**D**), and the isatin sulfonamide derivatives **3a**, **4b**, and **4c** (**E–J**) inside the EGFR (PDB:1M17) active site.

Table 5. Molecular docking study results of the isatin sulfonamide derivatives **3a**, **4b**, and **4c**, doxorubicin, and the co-crystalized ligand (Erlotinib), with binding energy and inter-acting groups with specific amino acids in residue inside the active site of the EGFR (PDB: 1M17).

Compound no.	S (Kcal/mol)	Amino Acids Residues	Ligand Atoms	Distance (Å A)	Strength (%)
3a	−21.74	Met769	C=O of isatin	2.86	15
4b	−19.21	Met769	C=O of isatin	2.71	23
4c	−20.80	Met769	C=O of isatin	2.85	33
		Asp831	NH of acetanilide	2.44	54
Doxorubicin	−22.82	Met769 Lys721 Asp831 Asn818 Arg817	C=O pf anthraquinone C=O of ethenone Hydroxy group NH2 group NH2 group	2.74 2.61 2.76 2.56 2.60	32 19 42 37 28
Erlotinib	−25.65	Met769	C=O of quinazoline	3.04	27

4. Discussion

Molecular hybridization drug design based on combining different bioactive moieties into one compound is an effective strategy to produce promising efficient anticancer agents [17,46]. In the present study, molecular hybridization approach was used to synthesize ten new isatin sulfonamide hybrids. The anticancer antiproliferative activities of these derivatives, characterized using two HCC cell lines, HepG2 and Huh7, revealed that six compounds (**2a**, **3a**, **4a**, **4b**, **4c**, and **4d**) provided the most cytotoxic effects. The safety of these six compounds was investigated through in vitro and *in vivo* studies that showed that three derivatives (**3a**, **4b**, and **4c**) exhibited the least toxicity and higher safety margins.

To shed more light on the mechanistic pathways underlying the anticancer and antiproliferative activities of compounds **3a**, **4b**, and **4c**, we evaluated their effect on cell death tendency through studying cell cycle apoptosis. Our results showed that the compound "**3a**", of all the tested compounds, has the highest cell cycle arresting potential during the cell cycle critical phases (M and pre-G1), rather than the other two new isatin hybrids (**4b** and **4c**) and the classically used cytotoxic positive control drug (Dox), in comparison to non-treated HepG2 cells (negative control). Cell cycle arrest percentage is now documented as a marker for new anti-cancer drug efficiency [47,48]. Our observations come in agreement with other studies reporting that some isatin derivatives arrested the cell cycle during the mitotic phase, in comparison to the positive cytotoxic control drugs cisplatin [49] and Dox [50].

One of the mechanisms by which anticancer agents could affect cell cycle machinery is through targeting various cyclin-dependent kinases (CDK_S). Depending on the type of targeted CDK, cell cycle machinery could be arrested at different phases. Targeting cyclinA-CDK2 causes apoptosis induction at the S phase, while inappropriate activation of cyclinB-CDK1 targets the cell cycle at the G2-M boundary [48,51]. This could explain why isatin sulfonamide derivatives 3a and 4c arrested the cell cycle at the G2-M and pre-G1 phases, while 4b accumulated the cells at the G0-G1 and S phases. However, this needs more investigation by evaluating the effects of these derivatives on CDKs and the other enzymes regulating cell cycle machinery.

Regarding the mode of cell death analysis using Annexin V-FITC/PI double staining, it was found that compound **3a** had the highest induction of both total and late apoptosis in comparison to the **3b** and **4c** compounds, and it was even more potent than Dox. This suggests that compound **3a** would modulate cell death regulators and may play a key role in the decision of cell death having an apoptotic effect that would also be related to other cancer hallmarks [52,53].

This apoptotic efficacy raised by the **3a** compound was further supported by the significantly decreased expression of the survival anti-apoptotic protein (Bcl-2) compared to untreated HepG2. In addition, treating HepG2 with the **4b** and **4c** compounds caused a significant reduction in the Bcl-2 levels. Similarly, to our findings, different isatin derivatives could induce HepG2 cells death through targeting cell cycle proteins, apoptotic induction [50], and down-regulation of Bcl-2 with autophagy promotion [54].

Although **3a** had a much more potent apoptotic efficacy than the positive chemotherapeutic cytotoxic control Dox, the latter depressed the Bcl-2 level more than compound **3a**; this could possibly be due to the confined Dox effect against Bcl-2 solely [55].

Compounds **3a**, **4b**, and **4c** significantly decreased angiogenesis measured as EGFR protein, and the greatest reduction was observed when HepG2 cells were treated with **3a**. The potential of isatin sulfonamide derivatives to depress angiogenesis was also previously reported with apoptosis induction and arresting the cell cycle at the G2/M phases [56]. This inhibitory effect of the newly synthesized isatin sulfonamide derivatives (**3a**, **4b**, and **4c**) was confirmed by the molecular docking studies, as they displayed low free binding energy in comparison to the co-crystalized ligand (Erlotinib) and to the positive control (Dox) and inside the EGFR active site, meaning better potential anti-angiogenic drugs with tyrosine kinase inhibitory activity.

uPA is a highly restricted serine protease that converts the zymogen plasminogen to active plasmin, a broad-spectrum serine proteinase capable of degrading most protein components of the extracellular matrix, facilitating cancer invasion and metastasis (one of the cancer hallmarks) [57,58]. The three superior tested isatin sulfonamide-derivative compounds, **3a**, **4b**, and **4c**, showed possible anti-invasive activity measured as decreased uPA levels. Compound 3a showed the most significant efficacy that is comparable to the Dox effect. This observation is in accordance with a previously reported work, showing some isatin derivatives to decrease the cancer invasiveness in treated cancer cell lines via down-regulating urokinase [59].

The three currently investigated potential isatin sulfonamide derivatives, **3a** (the most superior), **4b**, and **4c**, significantly decreased the invasive potential of liver cancer cells through decreasing heparanase expression in the HepG2 cell lysate by a comparable degree to that shown in the Dox-treated HepG2 cells. Heparanase is an endo-β-D glucuronidase that cleaves an extracellular component, heparan sulfate. High expression levels of heparanase have been reported in several tumors [60,61] and associated with tumor growth, invasiveness, metastasis, and poor prognosis. In addition, there is crosstalk between heparanase and many other proteases such as matrix metalloproteinase-9 and uPA [62,63]. Therefore, downregulation of both uPA and heparanase by isatin derivatives (**3a**, **4b**, and **4c**) highlights their promising anticancer activities via targeting tumor invasion and metastasis.

It is worth mentioning here the role of heparanase in augmenting angiogenesis, another hallmark of cancer, through both its enzymatic and non-enzymatic activities. Release of heparan sulfate induces the angiogenesis via the upregulation of various growth factors and other proangiogenic factors [64,65]. Moreover, heparanase could promote angiogenesis non-enzymatically through activation of hypoxia-inducible factors, heat shock proteins, and proangiogenic factors [63,66]. Hence, the inhibitory effect of our newly synthesized isatin sulfonamide molecular hybrids on heparanase and EGFRs suggests a possible role in targeting angiogenesis.

Alongside this, we tried to explore the contribution of oxidative stress in the whole anticancer or apoptotic induction, through measuring both MDA and GSH levels. The tested isatin sulfonamide-derivative compounds did not affect the oxidative stress markers measured currently, and, together with their cytotoxic efficacy, did not rely on this stressful pathway.

In summary, the anticancer molecular or biochemical analysis revealed, mechanistically, that isatin sulfonamide derivatives' cytotoxicity is possibly due to an effect on the apoptotic and the angiogenic machineries as well as targeting tumor invasiveness studied using the HepG2 cell line. The isatin sulfonamide derivatives showed an ability to depress the action of the survival mechanism of Bcl-2, inhibiting angiogenesis, and further hindering cancer invasion through inhibition of uPA and heparanase expression.

In future, more in vivo extrapolation for the compound **3a** will be conducted to introduce more insight into the anti-cancer autophagic/mitochondrial efficacy of the new molecularly hybridized isatin sulfonamide derivative, or, moreover, it will be engineered by nanotechnology for better drug targeting.

5. Conclusions

A schematic diagram is outlined in the graphical abstract, summarizing the present research's findings on the newly synthesized isatin sulfonamide derivatives **3a, 4b**, and **4c,** which have promising anticancer activities on HCC cell lines. Moreover, the compound (**3a**) presents, hopefully, a promising new anticancer agent to suppress HepG2 cell growth in vitro through cell cycle arrest, apoptosis induction, and possible anti-angiogenic and anti-metastatic activities.

Supplementary Materials: The following are available online at https://www.mdpi.com/article/10.3390/biomedicines10030722/s1. Synthesis and Identification of the Tested Isatin Sulfonamides Derivatives (10 Compounds).

Author Contributions: M.E. and E.F.S. contributed equally during the research conduction, data analysis, and drafting of the first manuscript and share the "First Authorship"; A.R. and Y.A.A. contributed equally to the new heterocyclic derivatives' design; A.R. performed the molecular docking study; K.M. contributed during the biochemical practical research conducted; M.M.A. and N.M.H. contributed to the study design and conceptualization and share the "Chief Supervisor Authorship", including revising the manuscript draft and ensuring partial funding; E.F.S. and N.M.H. contributed to writing and revising the manuscript until submission and full publication; N.M.H. being the corresponding author. All authors have read and agreed to the published version of the manuscript.

Funding: This research received no external funding.

Institutional Review Board Statement: Animal care and all the experimental protocols were approved and conducted in accordance with the ethical guidelines approved by the Institutional Review Board of Faculty of Pharmacy, Ain Shams University, Cairo, Egypt (Ethics Committee Approval 10/2019/1). The study was carried out in compliance with the ARRIVE guidelines.

Informed Consent Statement: Not applicable.

Data Availability Statement: Not applicable.

Acknowledgments: The authors are thankful to the partial fund provided by both the ABRL at the Biochemistry Dept., Faculty of Pharmacy, Ain Shams University (#2021NMH5) as well as the National Research Centre, Giza (Grant #11/2/7,2020).

Conflicts of Interest: The authors declare no conflict of interest.

References

1. Tsuchiya, K. Switching from Apoptosis to Pyroptosis: Gasdermin-Elicited Inflammation and Antitumor Immunity. *Int. J. Mol. Sci.* **2021**, *22*, 426. [CrossRef] [PubMed]
2. Hussien, A.G.; Borai, I.H.; Said, M.M.; Mahmoud, K.; Ali, M.M. Chemotherapeutic effect of Ulmus pumila leaves methanolic extract against N-methyl-N-nitrosourea-induced mammary carcinoma in female rats: An in vitro and in vivo study. *J. Appl. Pharm. Sci.* **2019**, *9*, 57–68. [CrossRef]
3. Gong, Y.; Fan, Z.; Luo, G.; Yang, C.; Huang, Q.; Fan, K.; Cheng, H.; Jin, K.; Ni, Q.; Yu, X.; et al. The role of necroptosis in cancer biology and therapy. *Mol. Cancer* **2019**, *18*, 1–17. [CrossRef] [PubMed]
4. Ke, S.; Shi, L.; Yang, Z. Discovery of novel isatin–dehydroepiandrosterone conjugates as potential anticancer agents. *Bioorg. Med. Chem. Lett.* **2015**, *25*, 4628–4631. [CrossRef] [PubMed]
5. Kumar, K.; Bhargava, G.; Land, K.M.; Chang, K.-H.; Arora, R.; Sen, S.; Kumar, V. N-Propargylated isatin-Mannich mono-and bis-adducts: Synthesis and preliminary analysis of in vitro activity against Tritrichomonas foetus. *Eur. J. Med. Chem.* **2014**, *74*, 657–663. [CrossRef]
6. Lin, H.-H.; Wu, W.-Y.; Cao, S.-L.; Liao, J.; Ma, L.; Gao, M.; Li, Z.-F.; Xu, X. Synthesis and antiproliferative evaluation of piperazine-1-carbothiohydrazide derivatives of indolin-2-one. *Bioorg. Med. Chem. Lett.* **2013**, *23*, 3304–3307. [CrossRef]
7. Krause-Heuer, A.M.; Howell, N.R.; Matesic, L.; Dhand, G.; Young, E.L.; Burgess, L.; Jiang, C.D.; Lengkeek, N.A.; Fookes, C.J.R.; Pham, T.Q. A new class of fluorinated 5-pyrrolidinylsulfonyl isatin caspase inhibitors for PET imaging of apoptosis. *MedChemComm* **2013**, *4*, 347–352. [CrossRef]
8. Salem, M.A.; Ragab, A.; El-Khalafawy, A.; Makhlouf, A.H.; Askar, A.A.; Ammar, Y.A. Design, synthesis, in vitro antimicrobial evaluation and molecular docking studies of indol-2-one tagged with morpholinosulfonyl moiety as DNA gyrase inhibitors. *Bioorg. Chem.* **2020**, *96*, 103619. [CrossRef] [PubMed]
9. Nath, R.; Pathania, S.; Grover, G.; Akhtar, M.J. Isatin containing heterocycles for different biological activities: Analysis of structure activity relationship. *J. Mol. Struct.* **2020**, *1222*, 128900. [CrossRef]
10. Wang, J.; Yun, D.; Yao, J.; Fu, W.; Huang, F.; Chen, L.; Wei, T.; Yu, C.; Xu, H.; Zhou, X. Design, synthesis and QSAR study of novel isatin analogues inspired Michael acceptor as potential anticancer compounds. *Eur. J. Med. Chem.* **2018**, *144*, 493–503. [CrossRef]
11. Gupta, A.K.; Tulsyan, S.; Bharadwaj, M.; Mehrotra, R. Systematic review on cytotoxic and anticancer potential of n-substituted isatins as novel class of compounds useful in multidrug-resistant cancer therapy: In silico and in vitro analysis. *Top. Curr. Chem.* **2019**, *377*, 15. [CrossRef] [PubMed]
12. Shin, J.-W.; Ohnishi, K.; Murakami, A.; Lee, J.-S.; Kundu, J.K.; Na, H.-K.; Ohigashi, H.; Surh, Y.-J. Zerumbone induces heme oxygenase-1 expression in mouse skin and cultured murine epidermal cells through activation of Nrf2. *Cancer Prev. Res.* **2011**, *4*, 860–870. [CrossRef] [PubMed]
13. Heller, L.; Schwarz, S.; Perl, V.; Köwitsch, A.; Siewert, B.; Csuk, R. Incorporation of a Michael acceptor enhances the antitumor activity of triterpenoic acids. *Eur. J. Med. Chem.* **2015**, *101*, 391–399. [CrossRef] [PubMed]
14. Mishra, S.; Singh, P. Hybrid molecules: The privileged scaffolds for various pharmaceuticals. *Eur. J. Med. Chem.* **2016**, *124*, 500–536. [CrossRef]
15. Fang, L.; Jumpertz, S.; Zhang, Y.; Appenroth, D.; Fleck, C.; Mohr, K.; Tränkle, C.; Decker, M. Hybrid molecules from xanomeline and tacrine: Enhanced tacrine actions on cholinesterases and muscarinic M1 receptors. *J. Med. Chem.* **2010**, *53*, 2094–2103. [CrossRef] [PubMed]
16. Abdel-Aziz, H.A.; Eldehna, W.M.; Keeton, A.B.; Piazza, G.A.; Kadi, A.A.; Attwa, M.W.; Abdelhameed, A.S.; Attia, M.I. Isatin-benzoazine molecular hybrids as potential antiproliferative agents: Synthesis and in vitro pharmacological profiling. *Drug Des. Dev. Ther.* **2017**, *11*, 2333. [CrossRef]
17. Ding, Z.; Zhou, M.; Zeng, C. Recent advances in isatin hybrids as potential anticancer agents. *Arch. Pharm.* **2020**, *353*, 1900367. [CrossRef] [PubMed]
18. Medvedev, A.; Buneeva, O.; Gnedenko, O.; Ershov, P.; Ivanov, A. Isatin, an endogenous nonpeptide biofactor: A review of its molecular targets, mechanisms of actions, and their biomedical implications. *BioFactors* **2018**, *44*, 95–108. [CrossRef] [PubMed]
19. Varun, V.; Sonam, S.; Kakkar, R. Isatin and its derivatives: A survey of recent syntheses, reactions, and applications. *Medchemcomm* **2019**, *10*, 351–368. [CrossRef]
20. Singh, S.N.; Regati, S.; Paul, A.K.; Layek, M.; Jayaprakash, S.; Reddy, K.V.; Deora, G.S.; Mukherjee, S.; Pal, M. Cu-mediated 1, 3-dipolar cycloaddition of azomethine ylides with dipolarophiles: A faster access to spirooxindoles of potential pharmacological interest. *Tetrahedron Lett.* **2013**, *54*, 5448–5452. [CrossRef]
21. Senwar, K.R.; Reddy, T.S.; Thummuri, D.; Sharma, P.; Naidu, V.G.M.; Srinivasulu, G.; Shankaraiah, N. Design, synthesis and apoptosis inducing effect of novel (Z)-3-(3′-methoxy-4′-(2-amino-2-oxoethoxy)-benzylidene) indolin-2-ones as potential antitumour agents. *Eur. J. Med. Chem.* **2016**, *118*, 34–46. [CrossRef] [PubMed]

22. Ammar, Y.A.; El-Sharief, A.M.S.; Belal, A.; Abbas, S.Y.; Mohamed, Y.A.; Mehany, A.B.M.; Ragab, A. Design, synthesis, antiproliferative activity, molecular docking and cell cycle analysis of some novel (morpholinosulfonyl) isatins with potential EGFR inhibitory activity. *Eur. J. Med. Chem.* **2018**, *156*, 918–932. [CrossRef] [PubMed]
23. Zhu, C.; Li, X.; Zhao, B.; Peng, W.; Li, W.; Fu, W. Discovery of aryl-piperidine derivatives as potential antipsychotic agents using molecular hybridization strategy. *Eur. J. Med. Chem.* **2020**, *193*, 112214. [CrossRef] [PubMed]
24. Maldonado-Santiago, M.; Santiago, Á.; Pastor, N.; Alvarez, L.; Razo-Hernández, R.S. Isatin derivatives as DNA minor groove-binding agents: A structural and theoretical study. *Struct. Chem.* **2020**, *31*, 1289–1307. [CrossRef]
25. Ali Mohamed, H.; Ammar, Y.A.; AM Elhagali, G.; Eyada, H.A.; Aboul-Magd, D.S.; Ragab, A. In Vitro Antimicrobial Evaluation, Single-Point Resistance Study, and Radiosterilization of Novel Pyrazole Incorporating Thiazol-4-one/Thiophene Derivatives as Dual DNA Gyrase and DHFR Inhibitors against MDR Pathogens. *ACS omega* **2022**, *7*, 4970–4990. [CrossRef] [PubMed]
26. Mosmann, T. Rapid colorimetric assay for cellular growth and survival: Application to proliferation and cytotoxicity assays. *J. Immunol. Methods* **1983**, *65*, 55–63. [CrossRef]
27. Denizot, F.; Lang, R. Rapid colorimetric assay for cell growth and survival: Modifications to the tetrazolium dye procedure giving improved sensitivity and reliability. *J. Immunol. Methods* **1986**, *89*, 271–277. [CrossRef]
28. Kamiloglu, S.; Sari, G.; Ozdal, T.; Capanoglu, E. Guidelines for cell viability assays. *Food Front.* **2020**, *1*, 332–349. [CrossRef]
29. Oecd, T.N. 425: Acute Oral Toxicity: Up-and-Down Procedure. *OECD Guidel. Test. Chem. Sect.* **2008**, *4*, 1–27.
30. Khan, M.A.; Aljarbou, A.N.; Aldebasi, Y.H.; Alorainy, M.S.; Khan, A. Combination of glycosphingosomes and liposomal doxorubicin shows increased activity against dimethyl-α-benzanthracene-induced fibrosarcoma in mice. *Int. J. Nanomed.* **2015**, *10*, 6331. [CrossRef]
31. Mahesh, B.U.; Shrivastava, S.; Kuncha, M.; Sahu, B.D.; Swamy, C.V.; Pragada, R.R.; Naidu, V.G.M.; Sistla, R. Ethanolic extract of Boswellia ovalifoliolata bark and leaf attenuates doxorubicin-induced cardiotoxicity in mice. *Environ. Toxicol. Pharmacol.* **2013**, *36*, 840–849. [CrossRef]
32. Akhila, J.S.; Shyamjith, D.; Alwar, M.C. Acute toxicity studies and determination of median lethal dose. *Curr. Sci.* **2007**, 917–920.
33. Suvarna, K.S.; Layton, C.; Bancroft, J.D. *Bancroft's theory and practice of histological techniques E-Book*; Elsevier Health Sciences, 2018; ISBN 0702068861.
34. Culling, C.F.A. *Handbook of Histopathological and Histochemical Techniques: Including Museum Techniques*; Butterworth-Heinemann: Cambridge, UK, 2013; ISBN 1483164799.
35. Karrowni, W.; Li, Y.; Jones, P.G.; Cresci, S.; Abdallah, M.S.; Lanfear, D.E.; Maddox, T.M.; McGuire, D.K.; Spertus, J.A.; Horwitz, P.A. Insulin resistance is associated with significant clinical atherosclerosis in nondiabetic patients with acute myocardial infarction. *Arterioscler. Thromb. Vasc. Biol.* **2013**, *33*, 2245–2251. [CrossRef] [PubMed]
36. Eldehna, W.M.; Almahli, H.; Al-Ansary, G.H.; Ghabbour, H.A.; Aly, M.H.; Ismael, O.E.; Al-Dhfyan, A.; Abdel-Aziz, H.A. Synthesis and in vitro anti-proliferative activity of some novel isatins conjugated with quinazoline/phthalazine hydrazines against triple-negative breast cancer MDA-MB-231 cells as apoptosis-inducing agents. *J. Enzyme Inhib. Med. Chem.* **2017**, *32*, 600–613. [CrossRef] [PubMed]
37. Rezki, N.; Almehmadi, M.A.; Ihmaid, S.; Shehata, A.M.; Omar, A.M.; Ahmed, H.E.A.; Aouad, M.R. Novel scaffold hopping of potent benzothiazole and isatin analogues linked to 1, 2, 3-triazole fragment that mimic quinazoline epidermal growth factor receptor inhibitors: Synthesis, antitumor and mechanistic analyses. *Bioorg. Chem.* **2020**, *103*, 104133. [CrossRef] [PubMed]
38. Treger, S.; Ackerman, S.; Kaplan, V.; Ghanem, S.; Nadir, Y. Progestin type affects the increase of heparanase level and procoagulant activity mediated by the estrogen receptor. *Hum. Reprod.* **2021**, *36*, 61–69. [CrossRef] [PubMed]
39. Shamroukh, A.H.; El-Shahat, M.; Drabowicz, J.; Ali, M.M.; Rashad, A.E.; Ali, H.S. Anticancer evaluation of some newly synthesized N-nicotinonitrile derivative. *Eur. J. Med. Chem.* **2013**, *69*, 521–526. [CrossRef] [PubMed]
40. Fayed, E.A.; Ammar, Y.A.; Saleh, M.A.; Bayoumi, A.H.; Belal, A.; Mehany, A.B.M.; Ragab, A. Design, synthesis, antiproliferative evaluation, and molecular docking study of new quinoxaline derivatives as apoptotic inducers and EGFR inhibitors. *J. Mol. Struct.* **2021**, *1236*, 130317. [CrossRef]
41. Beutler, E. Improved method for the determination of blood glutathione. *J. Lab. Clin. Med.* **1963**, *61*, 882–888.
42. Du, H.; Chen, B.; Jiao, N.-L.; Liu, Y.-H.; Sun, S.-Y.; Zhang, Y.-W. Elevated glutathione peroxidase 2 expression promotes cisplatin resistance in lung adenocarcinoma. *Oxid. Med. Cell. Longev.* **2020**, *2020*, 7370157. [CrossRef] [PubMed]
43. Gavino, V.C.; Miller, J.S.; Ikharebha, S.O.; Milo, G.E.; Cornwell, D.G. Effect of polyunsaturated fatty acids and antioxidants on lipid peroxidation in tissue cultures. *J. Lipid Res.* **1981**, *22*, 763–769. [CrossRef]
44. De Leon, J.A.D.; Borges, C.R. Evaluation of Oxidative Stress in Biological Samples Using the Thiobarbituric Acid Reactive Substances Assay. *JoVE J. Vis. Exp.* **2020**, *159*, e61122. [CrossRef]
45. El-Sharief, A.M.S.; Ammar, Y.A.; Belal, A.; El-Sharief, M.A.M.S.; Mohamed, Y.A.; Mehany, A.B.M.; Ali, G.A.M.E.; Ragab, A. Design, synthesis, molecular docking and biological activity evaluation of some novel indole derivatives as potent anticancer active agents and apoptosis inducers. *Bioorg. Chem.* **2019**, *85*, 399–412. [CrossRef] [PubMed]
46. Wu, Z.; Liao, X.; Yuan, L.; Wang, Y.; Zheng, Y.; Zuo, J.; Pan, Y. Visible-Light-Mediated Click Chemistry for Highly Regioselective Azide–Alkyne Cycloaddition by a Photoredox Electron-Transfer Strategy. *Chem. Eur. J.* **2020**, *26*, 5694–5700. [CrossRef] [PubMed]
47. Leal-Esteban, L.C.; Fajas, L. Cell cycle regulators in cancer cell metabolism. *Biochim. Biophys. Acta Mol. Basis Dis.* **2020**, *1866*, 165715. [CrossRef] [PubMed]

48. Bai, J.; Li, Y.; Zhang, G. Cell cycle regulation and anticancer drug discovery. *Cancer Biol. Med.* **2017**, *14*, 348–362. [CrossRef] [PubMed]
49. Balachandran, C.; Haribabu, J.; Jeyalakshmi, K.; Bhuvanesh, N.S.P.; Karvembu, R.; Emi, N.; Awale, S. Nickel (II) bis (isatin thiosemicarbazone) complexes induced apoptosis through mitochondrial signaling pathway and G0/G1 cell cycle arrest in IM-9 cells. *J. Inorg. Biochem.* **2018**, *182*, 208–221. [CrossRef] [PubMed]
50. Yousef, M.A.; Ali, A.M.; El-Sayed, W.M.; Qayed, W.S.; Farag, H.H.A.; Aboul-Fadl, T. Design and synthesis of novel isatin-based derivatives targeting cell cycle checkpoint pathways as potential anticancer agents. *Bioorg. Chem.* **2020**, *105*, 104366. [CrossRef] [PubMed]
51. Shapiro, G.I.; Harper, J.W. Anticancer drug targets: Cell cycle and checkpoint control. *J. Clin. Investig.* **1999**, *104*, 1645–1653. [CrossRef] [PubMed]
52. Pfeffer, C.M.; Singh, A.T.K. Apoptosis: A target for anticancer therapy. *Int. J. Mol. Sci.* **2018**, *19*, 448. [CrossRef] [PubMed]
53. Pistritto, G.; Trisciuoglio, D.; Ceci, C.; Garufi, A.; D'Orazi, G. Apoptosis as anticancer mechanism: Function and dysfunction of its modulators and targeted therapeutic strategies. *Aging* **2016**, *8*, 603. [CrossRef] [PubMed]
54. Cao, L.; Zhang, L.; Zhao, X.; Zhang, Y. A hybrid chalcone combining the trimethoxyphenyl and isatinyl groups targets multiple oncogenic proteins and pathways in hepatocellular carcinoma cells. *PLoS ONE* **2016**, *11*, e0161025. [CrossRef]
55. Lee, T.; Lau, T.; Ng, I. Doxorubicin-induced apoptosis and chemosensitivity in hepatoma cell lines. *Cancer Chemother. Pharmacol.* **2002**, *49*, 78–86. [CrossRef] [PubMed]
56. Zhang, Q.; Fu, Y.; Zhao, Y.; Cui, S.; Wang, J.; Liu, F.; Yuan, Y.; Galons, H.; Yu, P.; Teng, Y. 5-Acetamido-1-(methoxybenzyl) isatin inhibits tumor cell proliferation, migration, and angiogenesis. *RSC Adv.* **2019**, *9*, 36690–36698. [CrossRef]
57. Mekkawy, A.H.; Pourgholami, M.H.; Morris, D.L. Involvement of urokinase-type plasminogen activator system in cancer: An overview. *Med. Res. Rev.* **2014**, *34*, 918–956. [CrossRef]
58. Mahmood, N.; Mihalcioiu, C.; Rabbani, S.A. Multifaceted role of the urokinase-type plasminogen activator (uPA) and its receptor (uPAR): Diagnostic, prognostic, and therapeutic applications. *Front. Oncol.* **2018**, *8*, 24. [CrossRef] [PubMed]
59. Vine, K.L.; Indira Chandran, V.; Locke, J.M.; Matesic, L.; Lee, J.; Skropeta, D.; Bremner, J.B.; Ranson, M. Targeting urokinase and the transferrin receptor with novel, anti-mitotic N-alkylisatin cytotoxin conjugates causes selective cancer cell death and reduces tumor growth. *Curr. Cancer Drug Targets* **2012**, *12*, 64–73. [CrossRef] [PubMed]
60. Maxhimer, J.B.; Quiros, R.M.; Stewart, R.; Dowlatshahi, K.; Gattuso, P.; Fan, M.; Prinz, R.A.; Xu, X. Heparanase-1 expression is associated with the metastatic potential of breast cancer. *Surgery* **2002**, *132*, 326–333. [CrossRef]
61. El-Assal, O.N.; Yamanoi, A.; Ono, T.; Kohno, H.; Nagasue, N. The clinicopathological significance of heparanase and basic fibroblast growth factor expressions in hepatocellular carcinoma. *Clin. Cancer Res.* **2001**, *7*, 1299–1305. [PubMed]
62. Coombe, D.R.; Gandhi, N.S. Heparanase: A challenging cancer drug target. *Front. Oncol.* **2019**, 1316. [CrossRef] [PubMed]
63. Jayatilleke, K.M.; Hulett, M.D. Heparanase and the hallmarks of cancer. *J. Transl. Med.* **2020**, *18*, 1–25. [CrossRef] [PubMed]
64. Knelson, E.H.; Nee, J.C.; Blobe, G.C. Heparan sulfate signaling in cancer. *Trends Biochem. Sci.* **2014**, *39*, 277–288. [CrossRef] [PubMed]
65. Simons, M.; Gordon, E.; Claesson-Welsh, L. Mechanisms and regulation of endothelial VEGF receptor signalling. *Nat. Rev. Mol. Cell Biol.* **2016**, *17*, 611–625. [CrossRef] [PubMed]
66. Hu, X.; Zhang, L.; Jin, J.; Zhu, W.; Xu, Y.; Wu, Y.; Wang, Y.; Chen, H.; Webster, K.A.; Chen, H. Heparanase released from mesenchymal stem cells activates integrin beta1/HIF-2alpha/Flk-1 signaling and promotes endothelial cell migration and angiogenesis. *Stem Cells* **2015**, *33*, 1850–1862. [PubMed]

biomedicines

MDPI

Article

8-Hydroxydaidzein, an Isoflavone from Fermented Soybean, Induces Autophagy, Apoptosis, Differentiation, and Degradation of Oncoprotein BCR-ABL in K562 Cells

Pei-Shan Wu [1,†], Jui-Hung Yen [2,3,†], Chih-Yang Wang [4,5], Pei-Yi Chen [2,6], Jui-Hsiang Hung [7] and Ming-Jiuan Wu [1,*]

1 Department of Pharmacy, Chia Nan University of Pharmacy and Science, Tainan 717, Taiwan; dc7575@gmail.com
2 Department of Molecular Biology and Human Genetics, Tzu Chi University, Hualien 970, Taiwan; imyenjh@mail.tcu.edu.tw (J.-H.Y.); pyc571@gmail.com (P.-Y.C.)
3 Institute of Medical Sciences, Tzu Chi University, Hualien 970, Taiwan
4 Ph.D. Program for Cancer Molecular Biology and Drug Discovery, Taipei Medical University, Taipei 11031, Taiwan; chihyang@tmu.edu.tw
5 Graduate Institute of Cancer Biology and Drug Discovery, Taipei Medical University, Taipei 11031, Taiwan
6 Center of Medical Genetics, Buddhist Tzu Chi General Hospital, Hualien 970, Taiwan
7 Department of Biotechnology, Chia Nan University of Pharmacy and Science, Tainan 717, Taiwan; hung86@mail.cnu.edu.tw
* Correspondence: imwu@gm.cnu.edu.tw or mingjiuanwu@gmail.com; Tel.: +886-6-2664911 (ext. 2520)
† Both authors contributed equally to this work.

Received: 29 October 2020; Accepted: 13 November 2020; Published: 16 November 2020

Abstract: 8-Hydroxydaidzein (8-OHD, 7,8,4′-trihydoxyisoflavone) is a hydroxylated derivative of daidzein isolated from fermented soybean products. The aim of this study is to investigate the anti-proliferative effects and the underlying mechanisms of 8-OHD in K562 human chronic myeloid leukemia (CML) cells. We found that 8-OHD induced reactive oxygen species (ROS) overproduction and cell cycle arrest at the S phase by upregulating $p21^{Cip1}$ and downregulating cyclin D2 (CCND2) and cyclin-dependent kinase 6 (CDK6) expression. 8-OHD also induced autophagy, caspase-7-dependent apoptosis, and the degradation of BCR-ABL oncoprotein. 8-OHD promoted Early Growth Response 1 (EGR1)-mediated megakaryocytic differentiation as an increased expression of marker genes, *CD61* and *CD42b*, and the formation of multi-lobulated nuclei in enlarged K562 cells. A microarray-based transcriptome analysis revealed a total of 3174 differentially expressed genes (DEGs) after 8-OHD (100 μM) treatment for 48 h. Bioinformatics analysis of DEGs showed that hemopoiesis, cell cycle regulation, nuclear factor-κB (NF-κB), and mitogen-activated protein kinase (MAPK) and Janus kinase/signal transducers and activators of transcription (JAK-STAT)-mediated apoptosis/anti-apoptosis networks were significantly regulated by 8-OHD. Western blot analysis confirmed that 8-OHD significantly induced the activation of MAPK and NF-κB signaling pathways, both of which may be responsible, at least in part, for the stimulation of apoptosis, autophagy, and differentiation in K562 cells. This is the first report on the anti-CML effects of 8-OHD and the combination of experimental and in silico analyses could provide a better understanding for the development of 8-OHD on CML therapy.

Keywords: K562; 8-hydroxydaidzein; apoptosis; autophagy; BCR-ABL; MAPK; NF-κB

1. Introduction

Chronic myeloid leukemia (CML) is a hematopoietic disease characterized by the expression of the BCR-ABL fusion oncoprotein, which is caused by a reciprocal translocation between chromosomes 9 and 22 [t(9;22)(q34;q11)] [1]. BCR-ABL fusion protein exerts a deregulated tyrosine kinase activity that activates proliferative and anti-apoptotic signaling pathways and leads to the malignant expansion of pluripotent stem cells in bone marrow [2]. Historically, CML was treated with hematopoietic stem cell transplantation, chemotherapy, and interferon-α [3]. Small molecule tyrosine kinase inhibitors (TKIs), such as imatinib, bosutinib, dasatinib, nilotinib, radotinib, and ponatinib, have been developed to treat CML by blocking the kinase domain of the BCR-ABL oncoprotein. However, a variety of adverse effects, such as off-target and metabolic toxicities, resistance to TKI therapy due to BCR-ABL mutations, and progression to advanced disease, render the need to pursue novel therapeutic strategies [3,4].

Since the BCR-ABL protein plays a crucial role in the pathogenesis of CML, its degradation is a new strategy for overcoming the problem of TKI resistance [5,6]. Targeting approaches include ubiquitin–proteasome, ubiquitin–lysosome, autophagy–lysosome, and caspase-mediated degradation pathways [5]. For example, arsenic sulfide As_4S_4 and diterpenoid oridonin promote ubiquitin–proteasome degradation of BCR-ABL [7,8]. Andrographolide, geldanamycin (GA), and 17-allylamino-17-demethoxygeldanamycin (17-AAG) downregulate the BCR-ABL oncoprotein by inducing HSP90 cleavage [9,10]. Arsenic trioxide (As_2O_3) promotes p62/SQSTM1-mediated autophagic degradation of the BCR-ABL [11]. Platinum pyrithione and xanthohumol downregulate BCR-ABL level through caspase-mediated degradation [12,13].

Isoflavones are dietary phytoestrogens mainly produced by Fabaceae family [14]. They are able to compete with estrogen hormone, 17β-estradiol, for the ligand-binding domain of the estrogen receptors and are used in the prevention and/or treatment of cancers, cardiovascular disease, osteoporosis, and postmenopausal symptoms [15–17]. Furthermore, genistein can inhibit tyrosine kinase activity and displays cancer chemopreventive activity [18]. 8-Hydroxydaidzein (8-OHD, 7,8,4′-trihydroxyisoflavone, NSC-678112) is a hydroxylated derivative of daidzein commonly isolated from fermented soybean products. It has recently attracted research attention due to its strong anti-inflammatory, antioxidant, antitumor, anti-melanogenesis, and hepatoprotective activities [19–24]. It was found that 8-OHD had strong anti-proliferative activity toward HL-60 human promyelocytic leukemia cells [25].

In this study, we evaluated the anti-CML effects of 8-OHD with a focus on cell cycle arrest, apoptosis, autophagy, differentiation, and BCR-ABL levels in K562 cells, which is a model cell line derived from a female CML patient in blast crisis [26]. High-throughput microarray analysis is a valuable tool for early-stage drug discovery decision-making due to its potential to detect adverse effects at the transcriptomic level [27]. Thus, microarray analysis was employed to explore the differentially expressed genes (DEGs) in 8-OHD-treated K562 cells. Then, Gene Ontology (GO), (Kyoto Encyclopedia of Genes and Genomes) KEGG, and BioCarta pathways were analyzed [28–30]. We further established a protein–protein interaction (PPI) network of DEGs to screen out the hub genes with a high degree of connectivity in the biological process.

Our data showed that 8-OHD induces reactive oxygen species (ROS) overproduction and cell cycle arrest at the S phase by upregulating $p21^{Cip1}$ and downregulating cyclin-dependent kinase 6 (CDK6) and cyclin D2 (CCND2) expression. Caspase-7-dependent apoptosis and p62/UVRAV-dependent autophagy may be responsible for the decreased level of BCR-ABL. Furthermore, 8-OHD promoted Early Growth Response 1 (EGR1)-mediated megakaryocytic differentiation. 8-OHD also activated mitogen-activated protein kinase (MAPK) and nuclear factor-κB (NF-κB) signaling pathways, and both may be responsible for the stimulation of apoptosis, autophagy, and differentiation in K562 cells. This is the first transcriptomics report on the anti-leukemic effects of 8-OHD, and the obtained results can provide a more thorough understanding about potential development of 8-OHD in CML therapy.

2. Experimental Section

2.1. Preparation of 8-Hydroxydaidzein (8-OHD, 7,8,4′-trihydoxyisoflavone, NSC-678112)

8-OHD was isolated from soybean fermented by *Aspergillus oryzae*, and the NMR spectral data and purification of 8-OHD was reported previously [20,21,24]. The chemical structure of 8-OHD was shown in Figure 1a.

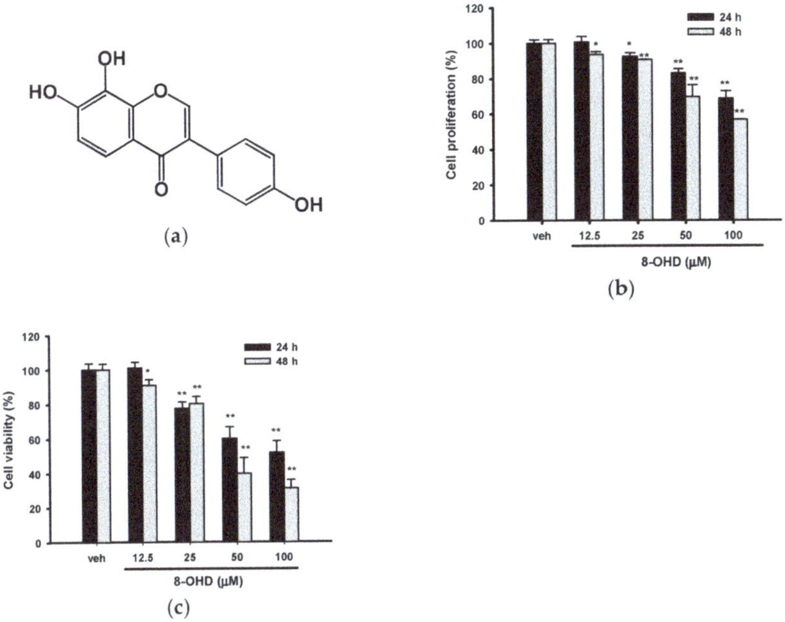

Figure 1. Effects of 8-hydroxydaidzein (8-OHD) on the cell viability of K562 cells. (**a**) Chemical structure of 8-OHD. K562 cells were treated with vehicle (0.1% DMSO) or 8-OHD (6.25–100 μM) for 24 and 48 h. (**b**) Cell viability was measured by 3-(4,5-dimethylthiazol-2-yl)-2,5-diphenyltetrazolium bromide (MTT) assay. (**c**) Cell viability was examined using trypan blue exclusion test. The experiments were repeated three times. These data represent the mean ± SD of three independent experiments. * $p < 0.05$ and ** $p < 0.01$ represent significant differences compared with the vehicle-treated cells.

2.2. Cell Culture

The K562 cells were provided by the Bioresource Collection and Research Center (Hsinchu, Taiwan) and cultured in RPMI-1640 medium supplemented with 10% fetal bovine serum (FBS), 1% nonessential amino acids (NEAA), 100 units/mL of penicillin, and 100 μg/mL of streptomycin (Thermo Fisher Scientific, Inc., Rockford, IL, USA) in a 5% CO_2 incubator at 37 °C.

2.3. Cell Proliferation and Viability Analysis

K562 cells (5×10^5/mL) were treated with vehicle (0.1% DMSO) or 8-OHD (12.5–100 μM) for 24 h or 48 h. Cell viability was analyzed by adding 1/10 volume of 0.5% MTT (3-(4,5-dimethylthiazol-2-yl)-2,5-diphenyltetrazolium bromide, in PBS) and incubating for another 3 h. Then, MTT solution was removed by centrifugation, and the formazan crystals produced inside cells were dissolved by DMSO, and the absorbance at 550 nm was measured spectrophotometrically [31]. The number of viable cells after treatment was further accessed by the trypan blue exclusion test as described in the literature [32].

2.4. Cell Cycle Analysis

K562 cells were synchronized by serum starvation overnight prior to shifting cells to 8-OHD-containing normal medium for 24 h. Then, cells were washed twice with PBS and fixed in ice-cold 70% ethanol overnight. The fixed cells were stained with 1 mL DNA-staining buffer (20 μg/mL of propidium iodide and 50 μg/mL of RNase in PBS) in the dark at 4 °C for 15 min before flow cytometry analysis (FACScan, BD Biosciences, San Jose, CA, USA). The singlet cell population was gated on the dot plot of FL2-A vs. FL2-W to exclude cell debris and aggregates. To evaluate the cell cycle, FL2-A histogram of gated population was analyzed by FlowJo software (FlowJo v7.6, LLC, Ashland, OR, USA) with Dean-Jett-Fox model.

2.5. Intracellular Reactive Oxygen Species (ROS) Assay

The fluorescence probe 2′,7′-dichlorodihydrofluorescein diacetate (H_2DCFDA) (Invitrogen) was used to measure ROS production. K562 cells were incubated with 8-OHDA (25–100 μM) for 24 h. Cells were collected by centrifugation and then washed with PBS. Subsequently, cells were loaded with 200 μL of 10 μM H_2DCFDA under dark for 30 min. Then, cells were washed twice with cold PBS and analyzed by fluorometer at Ex/Em: 495/530 nm. The relative ROS production from 10,000 cells was determined as the percentage of control after background subtraction [33].

2.6. Western Blot Analysis

Total cell lysate was prepared from cultured K562 cells using radioimmunoprecipitation assay buffer (RIPA buffer), while nuclear extracts were by a nuclear extraction kit (Cayman Chemical, Ann Arbor, Michigan, USA). Then, Bradford assay was employed to measure the protein concentration (Bio-Rad Laboratories, Hercules, CA, USA).

Equal amounts of protein were subjected to separate on 5–12% SDS-PAGE. Following electrophoretic separation, the proteins were transferred to a polyvinylidene difluoride (PVDF) membrane and then were blocked with freshly made buffer (5% skim milk in PBS with 0.05% Tween 20, pH 7.4). Then, the membrane was probed with specific primary antibody (Table 1) overnight at 4 °C. After rinsing, horseradish peroxidase (HRP)-conjugated secondary antibody (Jackson ImmunoResearch, West Grove, PA, USA) was then added and incubated for 1 h. The antigen–antibody reaction was detected using enhanced chemiluminescence detection (GE Healthcare, Wauwatosa, WI, USA).

Table 1. Primary antibodies used in Western blotting.

Antibody	Company	Catalog Number
α-Tubulin	Sigma	T6199
β-Actin	Genetex	GTX629630
c-ABL	Cell Signaling	2862
Caspase-3	Cell Signaling	9662
Caspase-7	Cell Signaling	9492
CDK6	Cell Signaling	13331
Cyclin D2	Cell Signaling	3741
JNK2	Cell Signaling	9258
Lamin A/C	Genetex	GTX101127
LC3B	Genetex	GTX127375
P21	Cell Signaling	2947
p38 MAPK	Cell Signaling	9212
p44/42 MAP Kinase	Cell Signaling	4695
p65 NF-κB	Cell Signaling	8242
PARP1	Santa Cruz	Sc-7150
Phoshpo-p38 MAPK	Cell Signaling	9215
Phoshpo-p44/42 MAPK	Cell Signaling	4370
Phospho-JNK1/2	Cell Signaling	4668
Phospho-p65 NF-κB	Cell Signaling	3033

2.7. RNA Extraction and Reverse Transcription Quantitative PCR

RNA was extracted from K562 cells with the Illustra RNA Spin Mini RNA Isolation Kit (GE Healthcare, Wauwatosa, WI, USA). High-Capacity cDNA Archive kit (Thermo Fisher Scientific, Inc., Rockford, IL, USA) was used for cDNA synthesis. Quantitative PCR was performed using a Power SYBR Green PCR Master Mix (Thermo Fisher Scientific, Inc., Rockford, IL, USA) in a total volume of 20 µL that contained 0.4 µM of each primer (Table 2). PCR program consisted of a pre-incubation for 2 min at 95 °C, followed by 40 cycles at 94 °C for 15 s and 60 °C for 60 s (ABI StepOne Real Time PCR System). Specificity verification was done by the melting curve. The relative mRNA expression was normalized with β-actin expression and then calculated by comparative Ct method.

Table 2. The primer pairs used in real-time PCR.

Gene	Primer Sequence (5′-3′)	Amplicon (bp)
β-actin	CATGTACGTTGCTATCCAGGC CTCCTTAATGTCACGCACGAT	250
ITGB3 (CD61)	TGTATGGGACTCAAGATTGGA AGCGATGGCTATTAGGTTCA	187
GP1BA (CD42b)	TCTGTATCAGAAGCCCTGTCTTCAC GCATCGGGAGCTTTGTCTTG	112
EGR1	CCGCAGAGTCTTTTCCTGAC TGGGTTGGTCATGCTCACTA	200
CD69	GCTGGACTTCAGCCCAAAATGC AGTCCAACCCAGTGTTCCTCTC	121
HECW2	GAGTATCCGCAGAACCATGACC GGTAGAGTGAGGCAGGATGTTC	133
SQSTM1 (P62)	AAGCCGGGTGGGAATGTTG CCTGAACAGTTATCCGACTCCAT	116

2.8. Analysis of Cell Morphology

K562 cells were incubated with vehicle (0.1% DMSO) or 8-OHDA (100 µM) for 24 and 48 h. Suspension cells (2.5×10^4) were added on Polysine™ slides and centrifuged at 200 rpm for 5 min using Cytospin™ 4 Cytocentrifuge (Thermo Fisher Scientific, Inc., Rockford, IL, USA). Slides were carefully removed and air-dried prior to staining with Wright-Giemsa dye (MilliporeSigma, St. Louis, MO, USA). The cell morphology and lobulation of the nucleus were analyzed as described previously [34].

2.9. Microarray Analysis

Cells were treated with vehicle (0.1% DMSO) or 8-OHD (100 µM) for 48 h. Microarray analysis was performed as previously described [35,36]. Briefly, fluorescent antisense RNA (aRNA) targets were prepared from 1 µg total RNA samples using Onearray Amino Allyl aRNA Amplification Kit (Phalanx Biotech Group, Hsinchu, Taiwan) and Cy5 dyes (Amersham Pharmacia, Piscataway, NJ, USA). Fluorescent targets were hybridized to the Human One Array Plus Version 7.1 (HOA 7.1, Phalanx) with phalanx hybridization buffer using Phalanx OneArray Plus Protocol. The signals of interest were scanned via an Agilent G2505C scanner using Agilent 0.1 XDR Protocol (Agilent Technologies, Santa Clara, CA, USA). The fluorescence intensity of each spot was analyzed and processed by GenePix 4.1 software (Molecular Devices, Sunnyvale, CA, USA) and a Rosetta Resolver System (Rosetta Biosoftware, Seattle, WA, USA), respectively. Differentially expressed genes (DEGs) related with the treatment of 8-OHD were screened, and those with fold change values larger than 2 or less than −2 and $p < 0.05$ were selected.

2.10. Gene Ontology, KEGG, and Biocarta Pathways and Protein–Protein Interaction Analysis

Gene Ontology (GO) term analysis [28] as well as KEGG [30] and Biocarta [29] Pathways of DEGs were further analyzed using the Database for Annotation, Visualization, and Integrated Discovery (DAVID, http://david.ncifcrf.gov) (version 6.8), an online biological information database, and $p < 0.05$ was used as the cut-off criterion [37]. The protein–protein interactions were analyzed using STRING version 11 [38] on 2 October 2020 (https://string-db.org/).

2.11. Pathway Enrichment and Process Network Analysis

Molecular functions and the pathway of Gene Ontology (GO) term in MetaCore (GeneGo, Inc., St. Joseph, MI, USA) were further used to screen and analyze the signaling pathway and process networks modulated by 8-OHD. The REVIGO web-based tool was used to summarize and remove redundant GO terms. Gene Set Enrichment Analysis (GSEA) [39] was used to determine the statistically significant molecular pathway with 8-OHD treatment.

2.12. Statistical Analysis

All experiments were repeated at least three times, and the values were expressed as the mean ± SD. The results were analyzed using one-way ANOVA with Dunnett's post hoc test, and a p value < 0.05 was considered statistically significant.

3. Results and Discussion

3.1. 8-OHD Reduces the Proliferation and Viability of K562 Cells

The K562 cell line is a well-known model of human chronic myeloid leukemia (CML). We first examined the effect of 8-OHD on K562 cell proliferation by MTT assay. Figure 1b shows that 8-OHD (12.5–100 μM) significantly decreased cell growth in a dose- and time-dependent manner. Cell proliferation decreased to 68.7% and 56.8% of vehicle control, 24 and 48 h after 100 μM 8-OHD treatment, respectively. It has been shown that MTT assay sensitivity decreases as cell concentration increases and causes an overestimation of viability [40]. Thus, we reconfirmed the cell viability by trypan blue exclusion assay. Figure 1c shows that 8-OHD exerted inhibitory effects on K562 cell viability in a dose- and time-dependent manner. The IC_{50} (half maximal cell viability inhibitory concentration) were 91.8 μM at 24 h and 49.4 μM at 48 h. Our results were in accordance with previous studies, which show that 8-OHD at high concentrations, 40 μM and 100 μM, exerted more than 40% cytotoxicity in B16 murine melanoma and Caco-2 cells, respectively [41,42].

3.2. 8-OHD Causes Cell Cycle Arrest at S phase

We further investigated how 8-OHD affects the cell population distribution in the cell cycle by staining the cellular DNA with propidium iodide (PI) in K562 cells. As shown in Figure 2a–c, 8-OHD caused a significant increase in S phase cell population, from about 32.2% of vehicle control to about 37.3% of 50 μM 8-OHD ($p < 0.05$) and 44.4% of 100 μM 8-OHD ($p < 0.01$). This was accompanied with decreases in the proportion of cells in the G_2/M phase of the cell cycle. Since cell proliferation is inhibited by 8-OHD (Figure 1), it seems unlikely that the increase in the S phase represents an increase in cells that are actively replicating DNA [43]. This rather indicates that K562 cells are arrested in the S phase upon treatment with 8-OHD (50–100 μM).

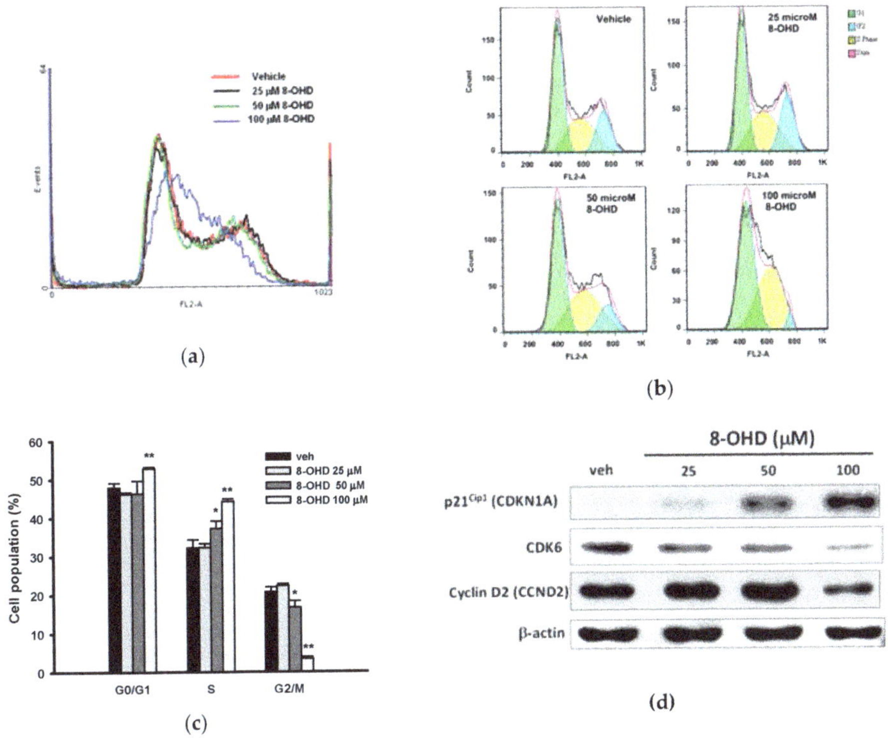

Figure 2. Effects of 8-OHD on the cell cycle distribution and regulatory gene expression. K562 cells were treated with vehicle or 8-OHD (25, 50, and 100 μM) for 24 h, and distribution of the cell cycle was measured by flow cytometric analysis. (**a**) FL2-A histogram overlay of K562 cells treated with various concentrations of 8-OHD. (**b**) FlowJo software analysis of DNA cell cycle. (**c**) The cell populations in G_0/G_1, S, and G_2/M phases were quantified as described in Section 2.4. The experiments were replicated three times. These data represent the mean ± SD of independent experiments. * $p < 0.05$ and ** $p < 0.01$ represent significant differences compared with the vehicle-treated cells. (**d**) Western blot analysis of expression of $p21^{Cip1}$, CDK6, cyclin D2, and β-actin of K562 cells treated with 8-OHD (25, 50, and 100 μM) for 24 h.

It has been shown that overexpression of the CDK inhibitor $p21^{Cip1}$ (CDKN1A, also known as $p21^{WAF1}$) induces S-phase arrest [43,44]. Figure 2d shows that the $p21^{Cip1}$ protein was upregulated by 8-OHD (25–100 μM) in a dose-dependent manner. It has been reported that $p21^{Cip1}$ is an inhibitor of the cyclin D-CDK complexes [45]. Thus, we further examined the expression of cyclin-dependent kinase 6 (CDK6) and cyclin D2 (CCND2) by Western blot analysis. As shown in Figure 2c, CDK6 and cyclin D2 were dose-dependently inhibited by 8-OHD (25–100 μM). Thus, 8-OHD induced K562 cell cycle arrest at the S phase by upregulating $p21^{Cip1}$ so as to downregulate the expression of CDK6 and cyclin D2.

3.3. *8-OHD Causes ROS Production and Apoptosis in K562 Cells*

It was found previously that 8-OHD induced ROS production and caused cell death in Caco-2 cells [42]. Therefore, we accessed the intracellular ROS production of K562 cells. Figure 3a shows that 8-OHD (25–100 μM) significantly increased ROS production in a dose-dependent manner in K562 cells, and 8-OHD (100 μM) enhanced about 400% ROS production as compared with vehicle control after 24 h treatment ($p < 0.01$).

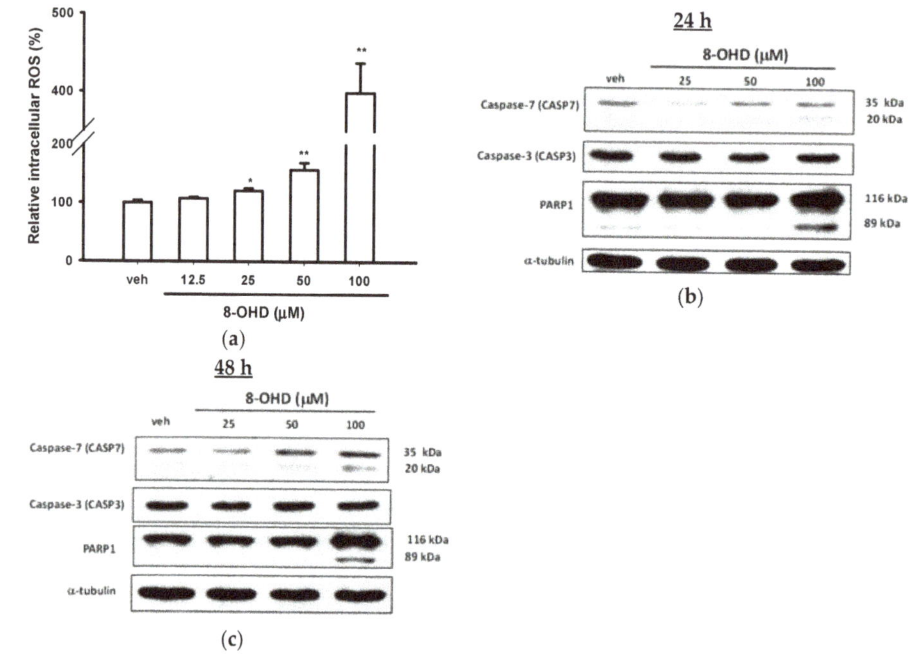

Figure 3. Effects of 8-OHD on reactive oxygen species (ROS) production and apoptosis in K562 cells. (**a**) K562 cells were exposed to various concentrations of 8-OHD for 24 h. Intracellular ROS was measured by DCF fluorescence as described in Materials and Methods. These data represent the mean ± SD of independent experiments. * $p < 0.05$ and ** $p < 0.01$ represent significant differences compared with the vehicle-treated cells. (**b**,**c**) Western blot analysis of apoptosis-related gene expression 24 and 48 h after 8-OHD treatment.

We further investigated whether K562 cells undergo apoptosis in response to 8-OHD. Caspase-3 (CASP3) and -7 (CASP7) are the most important executioner caspases involved in apoptotic cells [46]. Figure 3b shows that the treatment of K562 cells with 8-OHD (100 μM) for 24 h induced an increase in the expression and cleavage of caspase-7, which was more prominent after 48 h treatment (Figure 3c). In contrast, 8-OHD did not change the expression or activation of caspase-3 in K562 cells. It has been reported that caspase-3 and -7 are at the center of a fundamentally recursive apoptotic program, where they act as redundant signal amplifiers that are critical for activation of upstream and downstream apoptotic processes [47]. Our data indicate that caspase-7-dependent and caspase-3-independent apoptotic pathway is induced by 8-OHD.

Poly (ADP-ribose) polymerase-1 (PARP1) is the cellular substrate of caspases [48], caspase-7 is the most efficient caspase in PARP1 cleavage [49], and the proteolytic inactivation of PARP1 by caspases ensures the execution of the apoptotic process [50]. Figure 3b, c show that in parallel to caspase-7 activation, treatment of K562 cells with 100 μM 8-OHD for 24 and 48 h induced an increase in PARP1 cleavage.

It is well-known that natural compounds, such as vitamins and polyphenols, may play dual roles, antioxidant and pro-oxidant, in the prevention of cancer formation and cancer treatment [51–53]. It has been shown that 8-OHD, at lower μM concentrations, had strong free radical scavenging activities and decreased ROS production in cell-free and BV2 microglial cells [19,20,24,25]. On the other hand, 8-OHD could stimulate oxidative stress and caused Caco-2 cell death at higher concentrations [42]. Here, we reported that 8-OHD (50 and 100 μM) could stimulate ROS overproduction, which may cause caspase-7-dependent apoptosis in K562 cells.

3.4. 8-OHD Causes Autophagy and Decreases BCR-ABL Oncoprotein in K562 Cells

LC3s (MAP1LC3A, B, and C), the autophagosomal membrane proteins, are the most widely used markers of autophagosomes [54]. The conversion of LC3B from LC3BI (free form, ≈18 kDa) to LC3BII (phosphatidylethanolamine-conjugated form, ≈ 16 kDa) is an initiating step in autophagy in mammals [55]. Figure 4a,b show a dose-dependent increase in the LC3BII 24 and 48 h after 8-OHD (25–100 μM) treatment in K562 cells.

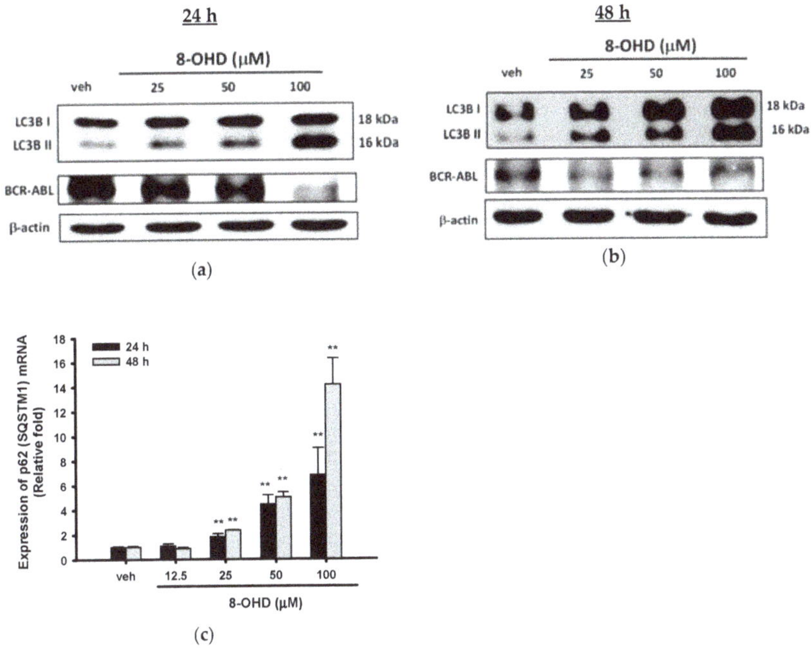

Figure 4. Effects of 8-OHD on autophagy and BCR-ABL protein level in K562 cells. (**a**,**b**) Western blot analysis of autophagy-related gene expression and BCR-ABL oncoprotein in K562 cells treated with 8-OHD for 24 and 48 h. (**c**) mRNA expression of p62/ SQSTM1 was analyzed by RT-Q-PCR as described in Materials and Methods. ** $p < 0.01$ represents significant differences compared with the vehicle-treated cells.

BCR-ABL tyrosine kinase is responsible for the proliferation and malignant transformation of CML. Therefore, we next investigated whether 8-OHD inhibits cell proliferation through decreasing the BCR-ABL protein level in K562 cells. As shown in Figure 4a, 8-OHD (100 μM) exerts a marked inhibitory effect on BCR-ABL expression in K562 cells after 24 h treatment. Furthermore, BCR-ABL levels were significantly attenuated by lower concentrations of 8-OHD (25–100 μM) after 48 h treatment (Figure 4b).

It was approved that ubiquitin-associated protein p62 (SQSTM1, autophagy receptor sequestosome 1), which binds to LC3, is an autophagy marker [56,57]. Increased expression of p62 (SQSTM1) was found in phorbol ester PMA- and the p38 MAPK inhibitor SB202190-induced autophagy in K562 cells [58]. Figure 4c shows that 8-OHD (25–100 μM) also stimulated *p62* (*SQSTM1*) mRNA upregulation in a dose- and time-dependent manner.

It has been reported that BCR-ABL protein degradation can be mediated via autophagy–lysosome pathway or caspase activation [5,11,59]. The above data indicate that 8-OHD causes cell death through caspase-7-dependent apoptosis and autophagy pathways, and both pathways may be responsible for, at least in part, the attenuation of the BCR-ABL protein level in K562 cells.

3.5. 8-OHD Promotes EGR1-Induced K562 Cell Differentiation

K562 cells represent stem cell precursors of the myeloid lineage and can differentiate into megakaryocyte/erythroid or granulocyte/macrophage lineages after exposure to various inducers [60,61]. It has been reported that autophagy is required for myeloid differentiation [58,62]. Furthermore, *p62* (*SQSTM1*) upregulation was found during megakaryocyte differentiation of K562 cells [58]. In addition, the BCR-ABL protein has been shown to confer on cells the characteristics of differentiation inhibition [63]. Since 8-OHD induced autophagy and *p62* expression, as well as decreased BCR-ABL protein level, we further investigated its effects on the expression of genes associated with megakaryocyte/erythroid differentiation. As shown in Figure 5a, 8-OHD (25–100 μM) significantly increased the mRNA expression of *CD61* (integrin β3, *ITGB3*), a megakaryocytic marker gene, 24 and 48 h after treatment. *CD61* mRNA expression was increased by 5.0-, 13.5- and 21.5-fold, 48 h after 25, 50, and 100 μM 8-OHD treatment ($p < 0.01$), respectively.

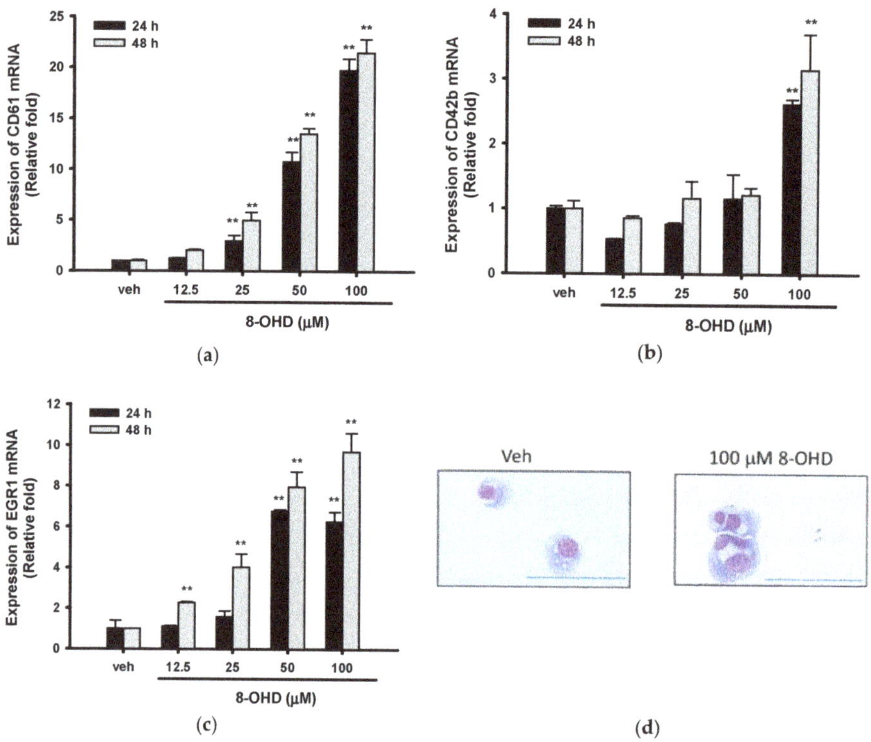

Figure 5. Effects of 8-OHD on the differentiation of K562 cells. (**a**–**c**) mRNA expression of CD61, CD42b, and EGR1 were analyzed by RT-Q-PCR as described in Materials and Methods. ** $p < 0.01$ represents significant differences compared with the vehicle-treated cells. (**d**) Cells were treated with vehicle (0.1% DMSO) or 8-OHD (100 μM) for 48 h. K562 cells were stained with Wright–Giemsa dye and the cell morphological changes and multi-lobulation of the nucleus were observed under microscopy. The scale bar is 100 μm.

8-OHD (100 μM) also induced *CD42b* (Glycoprotein Ib Platelet Subunit Alpha, *GP1BA*) expression by 2.6- and 3.1-fold, 24 and 48 h after treatment in K562 cells, respectively ($p < 0.01$) (Figure 5b). Early growth response 1 (EGR1) is a transcriptional regulator that is involved in promoting the differentiation of K562 cells and suppressing leukemia tumorigenesis [34,64,65]. 8-OHD (25–100 μM) exerted a dose- and time-dependent induction effect on *EGR1* mRNA expression (Figure 5c). 8-OHD

(100 μM) upregulated *EGR1* mRNA by 6.3- and 9.7-fold as compared with vehicle control, 24 and 48 h after treatment, respectively ($p < 0.01$).

It has been reported that multi-lobulation of the nucleus is a unique morphologic change during K562 megakaryocytic differentiation [34]. Figure 5d shows 100 μM 8-OHD promoted the formation of large cells with multi-lobulated nuclei 48 h after treatment. Microscope examination found that the ratio of cells with multi-lobulated nuclei increased from 0.83 ± 0.06% of untreated to 1.04 ± 0.13% and 3.75 ± 0.20% after 24 and 48 h 8-OHD (100 μM) treatment, respectively. This result indicates that a high concentration of 8-OHD (100 μM) induced K562 differentiation.

3.6. *Analysis of 8-OHD-Modulated Gene Expression*

From the above data, we discovered that a high concentration of 8-OHD (100 μM) induces apoptosis, autophagy, cell differentiation, and BCR-ABL downregulation in K562 cells. Recently, transcriptome analysis provides an ideal method to investigate molecular changes in response to stress or drug and is a valuable tool in early-stage drug discovery decision making [27]. Thus, we further investigated gene expression in 8-OHD-treated cells by human genome-wide microarray analysis. K562 cells were treated with vehicle or 8-OHD (100 μM) for 48 h, and transcriptome profiles were analyzed using the Human OneArray system, which contained 25,765 known genes. The general profile of transcriptome change is illustrated as a volcano plot (Figure 6a). Those with fold change values larger than 2 or less than −2, and $p < 0.05$ were selected as differentially expressed genes (DEGs). Among these, 1522 were upregulated and 1652 were downregulated by 100 μM 8-OHD. The most significantly down-related gene is *CRIP3* ($p = 2.7 \times 10^{-7}$), encoding a cysteine-rich protein with little known function. The most significantly upregulated gene is *SGMS2* ($p = 1.0 \times 10^{-7}$), encoding an enzyme sphingomyelin synthase 2, which is a regulator of cell surface levels of ceramide: an important mediator of signal transduction and apoptosis [66].

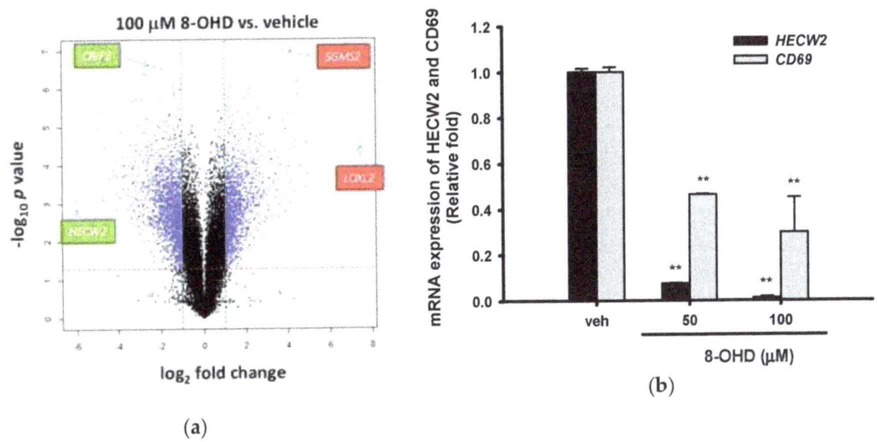

Figure 6. Microarray analysis of 8-OHD-treated K562 cells. (**a**) Volcano plots of total gene expression profiles of the K562 cells treated with 8-OHD (100 μM) for 48 h. Each dot represents the mean expression (n = 2) of the individual gene obtained from a microarray normalized dataset. The green and red dotted lines (cut-off values) were established according to the following parameters: fold= |2| and *p*-value = 5%, respectively. Genes above the cut-off lines have been considered as differentially expressed genes (DEG) and are shown as blue dots. (**b**) RT-Q-PCR analysis of HECW2 and CD69 expression in K562 cells 48 h after 8-OHD treatment (50 and 100 μM). ** $p < 0.01$ represents significant differences compared with the vehicle-treated cells.

The 30 DEGs with the most fold change (15 up- and 15 downregulated) are presented in Table S1. The most dramatic downregulated *HECW2* (downregulated by 64-fold) is a member of a family of E3 ubiquitin ligases, which play an important role in the proliferation, migration and differentiation. It has been reported that *HECW2* was one of the most downregulated genes by dasatinib, which is a selective tyrosine kinase receptor inhibitor that is used in the therapy of chronic myelogenous leukemia (CML) [67]. RT-Q-PCR in Figure 6b reconfirmed that 8-OHD (50 and 100 μM) repressed *HECW2* transcription to 0.075- and 0.011-fold of vehicle, respectively, 48 h after treatment ($p < 0.01$).

Microarray data also show that *CD69* expression was strongly downregulated by 8-OHD (down to 0.07-fold of vehicle). CD69, a member of the calcium-dependent lectin superfamily of type II transmembrane receptors, was reported as the marker for BCR-ABL activity in chronic myeloid leukemia [67]. In addition, CD69 was found to inhibit the apoptosis and differentiation in K562 cells [68]. RT-Q-PCR in Figure 6b validated that 8-OHD (50 and 100 μM) inhibits *CD69* mRNA expression after 48 h treatment. The above data support the notion that 8-OHD decreases BCR-ABL levels, which may cause *CD69* downregulation and in turn induce apoptosis and differentiation in K562 cells.

We also found that *LOXL2*, which encodes lysyl oxidase-like protein 2, was upregulated by more than 180-fold after 48 h treatment with 8-OHD (100 μM) (Figure 6a and Table S1). Both the down- and upregulation of *LOX2* in tumor tissues and cancer cell lines has been reported in the literature, suggesting a paradoxical role for *LOX2* as a tumor suppressor and a metastasis promoter gene [69]. The exact role of *LOXL2* overexpression on cell proliferation/differentiation in 8-OHD-treated K562 cells remains unclear at this stage.

3.7. *Analysis of 8-OHD-Modulated Biological Processes*

To determine which Gene Ontology (GO) term was modulated by 8-OHD (100 μM), we performed an analysis of the enrichment in the relevant ontological groups from the GO database. A whole set of DEGs consisting of 3174 genes was subjected to functional annotation and clusterization using the Database for Annotation, Visualization, and Integrated Discovery (DAVID) bioinformatics tools.

Table S2 presented the only GO BP (biological process) groups that fulfill the following criteria: $p < 0.05$ and the minimal number of genes per group ≥ 30. REVIGO was used for the visualization of a non-redundant GO term set in multiple ways to assist in interpretation. The scatterplot shows the cluster representatives in a two-dimensional space derived by applying multidimensional scaling to a matrix of the GO terms' semantic similarities [70]. It was found that small GTPase-mediated signal transduction, autophagy, transcription from RNA polymerase II promoter, and protein phosphorylation processes were significantly changed in the 8-OHD-treated cells (Figure 7a).

Furthermore, Over Representation Analysis (ORA) is a widely used approach to determine whether known biological functions or processes are enriched in DEGs, and GO-elite was a tool for ontology pruning [71]. It was found that biological processes related to oxidative stress, signal transduction, autophagy, cell death, and differentiation appeared significantly associated with upregulated DEGs in 8-OHD-treated groups. On the other hand, oxidative phosphorylation, cellular catabolic, and primary metabolic processes were significantly associated with downregulated DEGs in 8-OHD-treated groups (Figure 7b). These results were in good agreement with our experimental data that 8-OHD stimulates ROS, cell death, differentiation and changes of BCR-ABL tyrosine kinase signaling, as well as decreases normal cellular metabolism.

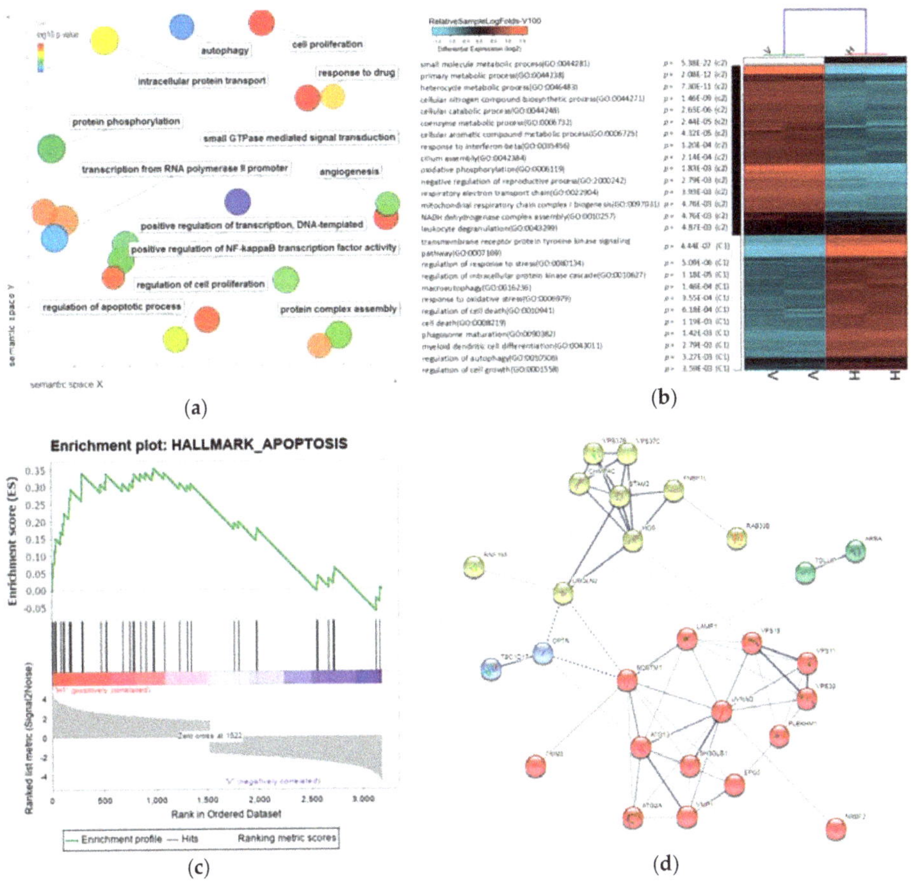

Figure 7. Enrichment analysis and visualization of Gene Ontology (GO), Gene Set Enrichment Analysis (GSEA), as well as predicted protein–protein interaction for DEGs in response to 8-OHD. (**a**) Significantly enriched GO BP (biological process) in 8-OHD (100 μM)- vs. vehicle (0.1% DMSO)-treated K562 cells was analyzed by REVIGO. The scatterplot represents functional clusters, the bubble color indicates the *p*-values of the GO analysis, and the bubble size indicates the frequency of the GO term in the underlying gene ontology database. (**b**) Heatmap of DEGs in response to 100 μM 8-OHD treatment. The GO listed in the upper half of the heatmap is gene ontology annotation BP of the downregulated DEGs. The lower processes are the BP of the upregulated DEGs. Only some of the representative GO BPs are shown. (**c**) Gene Set Enrichment Analysis (GSEA) demonstrates that the signature "Hallmark_Apoptosis" gene set is enriched in the DEGs of 8-OHD. The barcode plot indicates the position of the genes in each gene set. The horizontal bar in graded color from red to blue indicates positive and negative regulation by 8-OHD. The vertical axis in the lower plot indicates Ranked List Metric. (**d**) Predicted protein–protein interaction for 8-OHD-upregulated DEGs associated with autophagy (GO:0006914). The corresponding genes were uploaded to query the STRING interaction database (https://string-db.org/cgi/input.pl). Only the "Experiments", "Databases", and "Textmining" source options were selected, and the minimum interaction score was set to 0.4. For visual clarity, disconnected nodes were omitted from the interaction graph. Line thickness indicates the strength of data support. The identified clusters are colored differently. The solid and the dotted lines indicate connection within the same and different cluster, respectively.

To uncover dependencies of 8-OHD-treated apoptosis with DEGs, enrichment for genes in the signature "Hallmark_Apoptosis" using GSEA was employed [72]. We found that most of the genes associated with the signature "Hallmark_Apoptosis" were stimulated by 8-OHD (Figure 7c).

Table 3 shows 35 DEGs were associated with autophagy (GO: 0006914). Among them, 29 genes were upregulated, and six genes are downregulated by 100 μM 8-OHD. We further exploit protein interaction for those upregulated genes in this category by String analysis. Markov clustering algorithm (MCL) with inflation factor = 2 was employed to find cluster structure in graphs by a mathematical bootstrapping procedure [73]. As shown in Figure 7d, two clusters in the protein–protein interaction (PPI) network. One cluster with light green color proteins are mainly involved in intracellular signal transduction and vesicular trafficking process during autophagy. The other cluster with red color includes highly induced genes such as p62 (SQSTM1), autophagy-related proteins (ATG13 and ATG2A), and UVRAG. Among these, UVRAG has 10 edges within the cluster and is considered as a hub. UVRAG has been suggested to be involved in the maturation of autophagosomes and endocytosis during autophagy [74,75]. UVRAG upregulation also participates in the p62-mediated autophagic degradation of BCR/ABL in CML cells [76].

Table 3. Genes associated with autophagy categories (GO: 0006914) in response to 100 μM 8-OHD.

ID	Symbol	Description	log_2FC	p Value	q Value
8878	*SQSTM1*	autophagy receptor that interacts directly with both the cargo to become degraded and an autophagy modifier of the MAP1 LC3 family	3.48	1.79×10^{-4}	7.76×10^{-3}
54472	*TOLLIP*	component of the signaling pathway of IL-1 and Toll-like receptors.	2.37	7.55×10^{-4}	1.20×10^{-2}
55823	*VPS11*	vacuolar protein sorting-associated protein 11 homolog; plays a role in vesicle-mediated protein trafficking to lysosomal compartments including the endocytic membrane transport and autophagic pathway	2.3	3.22×10^{-4}	9.01×10^{-3}
10254	*STAM2*	signal transducing adapter molecule 2; Involved in intracellular signal transduction mediated by cytokines and growth factors.	2.15	1.61×10^{-3}	1.62×10^{-2}
3916	*LAMP1*	lysosomal associated membrane protein 1	1.98	7.97×10^{-4}	1.21×10^{-2}
51100	*SH3GLB1*	SH3 domain containing GRB2 like endophilin B1	1.87	5.95×10^{-3}	3.06×10^{-2}
81671	*VMP1*	vacuole membrane protein 1	1.8	8.26×10^{-5}	6.26×10^{-3}
85363	*TRIM5*	tripartite motif containing 5	1.75	1.77×10^{-2}	6.08×10^{-2}
23130	*ATG2A*	autophagy related 2A	1.74	1.07×10^{-3}	1.36×10^{-2}
91445	*RNF185*	ring finger protein 185	1.64	2.70×10^{-4}	8.72×10^{-3}
9146	*HGS*	hepatocyte growth factor-regulated tyrosine kinase substrate	1.62	1.18×10^{-3}	1.42×10^{-2}
9842	*PLEKHM1*	pleckstrin homology and RUN domain containing M1	1.61	1.70×10^{-3}	1.67×10^{-2}
79735	*TBC1D17*	TBC1 domain family member 17	1.57	1.47×10^{-4}	7.34×10^{-3}
83452	*RAB33B*	RAB33B, member RAS oncogene family	1.54	3.33×10^{-3}	2.26×10^{-2}
7405	*UVRAG*	UV radiation resistance associated	1.47	8.62×10^{-3}	6.33×10^{-3}
9776	*ATG13*	autophagy related 13	1.41	4.72×10^{-3}	2.70×10^{-2}
3428	*IFI16*	interferon gamma inducible protein 16	1.35	9.18×10^{-3}	3.97×10^{-2}
55048	*VPS37C*	VPS37C, ESCRT-I subunit	1.31	5.69×10^{-3}	2.99×10^{-2}
57617	*VPS18*	VPS18, CORVET/HOPS core subunit	1.31	2.87×10^{-4}	8.81×10^{-3}
79142	*PHF23*	PHD finger protein 23	1.3	2.16×10^{-3}	1.85×10^{-2}
54874	*FNBP1L*	formin binding protein 1 like	1.3	6.60×10^{-3}	3.25×10^{-2}
57724	*EPG5*	ectopic P-granules autophagy protein 5 homolog	1.26	1.06×10^{-3}	1.35×10^{-2}
79720	*VPS37B*	VPS37B, ESCRT-I subunit	1.25	3.11×10^{-6}	3.66×10^{-3}
29978	*UBQLN2*	ubiquilin 2	1.23	4.25×10^{-3}	2.55×10^{-2}
23339	*VPS39*	VPS39, HOPS complex subunit	1.23	1.29×10^{-3}	1.47×10^{-2}
92421	*CHMP4C*	charged multivesicular body protein 4C	1.18	1.76×10^{-2}	6.05×10^{-2}
410	*ARSA*	arylsulfatase A	1.18	9.04×10^{-3}	3.94×10^{-2}
10133	*OPTN*	optineurin	1.04	2.95×10^{-3}	2.14×10^{-2}
29982	*NRBF2*	nuclear receptor binding factor 2	1.01	5.32×10^{-4}	1.07×10^{-2}
7957	*EPM2A*	epilepsy, progressive myoclonus type 2A, Lafora disease (laforin)	−1.02	4.22×10^{-4}	9.84×10^{-3}
83734	*ATG10*	autophagy related 10	−1.29	8.40×10^{-5}	6.28×10^{-3}
51160	*VPS28*	VPS28, ESCRT-I subunit	−1.46	7.59×10^{-3}	3.52×10^{-2}
51028	*VPS36*	vacuolar protein sorting 36 homolog	−1.53	1.26×10^{-2}	4.84×10^{-2}
93343	*MVB12A*	multivesicular body subunit 12A	−1.86	1.28×10^{-4}	7.06×10^{-3}
84938	*ATG4C*	autophagy related 4C cysteine peptidase	−2.05	8.90×10^{-4}	1.26×10^{-2}

p62 (SQSTM1) has 10 edges (seven connect to the same cluster, and three connect to different clusters) and is considered as another hub. It has been reported that p62 (SQSTM1) inactivates the pro-oncogenic signaling through interaction with BCR-ABL and the subsequent autophagic degradation of BCR-ABL in CML [11]. In conclusion, the bioinformatics and Western blot (Figure 4) analyses suggest that 8-OHD decreased BCR-ABL possibly via autophagic degradation post-translationally.

3.8. Analysis of Signaling Pathways and Network Processes Using KEGG, BioCarta, and MetaCore

Twenty-one KEGG (Kyoto Encyclopedia of Genes and Genomes) and 25 BioCarta pathways were significantly associated with DEGs of 100 μM 8-OHD treatment (Supplemental Tables S3 and S4). Among these, the MAPK signal pathway was present in both KEGG and BioCarta pathways. 8-OHD-upregulated and downregulated genes involved in the KEGG MAPK signal pathway (hsa04010) were marked as red and yellow, respectively, in Figure 8a. Gene Set Enrichment Analysis (GSEA) reveals that the BioCarta MAPK pathway was predominantly correlated with upregulated DEGs of 8-OHD-treated K562 cells (Figure 8b).

Furthermore, we used the commercial MetaCore package to analyze the pathway enrichment by DEGs, and the top 15 significantly enriched pathways were shown in Figure 8c. Two MAPK-related and one oxidative stress-related signaling pathways were significantly regulated by 8-OHD. The canonical pathway map, which represents the neurogenesis Nerve growth factor/ Tropomyosin receptor kinase A (NGF/TrkA) MAPK signaling pathway in response to 8-OHD treatment, is shown in Figure S1. It has been reported that neurotrophin signaling endosomes are formed in CML cells as in neuronal cells, and imatinib treatment enhances NGF/TrkA-mediated signaling in K562 cells [77]. Current result indicates that 8-OHD enhanced NGF/TrkA MAPK signaling pathway, which may be due to its BCR-ABL lowered effect in K562 cells. Furthermore, NGF/TrkA MAPK signaling may be involved in 8-OHD-mediated K562 differentiation.

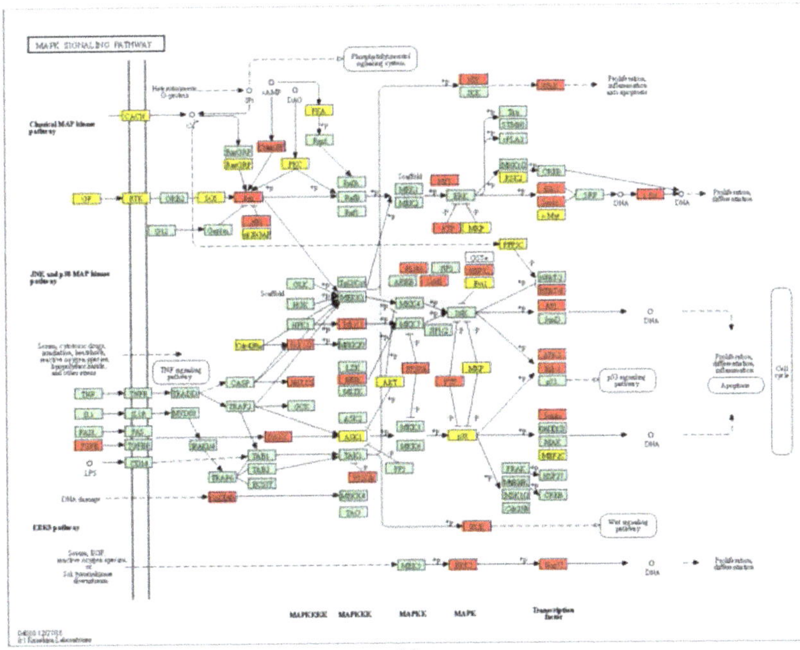

(a)

Figure 8. *Cont.*

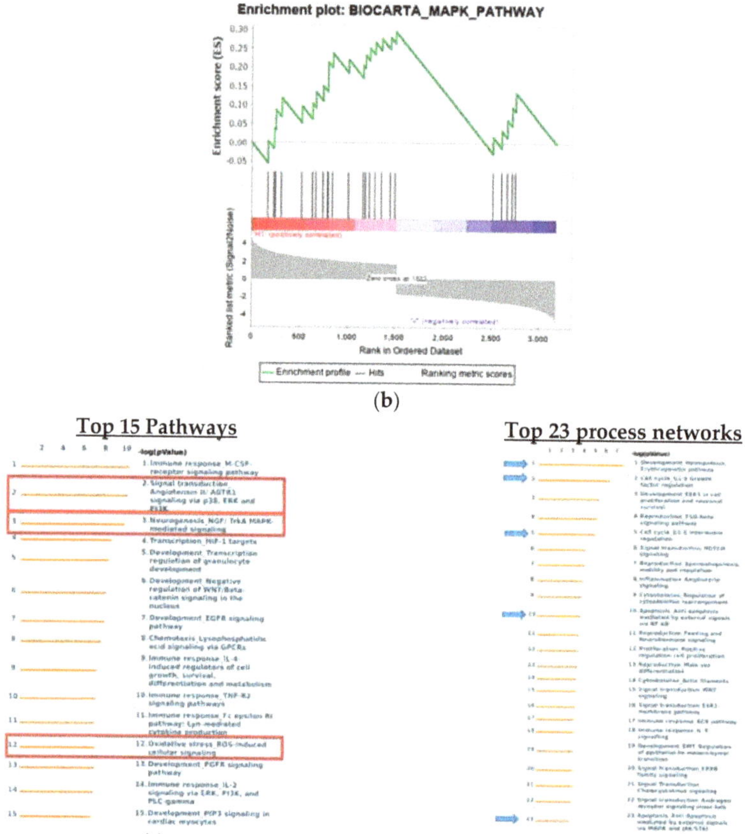

(b)

(c)

(d)

Figure 8. Signaling pathways and process networks associated with 8-OHD treatment. (**a**) KEGG MAPK signaling pathway (hsa04010) in K562 cells in response to 8-OHD (100 μM) treatment for 48 h. Expression changes of target genes are mapped by colors; red color: statistically significant increase in expression, yellow color: statistically significant decrease in expression, green color: expression statistically insignificant. (**b**) Gene Set Enrichment Analysis (GSEA) demonstrated that BioCarta MAPK pathway gene set was enriched in the DEGs of 8-OHD. (**c**) Top 15 significantly enriched pathways according to the MetaCore database in response to 8-OHD treatment. (**d**) Top 23 significantly enriched process networks according to the MetaCore database in response to 8-OHD treatment.

Comparative enrichment analysis of the DEG dataset by MetaCore identified top 23 process networks (Figure 8d). It was found that hemopoiesis, cell cycle regulation, NF-κB-, and MAPK-, and JAK/STAT-mediated apoptosis/anti-apoptosis networks are significantly regulated by 8-OHD. The JAK/STAT pathway is very important in transferring extracellular signals to the cell and initiates gene expression involved in cell proliferation, differentiation, survival, and developmental processes [78]. The network map of MAPK and JAK/STAT-mediated apoptosis/anti-apoptosis is shown in Figure S2. It has been reported that 8-OHD triggered apoptosis in breast cancer stem-like cells (BCSCs) through blocking of the JAK/STAT signaling pathway [79]. Here, we found that JAK/STAT-related genes, *SYK*, *STAT3*, *STAT5*, and *BIRC5*, were all downregulated by 8-OHD in K562 cells, indicating that the JAK/STAT signaling pathway was attenuated.

3.9. 8-OHD Modulates MAPK Pathways and NF-kB Activation

Since DAVID, Metacore, and GSEA analyses indicated the potential regulation of MAPK signaling pathways by 8-OHD, we continued to analyze MAPK activation by Western blot. It was found that 8-OHD (25–100 μM) significantly increased ERK (extracellular signal-regulated kinases), JNK (c-Jun N-terminal kinase), and p38 phosphorylation in a dose-dependent manner after 24 h treatment (Figure 9a). The activation of the ERK signaling pathway has been reportedly associated with K562 differentiation [34], while JNK activation is involved in ROS-mediated autophagy [80]. Furthermore, JNK and p38 phosphorylation have been implicated in apoptosis in K562 cells [81,82]. Here, we found that 8-OHD activated all three MAPK signaling pathways, which may be responsible for the differentiation and cell death in K562 cells.

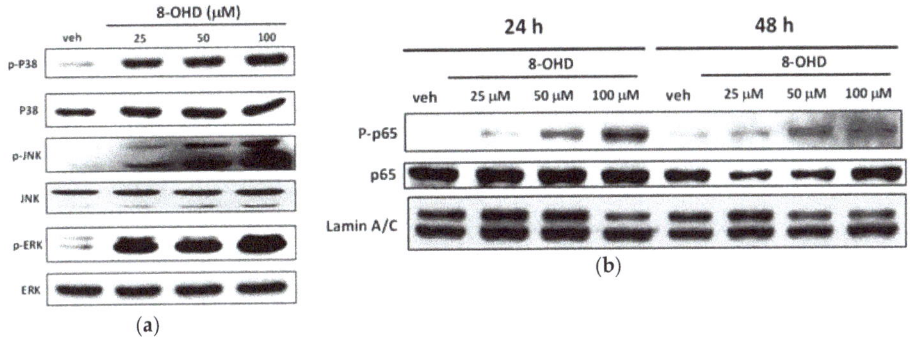

Figure 9. Western blotting analysis of MAPK and NF-kB activation. (**a**) Western blotting analysis of MAPK activation from total cell lysate of K562 cells treated with 8-OHD (25–100 μM) for 24 h. (**b**) Western blotting analysis of NF-κB activation from nuclear extract of K562 cells treated with 8-OHD (25–100 μM) for 24 and 48 h.

We also found that *RelA* (NF-κB p65) mRNA expression was induced by 8-OHD (Figure 8a). Thus, we continued investigating the effect of 8-OHD on NF-κB activation by analyzing the level of phosphorylation of p65 in nuclear extracts of K562 cells. It was found that the phosphorylation of nuclear p65 increased dose-dependently 24 and 48 h after 8-OHD (25–100 μM) treatments (Figure 9b). NF-κB is a multifunctional transcription factor that mediates pro- and anti-apoptotic as well as pro-autophagic signals [83,84]. In addition, NF-κB and MAP kinase pathways are involved in PMA-induced megakaryocytic differentiation of K562 cells [85]. NF-κB is a positive regulator for $p21^{Cip1}$ expression and induces cell cycle arrest and apoptosis [86]. Therefore, 8-OHD activated NF-κB, which may be responsible for the differentiation, cell cycle arrest, and apoptosis in K562 cells.

4. Conclusions

Our study provides the first evidence for the anti-CML activities and underlying molecular mechanisms of 8-OHD. Our results demonstrated that K562 cell proliferation was significantly inhibited by 8-OHD in a dose- and time-dependent manner. 8-OHD could induce K562 cell cycle arrest at the S phase by the upregulation of $p21^{Cip1}$, a negative regulator of the cell cycle, accompanied with downregulation of cyclin D (CCND2) and cyclin D-dependent kinase (CDK6). The treatment of K562 cells with 8-OHD (100 μM) resulted in ROS overproduction, caspase-7-dependent apoptosis, autophagy, and decreased BCR-ABL. Furthermore, 8-OHD also promoted EGR1-induced K562 differentiation as megakaryocytic marker genes, *CD61* and *CD42b*, were upregulated and the formation of large cells with multi-lobulated nuclei in K562 cells was found (Figure 10).

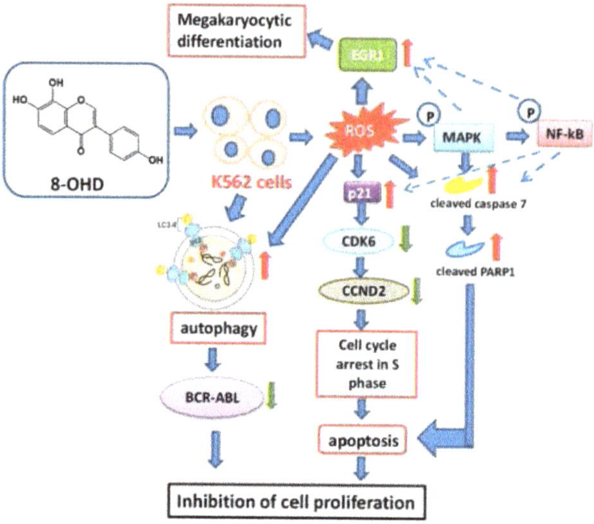

Figure 10. A hypothetical model for the anti-leukemic effects of 8-OHD in K562 cells. 8-OHD induces ROS overproduction, cell cycle arrest at the S phase, caspase 7-dependent apoptosis, and autophagy, as well as a decreased BCR-ABL oncoprotein level. It also promotes megakaryocytic differentiation in human CML K562 cells. MAPK and NF-κB signaling pathways may be responsible for the stimulation of apoptosis, autophagy, and differentiation in K562 cells.

A microarray-based transcriptome profiling was performed to offer a better understanding of the effects of 8-OHD in K562 cells. A total of 3174 differentially expressed genes (DEGs) (1522 upregulated and 1652 downregulated) were identified from K562 cells 48 h after 8-OHD (100 μM) incubation. The enrichment analysis showed that GO biological processes, such as small GTPase-mediated signal transduction, autophagy, transcription from RNA polymerase II promoter, and protein phosphorylation, were significantly associated with DEGs. Gene Set Enrichment Analysis (GSEA) reconfirms that most of the genes involved in the regulation of apoptosis were significantly stimulated by 8-OHD. Protein–protein interaction (PPI) network was constructed for upregulated DEGs associated with autophagy (GO: 0006914) using String. It was found that *p62/UVRAV*-mediated autophagy may be responsible for the decreased level of BCR-ABL.

The 8-OHD-modulated MAPK signaling pathway was confirmed using KEGG Pathview library, BioCarta and MetaCore package. The results of Western blot analysis reveal that 8-OHD (25–100 μM) significantly increased phosphorylation of ERK, JNK, and p38 MAPK in a dose-dependent manner in K562 cells. The phosphorylation of nuclear NF-κB-p65 was also increased by 8-OHD treatment dose-dependently. The combination of experimental and in silico analyses could provide a more thorough understanding about potential development of 8-OHD in CML therapy.

Supplementary Materials: The following are available online at http://www.mdpi.com/2227-9059/8/11/506/s1, Figure S1: MetaCore canonical pathway map showing the neurogenesis_NGF/TrkA MAPK signaling pathway in response to 8-OHD; Figure S2: MetaCore process network map showing apoptosis_anti-Apoptosis mediated by external signals via MAPK and JAK/STAT in response to 8-OHD; Table S1: The most significant regulated genes by 8-OHD (100 μM) in K562 cells; Table S2: Overrepresented gene sets in DAVID GO BP DIRECT annotations database obtained from comparisons in gene expression profiles between 100 μM 8-OHD treatment vs. vehicle control ($p < 0.05$ and minimal number of genes per group ≥ 30); Table S3: The KEGG pathways associated with 100 μM 8-OHD treatment in K562 cells; Table S4: The BioCarta pathways associated with 100 μM 8-OHD treatment in K562 cells.

Author Contributions: Conceptualization, M.-J.W. and J.-H.Y.; Data curation, P.-S.W. and C.-Y.W.; Formal analysis, P.-S.W. and C.-Y.W.; Funding acquisition, M.-J.W. and J.-H.Y.; Investigation, M.-J.W.; Methodology, P.-S.W., C.-Y.W. and P.-Y.C.; Project administration, M.-J.W.; Resources, J.-H.H.; Supervision, M.-J.W.; Validation, M.-J.W.;

Visualization, C.-Y.W. and M.-J.W.; Writing—original draft, P.-S.W.; Writing—review and editing, J.-H.Y. and M.-J.W. All authors have read and agreed to the published version of the manuscript.

Funding: This research was supported by research grants MOST 106-2320-B-041-006-MY3 and MOST-108-2320-B-320-002-MY3 from the Ministry of Science and Technology, Taiwan.

Acknowledgments: We thank Shiou-Yu Ding for technical support in supplying purified isoflavones.

Conflicts of Interest: The authors declare no conflict of interest.

References

1. Rowley, J.D. Letter: A new consistent chromosomal abnormality in chronic myelogenous leukaemia identified by quinacrine fluorescence and Giemsa staining. *Nature* **1973**, *243*, 290–293. [CrossRef]
2. Jabbour, E.; Cortes, J.E.; Giles, F.J.; O'Brien, S.; Kantarjian, H.M. Current and emerging treatment options in chronic myeloid leukemia. *Cancer* **2007**, *109*, 2171–2181. [CrossRef]
3. Westerweel, P.E.; Te Boekhorst, P.A.W.; Levin, M.D.; Cornelissen, J.J. New Approaches and Treatment Combinations for the Management of Chronic Myeloid Leukemia. *Front. Oncol.* **2019**, *9*, 665. [CrossRef] [PubMed]
4. Medeiros, B.C.; Possick, J.; Fradley, M. Cardiovascular, pulmonary, and metabolic toxicities complicating tyrosine kinase inhibitor therapy in chronic myeloid leukemia: Strategies for monitoring, detecting, and managing. *Blood Rev.* **2018**, *32*, 289–299. [CrossRef] [PubMed]
5. Zhu, H.Q.; Gao, F.H. Regulatory Molecules and Corresponding Processes of BCR-ABL Protein Degradation. *J. Cancer* **2019**, *10*, 2488–2500. [CrossRef] [PubMed]
6. Ablain, J.; Nasr, R.; Bazarbachi, A.; de The, H. The drug-induced degradation of oncoproteins: An unexpected Achilles' heel of cancer cells? *Cancer Discov.* **2011**, *1*, 117–127. [CrossRef] [PubMed]
7. Mao, J.H.; Sun, X.Y.; Liu, J.X.; Zhang, Q.Y.; Liu, P.; Huang, Q.H.; Li, K.K.; Chen, Q.; Chen, Z.; Chen, S.J. As4S4 targets RING-type E3 ligase c-CBL to induce degradation of BCR-ABL in chronic myelogenous leukemia. *Proc. Natl. Acad. Sci. USA* **2010**, *107*, 21683–21688. [CrossRef] [PubMed]
8. Huang, H.; Weng, H.; Dong, B.; Zhao, P.; Zhou, H.; Qu, L. Oridonin Triggers Chaperon-mediated Proteasomal Degradation of BCR-ABL in Leukemia. *Sci. Rep.* **2017**, *7*, 41525. [CrossRef] [PubMed]
9. Liu, S.H.; Lin, C.H.; Liang, F.P.; Chen, P.F.; Kuo, C.D.; Alam, M.M.; Maiti, B.; Hung, S.K.; Chi, C.W.; Sun, C.M.; et al. Andrographolide downregulates the v-Src and Bcr-Abl oncoproteins and induces Hsp90 cleavage in the ROS-dependent suppression of cancer malignancy. *Biochem. Pharmacol.* **2014**, *87*, 229–242. [CrossRef]
10. Nimmanapalli, R.; O'Bryan, E.; Bhalla, K. Geldanamycin and its analogue 17-allylamino-17-demethoxygeldanamycin lowers Bcr-Abl levels and induces apoptosis and differentiation of Bcr-Abl-positive human leukemic blasts. *Cancer Res.* **2001**, *61*, 1799–1804.
11. Goussetis, D.J.; Gounaris, E.; Wu, E.J.; Vakana, E.; Sharma, B.; Bogyo, M.; Altman, J.K.; Platanias, L.C. Autophagic degradation of the BCR-ABL oncoprotein and generation of antileukemic responses by arsenic trioxide. *Blood* **2012**, *120*, 3555–3562. [CrossRef] [PubMed]
12. Lan, X.; Zhao, C.; Chen, X.; Zhang, P.; Zang, D.; Wu, J.; Chen, J.; Long, H.; Yang, L.; Huang, H.; et al. Platinum pyrithione induces apoptosis in chronic myeloid leukemia cells resistant to imatinib via DUB inhibition-dependent caspase activation and Bcr-Abl downregulation. *Cell Death Dis.* **2017**, *8*, e2913. [CrossRef] [PubMed]
13. Lu, X.; Geng, J.; Zhang, J.; Miao, J.; Liu, M. Xanthohumol, a Prenylated Flavonoid from Hops, Induces Caspase-Dependent Degradation of Oncoprotein BCR-ABL in K562 Cells. *Antioxidants* **2019**, *8*, 402. [CrossRef] [PubMed]
14. Tsao, R. Chemistry and biochemistry of dietary polyphenols. *Nutrients* **2010**, *2*, 1231–1246. [CrossRef]
15. Wang, Q.; Ge, X.; Tian, X.; Zhang, Y.; Zhang, J.; Zhang, P. Soy isoflavone: The multipurpose phytochemical (Review). *Biomed. Rep.* **2013**, *1*, 697–701. [CrossRef]
16. Mann, G.E.; Bonacasa, B.; Ishii, T.; Siow, R.C. Targeting the redox sensitive Nrf2-Keap1 defense pathway in cardiovascular disease: Protection afforded by dietary isoflavones. *Curr. Opin. Pharmacol.* **2009**, *9*, 139–145. [CrossRef]
17. Vitale, D.C.; Piazza, C.; Melilli, B.; Drago, F.; Salomone, S. Isoflavones: Estrogenic activity, biological effect and bioavailability. *Eur. J. Drug Metab. Pharmacokinet.* **2013**, *38*, 15–25. [CrossRef]

18. Yan, G.R.; Xiao, C.L.; He, G.W.; Yin, X.F.; Chen, N.P.; Cao, Y.; He, Q.Y. Global phosphoproteomic effects of natural tyrosine kinase inhibitor, genistein, on signaling pathways. *Proteomics* **2010**, *10*, 976–986. [CrossRef]
19. Chang, T.S. Isolation, bioactivity, and production of ortho-hydroxydaidzein and ortho-hydroxygenistein. *Int. J. Mol. Sci.* **2014**, *15*, 5699–5716. [CrossRef]
20. Esaki, H.; Onozaki, H.; Morimitsu, Y.; Kawakishi, S.; Osawa, T. Potent Antioxidative Isoflavones Isolated from Soybeans Fermented with Aspergillus saitoi. *Biosci. Biotechnol. Biochem.* **1998**, *62*, 740–746. [CrossRef]
21. Fujita, T.; Funako, T.; Hayashi, H. 8-Hydroxydaidzein, an aldose reductase inhibitor from okara fermented with Aspergillus sp. HK-388. *Biosci. Biotechnol. Biochem.* **2004**, *68*, 1588–1590. [CrossRef] [PubMed]
22. Park, J.S.; Park, H.Y.; Kim, D.H.; Kim, D.H.; Kim, H.K. ortho-dihydroxyisoflavone derivatives from aged Doenjang (Korean fermented soypaste) and its radical scavenging activity. *Bioorg. Med. Chem. Lett.* **2008**, *18*, 5006–5009. [CrossRef] [PubMed]
23. Park, J.S.; Kim, D.H.; Lee, J.K.; Lee, J.Y.; Kim, D.H.; Kim, H.K.; Lee, H.J.; Kim, H.C. Natural ortho-dihydroxyisoflavone derivatives from aged Korean fermented soybean paste as potent tyrosinase and melanin formation inhibitors. *Bioorg. Med. Chem. Lett.* **2010**, *20*, 1162–1164. [CrossRef] [PubMed]
24. Wu, P.S.; Ding, H.Y.; Yen, J.H.; Chen, S.F.; Lee, K.H.; Wu, M.J. Anti-inflammatory Activity of 8-Hydroxydaidzein in LPS-Stimulated BV2 Microglial Cells via Activation of Nrf2-Antioxidant and Attenuation of Akt/NF-kappaB-Inflammatory Signaling Pathways, as Well As Inhibition of COX-2 Activity. *J. Agric. Food Chem.* **2018**, *66*, 5790–5801. [CrossRef] [PubMed]
25. Hirota, A.; Taki, S.; Kawaii, S.; Yano, M.; Abe, N. 1,1-Diphenyl-2-picrylhydrazyl radical-scavenging compounds from soybean miso and antiproliferative activity of isoflavones from soybean miso toward the cancer cell lines. *Biosci. Biotechnol. Biochem.* **2000**, *64*, 1038–1040. [CrossRef]
26. Lozzio, C.B.; Lozzio, B.B. Human chronic myelogenous leukemia cell-line with positive Philadelphia chromosome. *Blood* **1975**, *45*, 321–334. [CrossRef] [PubMed]
27. Verbist, B.; Klambauer, G.; Vervoort, L.; Talloen, W.; Consortium, Q.; Shkedy, Z.; Thas, O.; Bender, A.; Gohlmann, H.W.; Hochreiter, S. Using transcriptomics to guide lead optimization in drug discovery projects: Lessons learned from the QSTAR project. *Drug Discov. Today* **2015**, *20*, 505–513. [CrossRef]
28. The Gene Ontology, C. The Gene Ontology Resource: 20 years and still GOing strong. *Nucleic Acids Res.* **2019**, *47*, D330–D338. [CrossRef]
29. Nishimura, D. BioCarta. *Biotech Softw. Internet Rep.* **2001**, *2*, 117–120. [CrossRef]
30. Kanehisa, M.; Goto, S. KEGG: Kyoto encyclopedia of genes and genomes. *Nucleic Acids Res.* **2000**, *28*, 27–30. [CrossRef]
31. Carmichael, J.; DeGraff, W.G.; Gazdar, A.F.; Minna, J.D.; Mitchell, J.B. Evaluation of a tetrazolium-based semiautomated colorimetric assay: Assessment of chemosensitivity testing. *Cancer Res.* **1987**, *47*, 936–942. [PubMed]
32. Strober, W. Trypan Blue Exclusion Test of Cell Viability. *Curr. Protoc. Immunol.* **2015**, *111*, A3 B 1–A3 B 3. [CrossRef] [PubMed]
33. Hao, G.; Zhai, J.; Jiang, H.; Zhang, Y.; Wu, M.; Qiu, Y.; Fan, C.; Yu, L.; Bai, S.; Sun, L.; et al. Acetylshikonin induces apoptosis of human leukemia cell line K562 by inducing S phase cell cycle arrest, modulating ROS accumulation, depleting Bcr-Abl and blocking NF-kappaB signaling. *Biomed. Pharmacother.* **2020**, *122*, 109677. [CrossRef] [PubMed]
34. Yen, J.H.; Lin, C.Y.; Chuang, C.H.; Chin, H.K.; Wu, M.J.; Chen, P.Y. Nobiletin Promotes Megakaryocytic Differentiation through the MAPK/ERK-Dependent EGR1 Expression and Exerts Anti-Leukemic Effects in Human Chronic Myeloid Leukemia (CML) K562 Cells. *Cells* **2020**, *9*, 877. [CrossRef] [PubMed]
35. Chen, P.Y.; Chen, Y.T.; Gao, W.Y.; Wu, M.J.; Yen, J.H. Nobiletin Down-Regulates c-KIT Gene Expression and Exerts Antileukemic Effects on Human Acute Myeloid Leukemia Cells. *J. Agric. Food Chem.* **2018**, *66*, 13423–13434. [CrossRef] [PubMed]
36. Wang, C.-C.; Lin, Y.-C.; Cheng, Y.-H.; Tung, C.-W. Profiling transcriptomes of human SH-SY5Y neuroblastoma cells exposed to maleic acid. *PeerJ* **2017**, *5*, e3175. [CrossRef]
37. Huang da, W.; Sherman, B.T.; Lempicki, R.A. Bioinformatics enrichment tools: Paths toward the comprehensive functional analysis of large gene lists. *Nucleic Acids Res.* **2009**, *37*, 1–13. [CrossRef]

38. Szklarczyk, D.; Gable, A.L.; Lyon, D.; Junge, A.; Wyder, S.; Huerta-Cepas, J.; Simonovic, M.; Doncheva, N.T.; Morris, J.H.; Bork, P.; et al. STRING v11: Protein-protein association networks with increased coverage, supporting functional discovery in genome-wide experimental datasets. *Nucleic Acids Res.* **2019**, *47*, D607–D613. [CrossRef]
39. Subramanian, A.; Tamayo, P.; Mootha, V.K.; Mukherjee, S.; Ebert, B.L.; Gillette, M.A.; Paulovich, A.; Pomeroy, S.L.; Golub, T.R.; Lander, E.S.; et al. Gene set enrichment analysis: A knowledge-based approach for interpreting genome-wide expression profiles. *Proc. Natl. Acad. Sci. USA* **2005**, *102*, 15545–15550. [CrossRef]
40. Chung, D.M.; Kim, J.H.; Kim, J.K. Evaluation of MTT and Trypan Blue assays for radiation-induced cell viability test in HepG2 cells. *Int. J. Radiat. Res.* **2015**, *13*, 331–335. [CrossRef]
41. Goh, M.J.; Park, J.S.; Bae, J.H.; Kim, D.H.; Kim, H.K.; Na, Y.J. Effects of ortho-dihydroxyisoflavone derivatives from Korean fermented soybean paste on melanogenesis in B16 melanoma cells and human skin equivalents. *Phytother. Res.* **2012**, *26*, 1107–1112. [CrossRef] [PubMed]
42. Lo, Y.L. A potential daidzein derivative enhances cytotoxicity of epirubicin on human colon adenocarcinoma Caco-2 cells. *Int. J. Mol. Sci.* **2013**, *14*, 158–176. [CrossRef] [PubMed]
43. Hu, L.W.; Yen, J.H.; Shen, Y.T.; Wu, K.Y.; Wu, M.J. Luteolin modulates 6-hydroxydopamine-induced transcriptional changes of stress response pathways in PC12 cells. *PLoS ONE* **2014**, *9*, e97880. [CrossRef] [PubMed]
44. Zhu, H.; Zhang, L.; Wu, S.; Teraishi, F.; Davis, J.J.; Jacob, D.; Fang, B. Induction of S-phase arrest and p21 overexpression by a small molecule 2[[3-(2,3-dichlorophenoxy)propyl] amino]ethanol in correlation with activation of ERK. *Oncogene* **2004**, *23*, 4984–4992. [CrossRef]
45. Parry, D.; Mahony, D.; Wills, K.; Lees, E. Cyclin D-CDK subunit arrangement is dependent on the availability of competing INK4 and p21 class inhibitors. *Mol. Cell. Biol.* **1999**, *19*, 1775–1783. [CrossRef]
46. Walsh, J.G.; Cullen, S.P.; Sheridan, C.; Luthi, A.U.; Gerner, C.; Martin, S.J. Executioner caspase-3 and caspase-7 are functionally distinct proteases. *Proc. Natl. Acad. Sci. USA* **2008**, *105*, 12815–12819. [CrossRef]
47. McComb, S.; Chan, P.K.; Guinot, A.; Hartmannsdottir, H.; Jenni, S.; Dobay, M.P.; Bourquin, J.P.; Bornhauser, B.C. Efficient apoptosis requires feedback amplification of upstream apoptotic signals by effector caspase-3 or -7. *Sci. Adv.* **2019**, *5*, eaau9433. [CrossRef]
48. D'Amours, D.; Sallmann, F.R.; Dixit, V.M.; Poirier, G.G. Gain-of-function of poly(ADP-ribose) polymerase-1 upon cleavage by apoptotic proteases: Implications for apoptosis. *J. Cell Sci.* **2001**, *114*, 3771–3778.
49. Germain, M.; Affar, E.B.; D'Amours, D.; Dixit, V.M.; Salvesen, G.S.; Poirier, G.G. Cleavage of automodified poly(ADP-ribose) polymerase during apoptosis. Evidence for involvement of caspase-7. *J. Biol. Chem.* **1999**, *274*, 28379–28384. [CrossRef]
50. Los, M.; Mozoluk, M.; Ferrari, D.; Stepczynska, A.; Stroh, C.; Renz, A.; Herceg, Z.; Wang, Z.Q.; Schulze-Osthoff, K. Activation and caspase-mediated inhibition of PARP: A molecular switch between fibroblast necrosis and apoptosis in death receptor signaling. *Mol. Biol. Cell* **2002**, *13*, 978–988. [CrossRef]
51. Bezerra, D.P.; Militao, G.C.G.; de Morais, M.C.; de Sousa, D.P. The Dual Antioxidant/Prooxidant Effect of Eugenol and Its Action in Cancer Development and Treatment. *Nutrients* **2017**, *9*, 1367. [CrossRef] [PubMed]
52. Husain Yar, K.; Haseeb, Z.; Mohd Fahad, U.; Aamir, A.; Sheikh Mumtaz, H. A Prooxidant Mechanism for the Anticancer and Chemopreventive Properties of Plant Polyphenols. *Curr. Drug Targets* **2012**, *13*, 1738–1749. [CrossRef]
53. Schwartz, J.L. The dual roles of nutrients as antioxidants and prooxidants: Their effects on tumor cell growth. *J. Nutr.* **1996**, *126*, 1221S–1227S. [CrossRef]
54. Koukourakis, M.I.; Kalamida, D.; Giatromanolaki, A.; Zois, C.E.; Sivridis, E.; Pouliliou, S.; Mitrakas, A.; Gatter, K.C.; Harris, A.L. Autophagosome Proteins LC3A, LC3B and LC3C Have Distinct Subcellular Distribution Kinetics and Expression in Cancer Cell Lines. *PLoS ONE* **2015**, *10*, e0137675. [CrossRef] [PubMed]
55. Ohsumi, Y. Molecular dissection of autophagy: Two ubiquitin-like systems. *Nat. Rev. Mol. Cell Biol.* **2001**, *2*, 211–216. [CrossRef] [PubMed]
56. Moscat, J.; Diaz-Meco, M.T. p62 at the crossroads of autophagy, apoptosis, and cancer. *Cell* **2009**, *137*, 1001–1004. [CrossRef]
57. Feng, L.; Ma, Y.; Sun, J.; Shen, Q.; Liu, L.; Lu, H.; Wang, F.; Yue, Y.; Li, J.; Zhang, S.; et al. YY1-MIR372-SQSTM1 regulatory axis in autophagy. *Autophagy* **2014**, *10*, 1442–1453. [CrossRef]

58. Colosetti, P.; Puissant, A.; Robert, G.; Luciano, F.; Jacquel, A.; Gounon, P.; Cassuto, J.P.; Auberger, P. Autophagy is an important event for megakaryocytic differentiation of the chronic myelogenous leukemia K562 cell line. *Autophagy* **2009**, *5*, 1092–1098. [CrossRef]
59. Elzinga, B.M.; Nyhan, M.J.; Crowley, L.C.; O'Donovan, T.R.; Cahill, M.R.; McKenna, S.L. Induction of autophagy by Imatinib sequesters Bcr-Abl in autophagosomes and down-regulates Bcr-Abl protein. *Am. J. Hematol.* **2013**, *88*, 455–462. [CrossRef]
60. Sutherland, J.A.; Turner, A.R.; Mannoni, P.; McGann, L.E.; Turc, J.M. Differentiation of K562 leukemia cells along erythroid, macrophage, and megakaryocyte lineages. *J. Biol. Response Mod.* **1986**, *5*, 250–262.
61. Wang, H.; Hu, H.; Zhang, Q.; Yang, Y.; Li, Y.; Hu, Y.; Ruan, X.; Yang, Y.; Zhang, Z.; Shu, C.; et al. Dynamic transcriptomes of human myeloid leukemia cells. *Genomics* **2013**, *102*, 250–256. [CrossRef] [PubMed]
62. Gomez-Puerto, M.C.; Folkerts, H.; Wierenga, A.T.; Schepers, K.; Schuringa, J.J.; Coffer, P.J.; Vellenga, E. Autophagy Proteins ATG5 and ATG7 Are Essential for the Maintenance of Human CD34(+) Hematopoietic Stem-Progenitor Cells. *Stem Cells* **2016**, *34*, 1651–1663. [CrossRef] [PubMed]
63. Fang, G.; Kim, C.N.; Perkins, C.L.; Ramadevi, N.; Winton, E.; Wittmann, S.; Bhalla, K.N. CGP57148B (STI-571) induces differentiation and apoptosis and sensitizes Bcr-Abl-positive human leukemia cells to apoptosis due to antileukemic drugs. *Blood* **2000**, *96*, 2246–2253. [CrossRef]
64. Gibbs, J.D.; Liebermann, D.A.; Hoffman, B. Egr-1 abrogates the E2F-1 block in terminal myeloid differentiation and suppresses leukemia. *Oncogene* **2008**, *27*, 98–106. [CrossRef] [PubMed]
65. Cheng, T.; Wang, Y.; Dai, W. Transcription factor egr-1 is involved in phorbol 12-myristate 13-acetate-induced megakaryocytic differentiation of K562 cells. *J. Biol. Chem.* **1994**, *269*, 30848–30853. [PubMed]
66. Kitatani, K.; Taniguchi, M.; Okazaki, T. Role of Sphingolipids and Metabolizing Enzymes in Hematological Malignancies. *Mol. Cells* **2015**, *38*, 482–495. [CrossRef] [PubMed]
67. Hantschel, O.; Gstoettenbauer, A.; Colinge, J.; Kaupe, I.; Bilban, M.; Burkard, T.R.; Valent, P.; Superti-Furga, G. The chemokine interleukin-8 and the surface activation protein CD69 are markers for Bcr-Abl activity in chronic myeloid leukemia. *Mol. Oncol.* **2008**, *2*, 272–281. [CrossRef]
68. Huang, S.Y.; Liu, Y.H.; Chen, Y.J.; Yeh, Y.Y.; Huang, H.M. CD69 partially inhibits apoptosis and erythroid differentiation via CD24, and their knockdown increase imatinib sensitivity in BCR-ABL-positive cells. *J. Cell. Physiol.* **2018**, *233*, 7467–7479. [CrossRef]
69. Payne, S.L.; Hendrix, M.J.; Kirschmann, D.A. Paradoxical roles for lysyl oxidases in cancer—A prospect. *J. Cell. Biochem.* **2007**, *101*, 1338–1354. [CrossRef]
70. Supek, F.; Bosnjak, M.; Skunca, N.; Smuc, T. REVIGO summarizes and visualizes long lists of gene ontology terms. *PLoS ONE* **2011**, *6*, e21800. [CrossRef]
71. Zambon, A.C.; Gaj, S.; Ho, I.; Hanspers, K.; Vranizan, K.; Evelo, C.T.; Conklin, B.R.; Pico, A.R.; Salomonis, N. GO-Elite: A flexible solution for pathway and ontology over-representation. *Bioinformatics* **2012**, *28*, 2209–2210. [CrossRef] [PubMed]
72. Liberzon, A.; Birger, C.; Thorvaldsdottir, H.; Ghandi, M.; Mesirov, J.P.; Tamayo, P. The Molecular Signatures Database (MSigDB) hallmark gene set collection. *Cell Syst.* **2015**, *1*, 417–425. [CrossRef] [PubMed]
73. Enright, A.J.; Van Dongen, S.; Ouzounis, C.A. An efficient algorithm for large-scale detection of protein families. *Nucleic Acids Res.* **2002**, *30*, 1575–1584. [CrossRef] [PubMed]
74. Liang, C.; Feng, P.; Ku, B.; Dotan, I.; Canaani, D.; Oh, B.H.; Jung, J.U. Autophagic and tumour suppressor activity of a novel Beclin1-binding protein UVRAG. *Nat. Cell Biol.* **2006**, *8*, 688–699. [CrossRef] [PubMed]
75. Wu, S.; He, Y.; Qiu, X.; Yang, W.; Liu, W.; Li, X.; Li, Y.; Shen, H.M.; Wang, R.; Yue, Z.; et al. Targeting the potent Beclin 1-UVRAG coiled-coil interaction with designed peptides enhances autophagy and endolysosomal trafficking. *Proc. Natl. Acad. Sci. USA* **2018**, *115*, E5669–E5678. [CrossRef] [PubMed]
76. Huang, X.; Li, Y.; Shou, L.; Li, L.; Chen, Z.; Ye, X.; Qian, W. The molecular mechanisms underlying BCR/ABL degradation in chronic myeloid leukemia cells promoted by Beclin1-mediated autophagy. *Cancer Manag. Res.* **2019**, *11*, 5197–5208. [CrossRef]
77. Koch, A.; Scherr, M.; Breyer, B.; Mancini, A.; Kardinal, C.; Battmer, K.; Eder, M.; Tamura, T. Inhibition of Abl tyrosine kinase enhances nerve growth factor-mediated signaling in Bcr-Abl transformed cells via the alteration of signaling complex and the receptor turnover. *Oncogene* **2008**, *27*, 4678–4689. [CrossRef]
78. Luo, N.; Balko, J.M. Role of JAK-STAT Pathway in Cancer Signaling. In *Predictive Biomarkers in Oncology: Applications in Precision Medicine*; Badve, S., Kumar, G.L., Eds.; Springer International Publishing: Cham, Switzerland, 2019; pp. 311–319. [CrossRef]

79. Rajabi, S.; Shojaee, M.; Malmir, A.; Rezaei Tavirani, M.; Noori, S. Anti-Breast Cancer Activities of 8-Hydroxydaidzein by Targeting Breast Cancer Stem-Like Cells. *J. Pharm. Pharm. Sci.* **2020**, *23*, 47–57. [CrossRef]
80. Liu, Y.; Ren, Z.; Li, X.; Zhong, J.; Bi, Y.; Li, R.; Zhao, Q.; Yu, X. Pristimerin Induces Autophagy-Mediated Cell Death in K562 Cells through the ROS/JNK Signaling Pathway. *Chem. Biodivers.* **2019**, *16*, e1900325. [CrossRef]
81. Nica, A.F.; Tsao, C.C.; Watt, J.C.; Jiffar, T.; Kurinna, S.; Jurasz, P.; Konopleva, M.; Andreeff, M.; Radomski, M.W.; Ruvolo, P.P. Ceramide promotes apoptosis in chronic myelogenous leukemia-derived K562 cells by a mechanism involving caspase-8 and JNK. *Cell Cycle* **2008**, *7*, 3362–3370. [CrossRef]
82. Wu, X.P.; Xiong, M.; Xu, C.S.; Duan, L.N.; Dong, Y.Q.; Luo, Y.; Niu, T.H.; Lu, C.R. Resveratrol induces apoptosis of human chronic myelogenous leukemia cells in vitro through p38 and JNK-regulated H2AX phosphorylation. *Acta Pharmacol. Sin.* **2015**, *36*, 353–361. [CrossRef] [PubMed]
83. Copetti, T.; Bertoli, C.; Dalla, E.; Demarchi, F.; Schneider, C. p65/RelA modulates BECN1 transcription and autophagy. *Mol. Cell. Biol.* **2009**, *29*, 2594–2608. [CrossRef] [PubMed]
84. Kaltschmidt, B.; Kaltschmidt, C.; Hofmann, T.G.; Hehner, S.P.; Droge, W.; Schmitz, M.L. The pro- or anti-apoptotic function of NF-kappaB is determined by the nature of the apoptotic stimulus. *Eur. J. Biochem.* **2000**, *267*, 3828–3835. [CrossRef] [PubMed]
85. Kang, C.-D.; Han, C.-S.; Kim, K.-W.; Do, I.-R.; Kim, C.-M.; Kim, S.-H.; Lee, E.-Y.; Chung, B.-S. Activation of NF-κB mediates the PMA-induced differentiation of K562 cells. *Cancer Lett.* **1998**, *132*, 99–106. [CrossRef]
86. Basile, J.R.; Eichten, A.; Zacny, V.; Münger, K. NF-κB-Mediated Induction of p21Cip1/Waf1 by Tumor Necrosis Factor α Induces Growth Arrest and Cytoprotection in Normal Human Keratinocytes1 1 R01 CA81135 (KM), K16DE00275 (JRB), DAAD Doktorandenstipendium im Rahmen des gemeinsamen HSP III von Bund und Ländern (AE) and a Senior Postdoctoral Fellowship from the New England Division of the American Cancer Society (VZ). *Mol. Cancer Res.* **2003**, *1*, 262–270.

Publisher's Note: MDPI stays neutral with regard to jurisdictional claims in published maps and institutional affiliations.

© 2020 by the authors. Licensee MDPI, Basel, Switzerland. This article is an open access article distributed under the terms and conditions of the Creative Commons Attribution (CC BY) license (http://creativecommons.org/licenses/by/4.0/).

biomedicines MDPI

Article

Cytocidal Antitumor Effects against Human Ovarian Cancer Cells Induced by B-Lactam Steroid Alkylators with Targeted Activity against Poly (ADP-Ribose) Polymerase (PARP) Enzymes in a Cell-Free Assay

Nikolaos Nikoleousakos [1,*], Panagiotis Dalezis [1], Aikaterini Polonifi [1], Elena G. Geromichalou [1], Sofia Sagredou [1], Constantinos E. Alifieris [1], Maria V. Deligiorgi [1], Vasiliki Sarli [2] and Dimitrios T. Trafalis [1]

1 Laboratory of Pharmacology, Faculty of Medicine, National and Kapodistrian University of Athens, 11527 Athens, Greece; pdalezis@med.uoa.gr (P.D.); kater_pol@yahoo.com (A.P.); elena_geromich@outlook.com (E.G.G.); ssagredou@med.uoa.gr (S.S.); kalifier@med.uoa.gr (C.E.A.); mdeligiorgi@yahoo.com (M.V.D.); dtrafal@med.uoa.gr (D.T.T.)
2 Department of Chemistry, Aristotle University of Thessaloniki, University Campus, 54124 Thessaloniki, Greece; sarli@chem.auth.gr
* Correspondence: nikosakos@hotmail.com or nnikoleous@med.uoa.gr

Citation: Nikoleousakos, N.; Dalezis, P.; Polonifi, A.; Geromichalou, E.G.; Sagredou, S.; Alifieris, C.E.; Deligiorgi, M.V.; Sarli, V.; Trafalis, D.T. Cytocidal Antitumor Effects against Human Ovarian Cancer Cells Induced by B-Lactam Steroid Alkylators with Targeted Activity against Poly (ADP-Ribose) Polymerase (PARP) Enzymes in a Cell-Free Assay. *Biomedicines* **2021**, *9*, 1028. https://doi.org/10.3390/biomedicines9081028

Academic Editor: Simon J. Allison

Received: 15 May 2021
Accepted: 8 August 2021
Published: 17 August 2021

Publisher's Note: MDPI stays neutral with regard to jurisdictional claims in published maps and institutional affiliations.

Copyright: © 2021 by the authors. Licensee MDPI, Basel, Switzerland. This article is an open access article distributed under the terms and conditions of the Creative Commons Attribution (CC BY) license (https://creativecommons.org/licenses/by/4.0/).

Abstract: We evaluated three newly synthesized B-lactam hybrid homo-aza-steroidal alkylators (ASA-A, ASA-B and ASA-C) for their PARP1/2 inhibition activity and their DNA damaging effect against human ovarian carcinoma cells. These agents are conjugated with an alkylating component (POPA), which also served as a reference molecule (positive control), and were tested against four human ovarian cell lines in vitro (UWB1.289 + BRCA1, UWB1.289, SKOV-3 and OVCAR-3). The studied compounds were thereafter compared to 3-AB, a known PARP inhibitor, as well as to Olaparib, a standard third-generation PARP inhibitor, on a PARP assay investigating their inhibitory potential. Finally, a PARP1 and PARP2 mRNA expression analysis by qRT-PCR was produced in order to measure the absolute and the relative gene expression (in mRNA transcripts) between treated and untreated cells. All the investigated hybrid steroid alkylators and POPA decreased in vitro cell growth differentially, according to the sensitivity and different gene characteristics of each cell line, while ASA-A and ASA-B presented the most significant anticancer activity. Both these compounds induced PARP1/2 enzyme inhibition, DNA damage (alkylation) and upregulation of PARP mRNA expression, for all tested cell lines. However, ASA-C underperformed on average in the above tasks, while the compound ASA-B induced synthetic lethality effects on the ovarian cancer cells. Nevertheless, the overall outcome, leading to a drug-like potential, provides strong evidence toward further evaluation.

Keywords: anticancer drug; B-lactam steroid alkylators; synthetic lethality; poly (ADP-ribose) polymerase inhibitors; ovarian cancer; hybrid steroidal alkylating agents

1. Introduction

The second most common malignancy after breast cancer in women over the age of 40 is ovarian cancer. As of 2018, ovarian cancer was the seventh most common cancer worldwide in women and the fifth leading cause of cancer-related death in the same group. Due to the lack of a definitive screening tool and the vague signs and symptoms that can "masquerade" as other non-malignant conditions, curative and survival trends have not changed significantly, as early diagnosis remains a challenge [1].

Currently, the most common treatment consists of surgery and a combination of platinum/taxane chemotherapy [2]. Nevertheless, the treatment is often ineffective, since a major percentage of patients acquire resistance to chemotherapy regimens. The Platinum-resistant disease constricts further known chemotherapy approaches; treatment protocols at this point are often selected individually, but recurrent disease is typically not curable [3].

Inherited mutations in the BRCA1 and BRCA2 genes and the Lynch syndrome (hereditary nonpolyposis colorectal cancer) have been closely associated with a highly increased risk of developing ovarian cancer [4]. The protein products of the aforementioned genes and their functions have been extensively investigated, in reference to the repair of DNA strand breaks via homologous recombination [5] and the detriment of this ability (referred to as "BRCAness") [6]. This "malfunction" renders the cell susceptible to further DNA alterations and increases the dependency on PARP-mediated DNA repair [7].

Poly (ADP-ribose) polymerase (PARP) is a superfamily of 18 proteins characterized by the PARP homology domain, the catalytic domain. This catalytic domain helps in the ADP-ribosylation of various acceptor proteins using nicotinamide adenine dinucleotide (NAD+) as a donor for ADP-ribose. PARP-1 and PARP-2 carry out 80% of the poly-ADP-ribosylation of cellular protein [8]. Poly-ADP-ribosylation of nuclear proteins, a post-translational event, occurs in response to DNA damage. Poly-(ADP-ribose) polymerase (PARP) catalyzes the NAD-dependent addition of poly-(ADP-ribose) to itself and adjacent nuclear proteins such as histones. The contribution of PARP to the events that occur during DNA base excision repair (BER) is important [9].

Nevertheless, PARP overactivation depletes its substrate, NAD+, slowing the rate of glycolysis, electron transport and ATP formation, eventually leading to the functional cell impairment or a PARP-mediated induction of cell and tissue necrosis by extensive depletion of the intracellular NAD pool [10,11]. The cleavage of PARP-1 promotes apoptosis by preventing DNA repair-induced survival and by blocking energy depletion-induced necrosis [12].

Moreover, PARP activation promotes the transcription of pro-inflammatory genes and modulates important inflammatory pathways, while PARP-1 transcription and activity can be modulated as well by several endogenous mechanisms, such as "suicide" enzymes, like caspases and hormonal factors, including steroid hormones. Consequently, as has been presented in various studies, PARP inhibition could lead to several therapeutic outcomes by downregulating multiple simultaneous pathways of inflammation and tissue injury, including beneficial effects on neuronal and myocardial ischemia/reperfusion injury, various forms of heart failure, cardiomyopathies, circulatory shock, cardiovascular aging, diabetes and diabetic cardiovascular complications, myocardial hypertrophy, atherosclerosis, vascular remodeling following injury, angiogenesis of myocardial ischemia, septic shock, vascular stroke, arthritis, colitis and neurodegenerative disorders [11,13–15].

The pharmacological effects of PARP inhibitors potentially enhance the cytotoxicity of certain DNA-damaging anticancer drugs and promote the chemosensitization and radiosensitization of tumors [15,16].

The aim of this report was to assess the potential antitumor effects of three novel B-lactam-steroid alkylating compounds on human ovarian cancer cells, together with their potential to inhibit PARP1/PARP2 activity.

The tested hybrid homo-aza- (lactam) steroidal alkylating esters exert their cytostatic and cytotoxic effects on cancer cells combining multiple pharmacological functions, including the induction of genotoxic effects and DNA damage through their alkylating moiety, as well as the inhibition of PARP activity by binding on its catalytic site [17–21].

2. Materials and Methods

2.1. Compounds

Three lactamic homo-aza-steroidal alkylators—the B-lactam-steroid alkylator (derived from dehydroepiandrosterone), namely 7a,9a-dimethyl-2,10-dioxo-1,2,4,5,6,7,7a,7b, 8,9,9a,10,11, 12,12a,12b-hexadecahydrobenzo[d]indeno [4,5-b]azepin-5-yl 3-(4-(bis(2-chloroethyl)amino) phenoxy)propanoate (ASA-A), the B,D-homo-aza (bilactam)-steroid alkylator (derived from dehydroepiandrosterone), namely 11a,13a-dimethyl-2,6-dioxo-2,3,4,4a,4b,5,6,8,9,10,11,11a, 11b,12,13,13a-hexadecahydro-1H-benzo[4,5]azepino [2,3-f]quinolin-9-yl 3-(4-(bis(2-chlor oethyl)amino)phenoxy)propanoate (ASA-B), and the B-homo-aza (lactam)-alkylator derived from 5-cholesten-3b-ol-7-one (ASA-C), namely 7a,9a-dimethyl-10-((R)-6-methylheptan-2-yl)-2-oxo-1,2,4,5,6,7,7a,7b,8,9,9a,10,11,12,12a,12b-hexadecahydrobenzo[d]indeno[4,5-b] azepin-5-yl 3-(4-(bis(2-chloro ethyl)amino)phenoxy)propanoate were synthesized and prepared according to previously described methods and procedures (Figure 1) [17,22–24]. All three alkylating B-lactam steroidal esters were esterified with the alkylating nitrogen mustard 3-(4-(bis(2-chloroethyl)amino)phenoxy)propanoic acid (POPA).

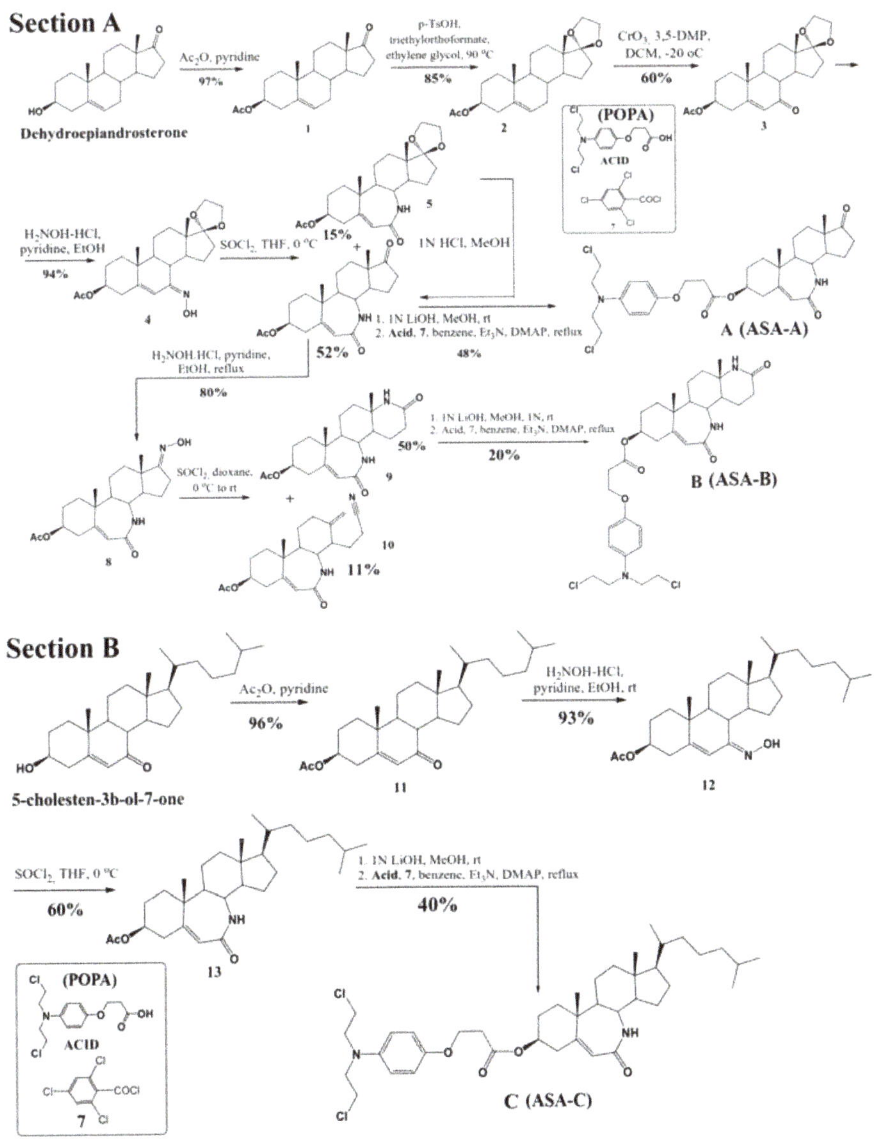

Figure 1. Schematic outlining the synthetic procedure and respective chemical reaction yields of B-homo-aza (lactam)-alkylator (ASA-A), B,D-homo-aza (bilactam)-steroid alkylator (ASA-B) derived from dehydroepiandrosterone (**Section A**) and B-homo-aza (lactam)-alkylator (ASA-C) derived from 5-cholesten-3b-ol-7-one (**Section B**), esterified with the alkylating acid 3-{4-[bis (2-chloroethyl) amino] phenoxy}propanoic acid (POPA).

Furthermore, the third-generation well-established PARP inhibitor Olaparib (AZD2281, AstraZeneca Pharmaceuticals) was tested and used as a positive control.

Stock solutions of the tested compounds were made immediately before use. The compounds were initially dissolved in a small volume of 10% dimethyl sulfoxide (DMSO). Afterwards, they were diluted in a culture medium to reach the final concentrations of 0.05–100 µM for in vitro testing. The solvent (DMSO) in the highest concentration used in the test did not reveal any cytotoxic activity.

2.2. Cell Lines

All human ovarian cancer cell lines described below and used for this study were obtained from the American Type Culture Collection (ATCC®).

UWB1.289 (CRL-2945™) and UWB1.289+BRCA1 (CRL 2946™) cell lines were cultured in 50% RPMI 1640 and 50% MEGM medium supplemented with 1% antibiotics (penicillin/streptomycin) and 6% fetal bovine serum. The mediums used for OVCAR-3 and SKOV-3 cells are RPMI 1640 and McCoy's 5A Medium, respectively, containing 10% *v/v* of fetal bovine serum (Gibco FBS, Thermo Fisher Scientific Inc., Waltham, MA, USA,) and a mixture of antibiotics (100 IU/mL penicillin and 100 µg/mL streptomycin). Moreover, the normal human lung fibroblasts MRC-5 cells (CCL-171™) were cultured in RPMI 1640 medium and tested in order to prove the compound selectivity versus the ovarian cancer cells. All mediums used were sterilized by filtration (Corning-Costar filter, diameter 0.2 µm). The cell cultures were maintained at 37 °C in a humid atmosphere saturated with 5% CO_2.

The in vitro experiments included the analysis of the compounds' biological activity and efficacy on four studied human ovarian cancer cell lines; these cell lines were selected based on their characteristics, such as their genetic background and chemotherapeutic resistance (Table 1). These selection criteria are considered essential in order to facilitate the investigation of ovarian cancer drug resistance and the evaluation of potential therapeutic protocols.

Table 1. Human ovarian cancer cell lines tested for the in vitro analysis of the compounds' biological activity.

Cancer Type	Human Cell Line Designation	Mutated Genes	Special Characteristics
Ovarian Carcinoma	UWB1.289	p53 (625delAG) BRCA1-null (2594delC)	p53, cytokeratin 7 (CK-7) positive; calretinin positive; Wilms' tumor protein (WT) positive; Estrogen/Progesterone receptor negative; MSS **
Ovarian Carcinoma	UWB1.289 + BRCA1	p53 (625delAG)	Estrogen/Progesterone receptor negative; MSS **
Ovarian Adenocarcinoma	SK-OV-3 (SKOV-3)	TP53 (p.S90fs) PIK3CA (p.H1047R)	Tumor Necrosis Factor; Diphtheria Toxin; Cisplatin and Adriamycin not sensitive; Estrogen receptor α (ERα) positive/Progesterone receptor negative; MSI-High *
Epithelial Ovarian Adenocarcinoma	NIH:OVCAR-3	MEK1 (p.G159R; p.R160K; p.A158) TP53 (p.R248Q)	Androgen/Estrogen/Progesterone receptor positive; Cisplatin, Adriamycin not sensitive; Melphalan resistant; MSS **

* MSI-high: High Microsatellite Instability, ** MSS: Microsatellite stable.

Both UWB1.289 and UWB1.289 + BRCA1 carry mutations in the tumor suppressor gene TP53 similar to the BRCA1-null breast cancer cell line HCC1937. UWB1.289 and UWB1.289 + BRCA1 have a negative expression of progesterone, estrogen and androgen receptors. UWB1.289 carries a germline BRCA1 mutation within exon 11 and has a deletion of the wild-type allele. In the UWB1.289 + BRCA1 cell line, wild-type BRCA1 is restored [25,26].

2.3. MTT Assay

The cells were plated in 96-well plates at a density of 1×10^4 cells/mL per well. They were grown as monolayers for 24 h before treatment with 0.1–100 μM of the compounds ASA-A, ASA-B, ASA-C, POPA, 3-AB and Olaparib for 48 h. The whole duration of the procedure was 72 h, during which the cells were maintained in an incubator. The viability of the cultured cells was estimated by (3-(4,5-dimethylthiazol-2-yl)-2,5-diphenyltetrazolium bromide (MTT) metabolic assay as described previously [27–30]. MTT (Sigma, St Louis, MO, USA) was dissolved in PBS at a concentration of 5 mg/mL, filter sterilized, and stored at 4oC. MTT (0.2 mL of stock solution) was added to each culture (per mL) and incubated for 3 h at 37 °C to allow metabolism. Formazan crystals were solubilized by acidic isopropanol (0.04 N HCl in absolute isopropanol in a ratio 1:3 *v*/*v*). The absorbance of the converted dye was measured at a 540 nm wavelength on an ELISA reader (Versamax, Orleans, MA, USA). The mean concentrations of each drug that generated 50% or total (100%) growth inhibition (GI50 and TGI, respectively) and the drug concentrations that produced cytotoxicity against 50% of the cultured cells [(half maximal inhibitory concentration (IC50)] were calculated using the linear regression method [27]. Using seven absorbance measurements [time 24 h(Ct24), control growth 72 h(Ct72) and test growth in the presence of the drug at 5 concentration levels (Tt72x)], the percentage of growth was calculated at each level of the drug concentrations. The percentage growth inhibition was calculated according to the National Cancer Institute (NCI) as: $[(Tt72x) - (Ct24)/(Ct72) - (Ct24)] \times 100$ for concentrations for which $Tt72x > Ct24$ and $[(Tt72x)\text{-}(Ct24)/Ct24] \times 100$ for concentrations for which $Tt72x < Ct24$. GI50 was calculated from $[(Tt72x) - (Ct24)/(Ct72)\text{-}(Ct24)] \times 100 = 50$, TGI from $[(Tt72x) - (Ct24)/(Ct72) - (Ct24)] \times 100 = 0$, and IC50 from $[(Tt72x) - (Ct24)/Ct24] \times 100 = 50$. All the experiments were carried out in triplicate.

2.4. PARP-1/2 Activity Assay

The Universal PARP Colorimetric Assay Kit (Trevigen® Inc., Gaithersburg, MD, USA) was used in order to evaluate the inhibition effect of the tested chemical compounds on PARP1/2 activity. PARP enzymatic activity was assayed according to the manufacturer's instructions and as previously described [31,32]. The method is based on the incorporation of biotinylated poly (ADP-ribose) (PAR) onto histone proteins in a 96-stripwell microplate format. Colorimetric readouts at 450 nm absorbance of the PARP activity in a serial of PARP dilutions (units/well) were performed, and a standard curve was prepared (Figure 2). 3-Aminobenzamide (3-AB), a well-known inhibitor of PARP enzymes, was used comparatively as a positive control for testing the inhibitory activity on PARP. The compounds ASA-A, ASA-B, ASA-C and POPA were tested in comparison to the standard well-established PARP inhibitor 3-AB at concentrations of 1–300 μM, for the inhibition of PARP activity (0.6 Unit/well PARP-HSA Enzyme). The tested compounds were loaded in quadruplicate into a 96-well plate coated with histones and biotinylated poly ADP-ribose and allowed to incubate for 1 h. Then, they were treated with horseradish peroxidase (HRP) conjugated streptavidin (strep-HRP) and read in a spectrophotometer at 450 nm [25,33].

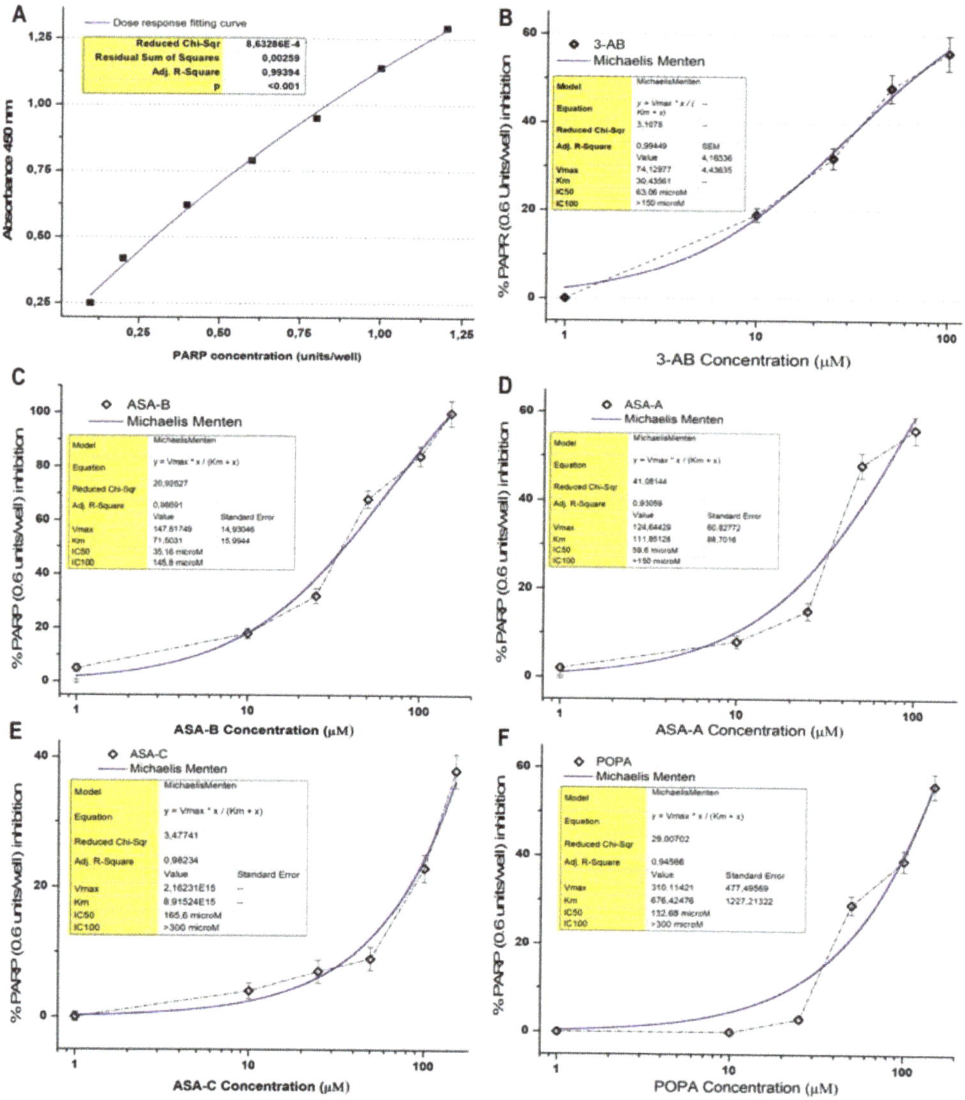

Figure 2. (**A**) Graphical representation of the colorimetric readout of the PARP dose response standard curve (Adj. R-square 0.994, $p < 0.001$); (**B**) PARP (0.6 units/well) inhibition curve and respective kinetics data analysis according to Michaelis Menten's model for 3-aminobenzamide (3-AB); (**C**) PARP (0.6 units/well) inhibition curve and respective kinetics data analysis according to Michaelis Menten's model induced by the B,D-bilactam, dehydroepiandrosterone-derived, steroid alkylator ASA-B; (**D**) PARP (0.6 units/well) inhibition curve and respective kinetics data analysis according to Michaelis Menten's model induced by the B-lactam, dehydroepiandrosterone-derived, steroid alkylator ASA-A; (**E**) PARP (0.6 units/well) inhibition curve and respective kinetics data analysis according to Michaelis Menten's model induced by the B-lactam, cholesten-derived, steroid alkylator ASA-C; (**F**) PARP (0.6 units/well) inhibition curve and respective kinetics' data analysis according to Michaelis Menten's model induced by the conjugant alkylator POPA.

2.5. *PARP1 and PARP2 mRNA Expression Analysis*

RNA isolation: Cells growing exponentially in culture were treated with 25 μM and 50 μM of each of the tested compounds for 24 h and then washed in phosphate buffered saline (PBS). The total RNA was extracted using the RNA extraction kit, according to the standard protocols. RNA integrity was assessed by denaturing formaldehyde 1.5% agarose gel electrophoresis and quantified by a spectrophotometer at A260/280 nm.

For each sample in duplicate, 1 μg of checked quality RNA was used to synthetize cDNA with M-MLV reverse transcriptase (RT) (Invitrogen, Thermo Fisher Scientific Inc., Waltham, MA, USA) and 100 pmol/μL hexamer random primers (Invitrogen). An RT reaction mix was incubated for 50 min at 37 °C. The cDNA yield was assessed after 1:10 dilution, by a regular PCR using β-actin (β-ACT) primers. No genomic contamination was detected.

Primers for the genes of interest, PARP1 and PARP2, were designed using Primer3 software (http://biotools.umassmed.edu/bioapps/primer3_www.cgi, accessed on 10 May 2021), based on the complete cDNA sequences deposited in GenBank. Both primers to each set were into different exons to avoid the amplification of contaminating genomic DNA and to eliminate mis-priming events generating a detectable signal. The specificity of the primers and the singularity of the amplified region were verified using http://www.ncbi.nlm.nih.gov/BLAST/, accessed on 10 May 2021 and http://genome.ucsc.edu/cgi-bin/hgBlat?command = start respectively, accessed on 10 May 2021. The amplified fragments were planned to correspond to common regions among different isoforms for each gene [34]. The sequences of primers as designed for real-time PCR were: β-actin (ACTB)/224 bp (BC014861): F 5′AGG ATG CAG AAG GAG ATC ACT G 3′/R 5′GGG TGT AAC GCA ACT AAG TCA TAG 3′; PARP1/94 bp (NM 001618.3): PARP1_F 5′CCTGATCCCCCACGACTTT 3′/PARP1_R 5′GCAGGTTGTC AAGCATTTC 3′; PARP2/229 bp (NM 005484.3): PARP2_10F-CTCCTCC ACTAATCCGGACA/PARP2_12 R-GTGTGGGAGCATGGGTAGAT.

The qRT-PCR was carried out in triplicate for each individual gene of each sample, and all reactions were performed twice by means of the SYBR-Green fluorescent dye detection system (Molecular Probes, Eugene, OR, USA) on a thermocycler- BIORAD CFX96. The reaction volume was 20 μL: 1 μL CDNA, 0.4 μL Forward primer (5 pmol/μL), 0.4 μL Reverse primer (5 pmol/μL), 10 μL buffer 2x (QIAGEN). Forty cycles of PCR amplification were run with 95 °C for 15 min of enzyme activation, 95 °C for 5 s for denaturation, at 59, and 15 s for annealing. The PCR products were directly monitored by measuring the increase of fluorescence due to the binding of the SYBR Green to double-stranded DNA. Melting curve experiments had previously established that the fluorescence signal for each amplicon was derived from the products only, and no primer dimers were found.

The relative gene expression between the treated and untreated samples was calculated using the 2-ΔCT method. The fractional cycle number (CT) expresses the fluorescent signal reached above the detection threshold (minimum detection level). The ΔCT was computed by calculating the difference of the average CT between target genes and the reference β-ACT (ACTB). Due to the exponential nature of PCR, the ΔCT is converted to a linear form by a 2-ΔCT or fourfold difference. RNA samples from treated cell lines versus untreated samples were analyzed. The values of the untreated cells' mRNA expression were set as 1 for all studied genes [17,34,35].

The absolute gene expression among untreated cell lines was carried out by calculating the ratio of the number of transcripts of the target gene to that of the reference gene, ACTB. This comparison provides the exact number of the PARP1 and PARP2 transcripts in each cell line. The real-time PCR efficiencies were calculated from the slope. The amplification efficiency was similar (1.984–1.986) between the targets and the reference. Finally, all cDNA samples were made from the same amount of RNA (1 μg) and diluted to the same volume.

The PARP1, PARP2 and ACTB absolute mRNA cellular content as well as the relative PARP1/ACTB and PARP2/ACTB mRNA cellular content (± Standard Error, SE) were assessed, as they were induced by the treatment of UWB1.289, UWB1.289 + BRCA1, SKOV-

3 and OVCAR-3 human ovarian cancer cells with ASA-A, ASA-B, ASA-C and POPA at 20 and 40 μM concentrations, respectively, for 72 h and compared to the corresponding control values (untreated cells).

The data are presented as mean value (MV) ± SE. A statistical analysis was conducted with a one-paired *t*-student test ($p < 0.01$). Group differences were analyzed using the standard Student *t*-test. $p < 0.01$ was considered statistically significant. All experiments were performed in triplicate.

2.6. Genotoxicity Assessment–Sister Chromatid Exchanges (SCEs) Assay

Heparinized peripheral blood samples from 5 healthy donors were obtained. Peripheral blood mononuclear cells (PBMCs) were isolated using density gradient centrifugation through Ficoll-Hypaque. The PBMCs were cultured at a density of 1.5×10^5 cells/mL in McCoy's 5A medium, containing L-glutamine 1% and antibiotics (penicillin 100 U/mL–streptomycin 100 μg/mL), supplemented with 10% heat-inactivated fetal calf serum (FCS) at 37 °C in a humidified 5% CO2 cell culture incubator. T-lymphocytes were stimulated to proliferate with 20 μg/mL phytohaemagglutinin (PHA-L). Moreover, UWB1.289 and UWB1.289 + BRCA1 cells were cultured and maintained as described above. The cultures were set up by adding 1.5×10^5 of cancer cells per mL of the cell culture medium. The PBMC cultures were maintained for 72 h, while UWB1.289 and UWB1.289 were maintained for 144 h. The normal PBMCs and ovarian cancer cells were treated with ASA-A, ASA-B, ASA-C, POPA, 3-AB and olaparib at a 2.5 μM drug concentration. The SCEs method was performed as previously described [17,20,21,36]. Briefly, 10 μM of 5- bromodeoxyuridine (5-BrdU) was added to PBMCs cultures at 24 h and to ovarian cancer cell cultures at 48 h for two cell cycles. The cell growth in the presence of 5-BrdU for two rounds of DNA replication was followed by collecting metaphase spreads on glass slides. To prepare the metaphase spreads, exponentially growing cells were treated with 0.2 μg/mL of colcemid for 3–5 h. The cells were collected and incubated in a hypotonic solution (0.56% KCl), fixed in methanol:acetic acid (3:1), spotted to slides and air-dried. To measure chromosomal aberrations, the slides were stained with a 2% Giemsa/Sorensen's buffer. Differential chromosome staining was achieved by treatment with the UV-sensitive dye Hoechst 33,258 (10 μg/mL), long-wave UV light exposure (for about 1 h and 30 min to long-wave ultra-violet light emitted from a lamp located at a distance of 10 cm from the slides) and Giemsa staining, which gives a permanent record of the exchanges. More than 30 suitably spread second-division cells from each culture were measured on coded slides to establish mean SCE values. The preparations were scored for cells in their first, second and subsequent divisions, with criteria that were previously described, and the cells that were suitably spread were scored blindly for SCEs and cancer cell proliferation kinetics. For the assessment of the Proliferation Rate Index (PRI), at least 100 metaphases were evaluated. The PRI was calculated according to the formula PRI = (M1 + 2M2 + 3M3)/N × 100, where M1, M2 and M3+ indicate the number of metaphases corresponding to the first, second and third or subsequent divisions, respectively, and N is the total number of metaphases scored (M1 + M2 + M3).

3. Results

3.1. In Vitro Anticancer Activity Testing

The results were expressed as GI50 (μM) (drug concentration that induces a 50% inhibition of cell growth) TGI (μM) (drug concentration that induces a 100% cell growth inhibition) indicating the cytostatic effects, and IC50 (μM) (drug concentration that induces 50% cytocidal effects) indicating the cytotoxic effects of the tested compounds on UWB1.289, UWB1.289 + BRCA1, OVCAR-3 and SKOV-3 human ovarian cancer cells. The tested compounds inhibited the cell growth in a dose-dependent manner. The results are summarized in Table 2. The B- lactam steroid alkylator, ASA-A, and the B,D-bilactam steroid alkylator, ASA-B, were shown to be effective with significantly elevated cytostatic and cytocidal activity, considerably higher (*t*-test, $p < 0.001$) than the aza-steroid alkylator

ASA-C and the alkylating component POPA. ASA-C and the alkylating component 3-{4-[bis (2-chloroethyl) amino] phenoxy} propanoic acid (POPA) appeared almost inactive as far as their in vitro cytostatic and cytotoxic effects are concerned, at varying doses (with a maximum of 100 μM) on human ovarian cancer cell lines. At the tested drug concentrations, 3-AB showed no significant cytostatic or cytotoxic effects, while olaparib generated cytostatic and no cytocidal effects against the BRCA1 mutated UWB1.289 and a lower cytostatic potential against the OVCAR-3 ovarian cancer cells. In total, the B-lactam steroidal alkylator ASA-A appeared to be more potent than the B,D-bilactam steroid alkylator ASA-B against the human ovarian cancer cells, with the exception of UWB1.289 cells, where ASA-B had a better cytostatic and cytotoxic activity, presenting a synergistic lethality effect. The tested aza-steroidal alkylators ASA-A and ASA-B, in terms of their cytostatic and cytotoxic activity, presented a significant selectivity against the ovarian human cancer cells, since they generated significantly lower cytostatic (IG50 and TGI) and cytotoxic (IC50) effects on normal human lung fibroblasts MRC-5.

Table 2. In vitro cytostatic (IG50 and TGI ±SEM) (μM) and cytotoxic (IC50 ±SEM) (μM) effects of compounds ASA-A, ASA-B, ASA-C, POPA, 3-AB and Olaparib on all four studied human ovarian cancer cell lines in vitro. All values are representative of experiments performed in triplicate (Significance level, $p < 0.01$).

Human Cell Line	ASA-A			ASA-B		
	GI50	**TGI**	**IC50**	**GI50**	**TGI**	**IC50**
UWB1.289	19 ± 1.83	52 ± 4.61	71 ± 7.01	10 ± 0.81	35 ± 3.63	54 ± 5.11
UWB1.289 + BRCA1	22 ± 2.01	54 ± 4.33	76 ± 6.66	20 ± 1.86	72 ± 6.71	95 ± 9.03
SKOV-3	15 ± 1.26	33 ± 2.79	67 ± 7.11	24 ± 2.42	51 ± 4.46	98 ± 8.97
OVCAR-3	29 ± 3.04	38 ± 2.95	47 ± 4.82	44 ± 5.04	81 ± 7.82	>100
MRC-5	69 ± 7.13	>100	>100	86 ± 9.95	>100	>100
	ASA-C			**POPA**		
	GI50	**TGI**	**IC50**	**GI50**	**TGI**	**IC50**
UWB1.289	73 ± 7.58	>100	>100	64 ± 6.72	>100	>100
UWB1.289 + BRCA1	>100	>100	>100	>100	>100	>100
SKOV-3	>100	>100	>100	77 ± 7.54	>100	>100
OVCAR-3	>100	>100	>100	>100	>100	>100
MRC-5	>100	>100	>100	>100	>100	>100
	3-AB			**Olaparib**		
	GI50	**TGI**	**IC50**	**GI50**	**TGI**	**IC50**
UWB1.289	92 ± 12.98	>100	>100	14 ± 1.78	85 ± 9.34	>100
UWB1.289 + BRCA1	>100	>100	>100	>100	>100	>100
SKOV-3	>100	>100	>100	>100	>100	>100
OVCAR-3	>100	>100	>100	32 ± 4.64	>100	>100
MRC-5	>100	>100	>100	>100	>100	>100

The data showed that the compounds under investigation influenced tumor cell growth differentially, demonstrating the BRCA1-null ovarian cancer cell line UWB1.289 to be significantly more sensitive to the antiproliferative effects of the investigated compounds than the respective BRCA1-wild type UWB1.289 + BRCA1 ovarian cancer cell line (t-test, $p < 0.001$). The BRCA1 mutated ovarian cell line UWB1.289 appeared to be particularly more sensitive to the cytostatic and cytocidal effects of the tested alkylating agents than the other ovarian cancer cell lines tested in consistence to its BRCAness phenotype (t-test, $p < 0.001$). ASA-B showed the greatest anticancer activity against UWB1.289 cells, in vitro (t-test, $p < 0.001$), as already mentioned. The second most significant activity against UWB1.289 cancer cells was generated by the B-lactam steroidal alkylator, ASA-A (t-test, $p < 0.001$).

The SKOV-3 and OVCAR-3 ovarian cancer cells were shown to be irresponsive to the anticancer activity of alkylators such as Cisplatin and Melphalan, as well as to the chemotherapeutic cytotoxic agent Adriamycin. Accordingly, our experimental data show SKOV-3 and OVCAR-3 cells to be relatively resistant to the anticancer activity of the tested nitrogen mustard alkylating agent POPA. However, this apparent phenotype of the insensitivity of SKOV-3 and OVCAR-3 human ovarian cancer cells is overcome by the lactam steroid alkylators, ASA-A and ASA-B, that produced excellent cytostatic and cytotoxic anticancer effects on these cells (t-test, $p < 0.001$).

All four tested cancer cell lines bear different active mutations of the onco-suppressor gene TP53. Both ASA-A and ASA-B appear to procure a remarkable anticancer activity independently of the presence of these mutations on the TP53 gene (t-test, $p < 0.001$).

Moreover, SKOV-3 ovarian cancer cells that bear an active mutation on the PIK3CA gene and present a high microsatellite instability (MSI-high), in contrast to the OVCAR-3 and UWB1.289 + BRCA1 ovarian cancer cells that are microsatellite-stable (MSS), demonstrate a clearly higher susceptibility to the anticancer activity of ASA-A and ASA-B as far as their cytostatic effects and the 100% cell growth inhibition (TGI) are concerned (t-test, $p < 0.01$).

Finally, the cytostatic or cytotoxic antitumor effects of the homo-aza-steroidal alkylating esters, ASA-A and ASA-B, were observed in both steroid receptors (Estrogen/Progesterone/Androgen receptors), expressing and non-expressing ovarian cancer cell lines (t-test, $p < 0.001$).

3.2. PARP Enzymatic Activity Inhibition

The B-lactam steroid alkylators ASA-A, ASA-B and ASA-C, as well as the alkylating moiety POPA (nitrogen mustard) were tested for their direct inhibitory effect against the activity of PARP enzymes in a cell-free experimental biochemical assay in comparison to the well-known PARP inhibitor 3-AB. The inhibition curves and the respective kinetics data analysis according to Michaelis Menten's model for the 3-aminobenzamide (3-AB), B-lactam (dehydroepiandrosterone-derived) steroid alkylator ASA-A, B,D-bilactam (dehydroepiandrosterone-derived) steroid alkylator ASA-B, B-lactam (cholesten-derived) steroid alkylator ASA-C and alkylating component POPA are presented in Figure 2 and Table 3. The B,D-bilactam steroid alkylator ASA-B and the B-lactam steroidal alkylator ASA-A showed significant inhibitory effects on PARP activity (t-test, $p < 0.0001$), ASA-B being the most potent inhibitor producing better inhibitory effects on PARP activity than 3-AB (t-test, $p < 0.01$). Thus, 3-AB, ASA-A, ASA-B, ASA-C and POPA induce a 50% inhibition of the PARP enzymatic activity (IC50) at concentrations of 63.06, 59.6, 35.16, 165.6 and 132.68 μM, respectively (Table 3). ASA-B and ASA-A exert their inhibitory effects on PARP activity, generating much higher kinetic parameters of Km and Vmax than 3-AB (t-test, $p < 0.01$). Contrarily, ASA-C appeared as a noticeably weak inhibitor of PARP activity, with high IC50 and very low Km and Vmax. Moreover, POPA acts like an inhibitor of PARP activity at significantly higher concentrations, with a relatively high IC50 and the highest Km and Vmax inhibition kinetic parameters. On the other hand, Olaparib generated by far the most potent inhibitory effect on PARP activity, demonstrating an IC50 at 0.95 μM and an IC100 at 13.17 μM, respectively (Table 3).

Table 3. Michaelis Menten's model kinetics data (± SEM) (of PARP (0.6 units/well) inhibition-induced by the 3-aminobenzamide (3-AB), B-lactam (dehydroepiandrosterone-derived) steroid alkylator ASA-A, B,D-bilactam (dehydroepiandrosterone-derived) steroid alkylator ASA-B, B-lactam (cholesten-derived) steroid alkylator ASA-C, alkylating component POPA and Olaparib.

Compounds	3-AB	ASA-A	ASA-B	ASA-C	POPA	Olaparib
Adj. R-Square	0.99449	0.93059	0.98691	0.98234	0.94566	0.97982
Vmax	74.130 ± 4.16	124.644 ± 60.83	147.817 ± 14.93	$2.162 \times 10^{15} \pm 4.18 \times 10^{15}$	310.114 ± 447.49	109.079 ± 10.23
Km	30.436 ± 4.43	111.851 ± 88.70	71.5031 ± 15.99	$8.915 \times 10^{15} \pm 13.01 \times 10^{15}$	676.425 ± 1227.41	1.117 ± 0.214
IC50 (μM)	63.06 ± 4.28	59.6 ± 6.42	35.16 ± 3.78	165.6 ± 19.77	132.68 ± 22.15	0.95 ± 0.12
IC100 (μM)	>150	>150	145.8 ± 14.29	>300	>300	13.17 ± 1.57

In conclusion, the potency of inhibition of the PARP activity induced by ASA-A and ASA-B is notably correlated to their cytostatic effects on UWB1.289, UWB1.289 + BRCA1, OVCAR-3 and SKOV-3 human ovarian cancer cells (Figure 3, Table 4). The respective

Pearson's correlation indexes (r) between cytostatic activity and PARP inhibitory effects for ASA-A and ASA-B are 0.95 and 0.93, respectively, at a significance level $p < 0.01$. On the other hand, the corresponding correlations for ASA-C and POPA are low and not significant (NS) (Figure 3, Table 4), while the 3-AB induced cytostatic effects were significantly correlated with PARP inhibition only on BRCA1 mutated (presenting "BRCAness") UWB1.289 human ovarian cancer cells (Table 4).

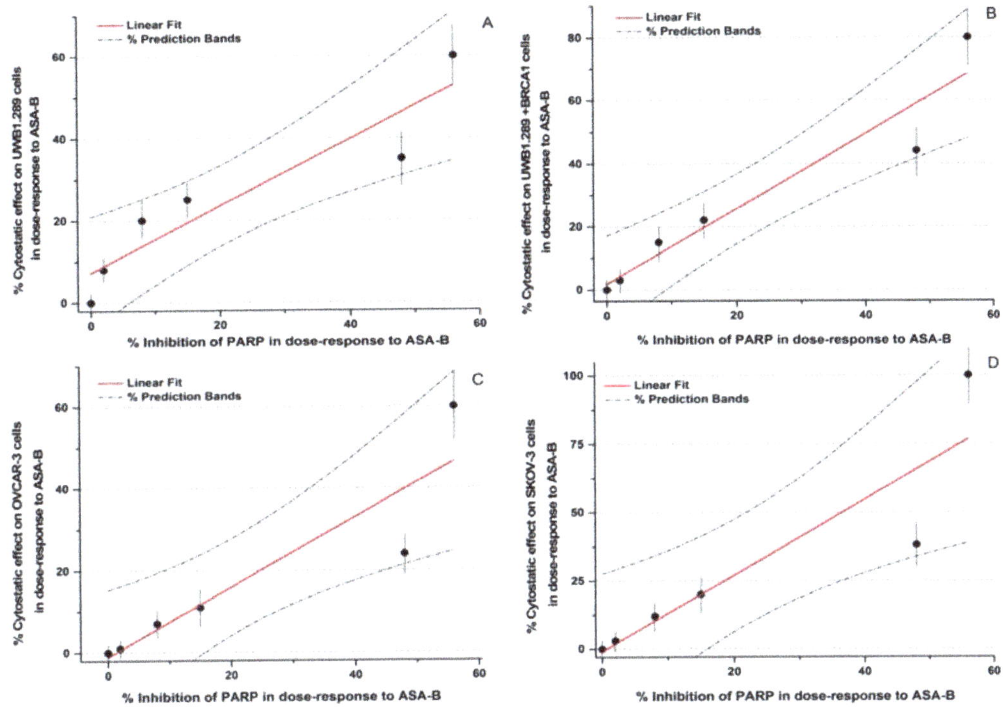

Figure 3. Correlation curves and 95% prediction bands between %PARP inhibition and %cytostatic effects in dose-response to the B,D-bilactam steroid alkylator ASA-B on UWB1.289 (**A**), UWB1.289+ BRCA1 (**B**), OVCAR-3 (**C**) and SKOV-3 (**D**) human ovarian cancer cells.

Table 4. Pearson's correlation indexes and respective statistical significance levels (p) of cytostatic effects and PARP inhibition induced by the homo-aza lactamic steroid alkylators ASA-B, ASA-A, ASA-C and the alkylating acid POPA, as well as 3-AB on UWB1.289, UWB1.289 + BRCA1, OVCAR-3 and SKOV-3 human ovarian cancer cells.

		UWB1.289	UWB1.289 + BRCA1	OVCAR-3	SKOV-3
ASA-B	Pearson's r	0.93	0.96	0.93	0.91
	Significance level (p)	<0.01	<0.01	<0.01	<0.01
ASA-A	Pearson's r	0.95	0.97	0.91	0.90
	Significance level (p)	<0.01	<0.01	<0.01	<0.01
ASA-C	Pearson's r	<0.25	<0.25	<0.25	<0.25
	Significance level (p)	NS *	NS	NS	NS
POPA	Pearson's r	<0.50	<0.50	<0.50	<0.50
	Significance level (p)	NS	NS	NS	NS
3-AB	Pearson's r	0.91	<0.50	<0.50	<0.50
	Significance level (p)	<0.01	NS	NS	NS

* NS: not significant ($p > 0.05$).

3.3. PARP1 and PARP2 Expression

Quantitative PCR analysis of PARP1 and PARP2 absolute mRNA content in UWB1.289 + BRCA1, UWB1.289, OVCAR-3 and SKOV-3 untreated human ovarian cancer cell lines is presented in Figure 4.

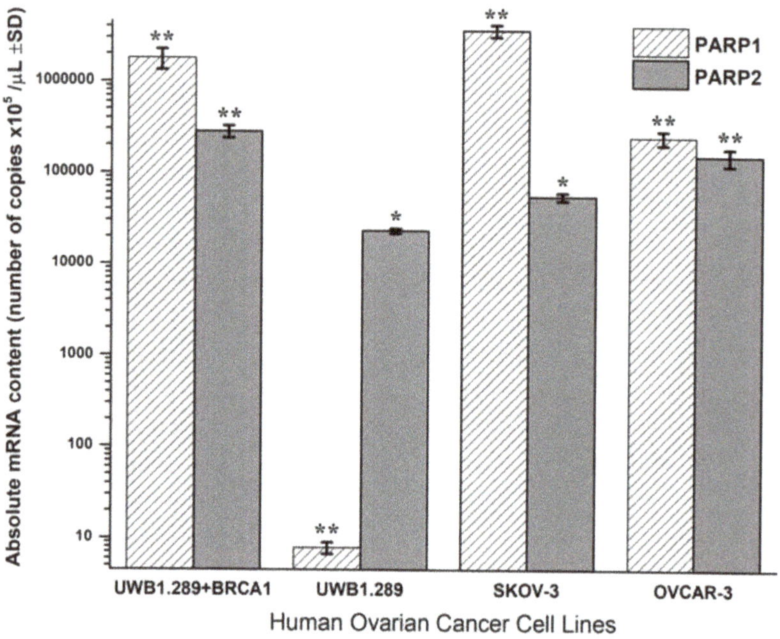

Figure 4. Absolute mRNA cellular content and transcription of PARP1 and PARP2 (± SD) in untreated UWB1.289 + BRCA1, UWB1.289, SKOV-3 and OVCAR-3 human ovarian cancer cell cultures at the exponential phase of cell growth. Significance levels in between PARP1 or PARP2 mRNA transcription in the tested ovarian cancer cell lines were denoted, (**) for $p < 0.0001$ and (*) for $p < 0.001$.

BRCA1-null UWB1.289 cancer cells transcribe a significantly higher PARP2 (*t*-test, $p < 0.001$) and a strikingly lower PARP1 mRNA content (*t*-test, $p < 0.001$) due to the BRCA1 loss of activity. Contrariwise, the wild-type UWB1.289 + BRCA1 cells presented normally a high PARP1 and a lower PARP2 expression (*t*-test, $p < 0.001$). In general, the UWB1.289 + BRCA1, OVCAR-3 and SKOV-3 cells showed a natural and significantly higher PARP1 and a lower PARP2 expression. Comparatively, the BRCA1 mutated UWB1.289 cancer cells demonstrated a much lower PARP1 and a reduced PARP2 mRNA transcription than the other ovarian cancer cell lines tested (*t*-test, $p < 0.001$).

The changes (%) in the relative PARP1/ACTB and PARP2/ACTB mRNA cellular content (± Standard Error, SE) as compared to the corresponding control values (untreated cells) induced by the treatment of UWB1.289, UWB1.289 + BRCA1, SKOV-3 and OVCAR-3 human ovarian cancer cells with ASA-A, ASA-B, ASA-C and POPA at 20 and 40 μM concentrations for 72 h, as demonstrated by the quantitative PCR analysis, are presented in Figure 5, Figure 7, respectively. Moreover, a similar expression analysis is described in Figure 6, regarding the treatment of the same tested human ovarian cancer lines with a 3-AB agent in combination with ASA-A, ASA-B or ASA-C, enclosing the respective alterations on the relative PARP1/ACTB mRNA cellular content.

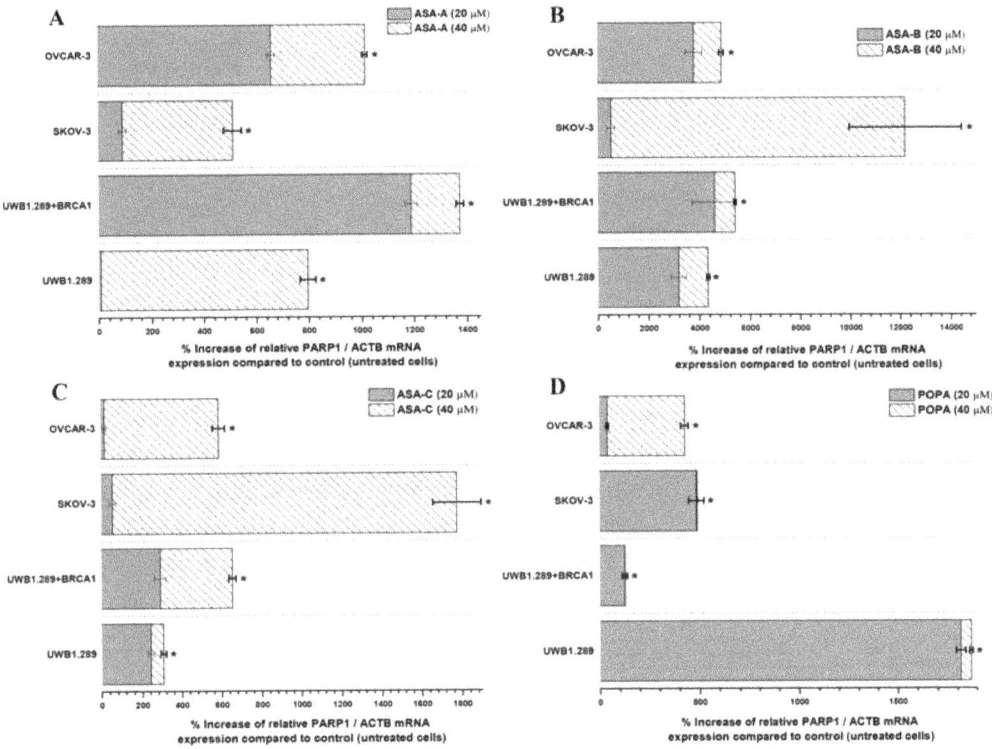

Figure 5. % Increase of the relative PARP1/ACTB mRNA cellular content (± Standard Error, SE) as compared to the corresponding control values (untreated cells) induced by the treatment of UWB1.289, UWB1.289 + BRCA1, SKOV-3 and OVCAR-3 human ovarian cancer cells with ASA-A (**A**), ASA-B (**B**), ASA-C (**C**) and POPA (**D**) at 20 and 40 μM concentrations, respectively, for 72 h. (*) Statistical significance for $p < 0.001$ (two-tail paired t-test).

In general, the results indicated that the PARP1/ACTB and PARP2/ACTB mRNA cellular content ratios are notably increased (t-test, $p < 0.001$) in most cases and for all tested cancer cell lines after treatment with the alkylators ASA-A, ASA-B, ASA-C and POPA. However, these rises were obviously differentiated depending on the variant cell types and biology, as well as due to the differential cell sensitivity in the tested compounds. Consequently, as was revealed, the more sensitive a cancer cell line to the cytostatic and cytotoxic effects of a tested compound, the higher the increase of the relative PARP1/ACTB and PARP2/ACTB mRNA ratios observed.

The highest increases of PARP1/ACTB mRNA ratios were induced by the most potent PARP inhibitor, the bilactamic steroid alkylator ASA-B, reaching at 120-fold increase in SKOV-3 and a 42–50-fold increase in UWB1.289, UWB1.289 + BRCA1 and OVCAR-3 human ovarian cancer cells. ASA-A appeared as the second most potent compound to produce a PARP1/ACTB mRNA ratio increment, while the POPA and ASE-C were the least potent, respectively. The aza-steroidal alkylators, ASA-A, ASA-B and ASA-C induced a conspicuously higher increase of the PARP1/ACTB mRNA ratio in the UWB1.289 + BRCA1 than in the BRCA1 mutated UWB1.289 cancer cells. On the other hand, POPA was the best augmenter of the PARP1/ACTB mRNA ratio in BRCA1-null UWB1.289 cancer cells (Figure 5).

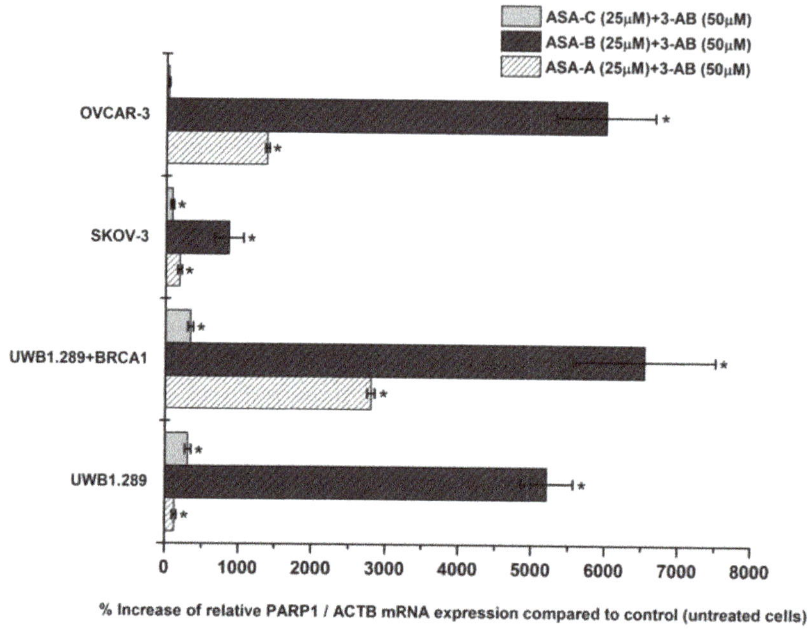

Figure 6. % Increase of the relative PARP1/ACTB mRNA cellular content (± Standard Error, SE) as compared to the corresponding control values (untreated cells) induced by the treatment of UWB1.289, UWB1.289 + BRCA1, SKOV-3 and OVCAR-3 human ovarian cancer cells with the combination of ASA-A, ASA-B or ASA-C at 25 μM concentration along with 3-AB at 50 μM concentration for 72 h. (*) Statistical significance for $p < 0.001$ (two-tail paired *t*-test).

The simultaneous treatment of the UWB1.289, UWB1.289 + BRCA1, SKOV-3 and OVCAR-3 human ovarian cancer cells with 3-AB in combination with ASA-A, ASA-B or ASA-C induced a significant ($p < 0.001$) synergistic effect on the % increase of PARP1/ACTB mRNA cellular content, mostly in ASA-B and less in ASA-A combinations with 3-AB. The treatment with combinations of 3-AB with ASA-C produced obviously lower increases, as well as synergistic effects (Figure 6). The treatment of the tested cancer cell lines with the 3-AB agent alone generated non-significant alterations in the PARP1/ACTB mRNA cellular content (data not shown).

As far as the inducible increases of PARP2/ACTB mRNA ratios are concerned, ASA-A, ASA-B and POPA were comparably more effective, whereas ASA-C was less effective against all tested cancer cell lines (Figure 7). ASA-A, ASA-B and POPA brought about a considerably higher increase of the PARP2/ACTB mRNA ratio in the BRCA1-null UWB1.289 than in the BRCA1 wild-type UWB1.289 + BRCA1 cancer cells, with the ASA-B begetting a 140- to150-fold increase of the PARP2/ACTB mRNA ratio in UWB1.289 cells.

ASA-A, ASA-B, ASA-C and POPA generated a statistically significant enhancement and increase of the PARP1 and PARP2 mRNA expression in all tested cancer cell lines. These material results on the regulation of PARP1/2 transcription were differential and depended on the drug's PARP inhibitory activity, the drug concentration, the cell type and the cancer cells' sensitivity for each tested compound.

All tested chemical derivatives produced a similar outcome onto UWB1.289 + BRCA1, UWB1.289, SKOV-3 and OVCAR-3 human ovarian cancer cells, with the activity of the lactam steroid alkylators ASA-A and ASA-B being more profound than the alkylating agent POPA. Furthermore, it is notable that the tested compounds induced a higher increase of the expression of PARP1 in the BRCA1 wild-type UWB1.289 + BRCA1 than in BRCA1-null

UWB1.289 cells, whilst the higher increase of the expression of PARP2 was observed in UWB1.289 cells.

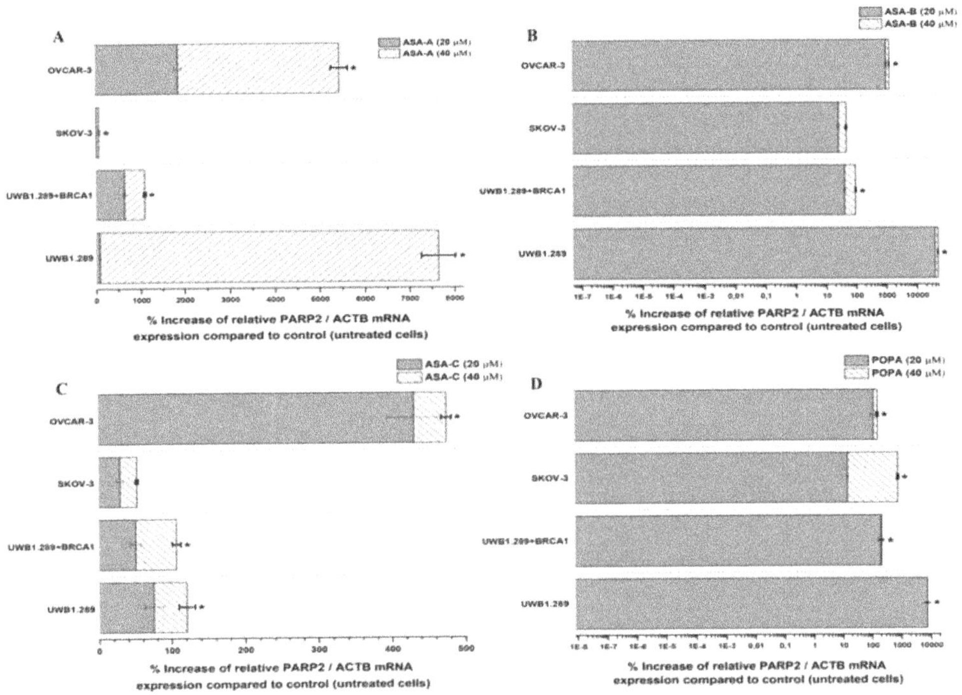

Figure 7. % Increase of the relative PARP2/ACTB mRNA cellular content (± Standard Error, SE) as compared to the corresponding control values (untreated cells) induced by the treatment of UWB1.289, UWB1.289 + BRCA1, SKOV-3 and OVCAR-3 human ovarian cancer cells with ASA-A (**A**), ASA-B (**B**), ASA-C (**C**) and POPA (**D**) at 20 and 40 μM concentrations, respectively, for 72 h. (*) Statistical significance for $p < 0.001$ (two-tail paired *t*-test).

3.4. Genotoxicity Assessment–Sister Chromatid Exchanges (SCEs)

The B-lactam- steroid alkylators ASA-A, ASA-B and ASA-C, as well as the alkylating moiety POPA and the PARP inhibitors 3-AB and Olaparib were tested for their genotoxic effects on PHA-stimulated T-lymphocytes obtained from five healthy donors, as well as on UWB1.289 and UWB1.289 +BRCA1 human ovarian cancer cells at a concentration of 3 μM.

ASA-A and ASA-B induced a significant ($p < 0.01$) increase of SCEs and a significant decrease of PRIs in all the cases tested. ASA-B was the most potent genotoxic agent. On the other hand, ASA-C and POPA showed no significant genotoxic or cytostatic effects at the tested drug concentration. The PARP inhibitor 3-AB induced a significant increase of SCEs in all cases but no cytostatic effects were observed. The well-established potent PARP inhibitor, Olaparib, exhibited a significantly high genotoxic activity in all cases, and in particular against the BRCA1 mutated UWB1.289 ovarian cancer cells. Moreover, Olaparib induced the attenuation of PRI and cytostatic effects only against UWB1.289 cells.

BRCA1-defective UWB1.289 cells showed a certain sensitivity to the active tested compounds, and in particular to B-lactam steroid alkylators ASA-B and ASA-A, and a significant but lower sensitivity to Olaparib at the tested drug concentration. ASA-B performed with the highest activity against the UWB1.289 cells. The results are shown in Table 5.

Table 5. Assessment of sister chromatid exchanges (SCEs) and respective PRIs induced in vitro by ASA-A, ASA-B, ASA-C, POPA, 3-AB and Olaparib at a drug concentration of 3 μM in PHA-stimulated PBMCs obtained from 5 healthy donors, as well as in UWB1.289 and UWB1.289 + BRCA1 human ovarian cancer cell metaphases. * Significance level $p < 0.01$, compared to control values; two-side paired *t*-test.

		Control	ASA-A	ASA-B	ASA-C	POPA	3-AB	Olaparib
Donor 1	SCEs/Metaphase ± SEM	8.23 ± 0.43	29.5 ± 1.4	36.5 ± 1.6	9.13 ± 0.45	8.4 ± 0.56	14.44 ± 0.92	20.43 ± 1.1
	SCE/Chromosome ± SEM	0.18 ± 0.01	0.64 ± 0.03	0.79 ± 0.03	0.2 ± 0.01	0.18 ± 0.01	0.31 ± 0.02	0.44 ± 0.02
	PRI	2.68	2.32	2.34	2.65	2.74	2.76	2.60
Donor 2	SCEs/Metaphase ± SEM	8.02 ± 0.46	30.7 ± 1.6	37.8 ± 1.9	8.46 ± 0.46	8.06 ± 0.43	13.26 ± 0.61	21.6 ± 1.4
	SCE/Chromosome ± SEM	0.17 ± 0.01	0.67 ± 0.03	0.82 ± 0.04	0.18 ± 0.01	0.18 ± 0.009	0.30 ± 0.013	0.47 ± 0.03
	PRI	2.72	2.26	2.29	2.74	2.78	2.70	2.56
Donor 3	SCEs/Metaphase ± SEM	8.43 ± 0.61	32.4 ± 1.7	41.7 ± 2.3	8.50 ± 0.55	8.54 ± 0.51	12.85 ± 0.73	23.87 ± 1.5
	SCE/Chromosome ± SEM	0.19 ± 0.013	0.70 ± 0.04	0.91 ± 0.05	0.18 ± 0.01	0.19 ± 0.011	0.28 ± 0.016	0.52 ± 0.03
	PRI	2.56	2.13	2.30	2.48	2.51	2.61	2.47
Donor 4	SCEs/Metaphase ± SEM	7.71 ± 0.35	27.2 ± 1.5	43.3 ± 2.4	8.61 ± 0.38	7.56 ± 0.31	13.21 ± 0.67	21.92 ± 1.2
	SCE/Chromosome ± SEM	0.17 ± 0.008	0.59 ± 0.03	0.94 ± 0.05	0.19 ± 0.01	0.16 ± 0.006	0.29 ± 0.015	0.48 ± 0.03
	PRI	2.84	2.21	2.18	2.72	2.73	2.86	2.76
Donor 5	SCEs/Metaphase ± SEM	9.12 ± 0.58	31.6 ± 1.7	38.4 ± 2.1	9.34 ± 0.64	9.02 ± 0.6	15.98 ± 0.96	25.32 ± 1.6
	SCE/Chromosome ± SEM	0.2 ± 0.012	0.69 ± 0.04	0.83 ± 0.05	0.2 ± 0.01	0.2 ± 0.013	0.35 ± 0.02	0.55 ± 0.04
	PRI	2.41	1.98	2.04	2.47	2.60	2.55	2.34
	Mean SCEs/metaphase Mean ± SD	*8.3 ± 0.53*	*30.28 ± 2.03**	*39.54 ± 2.84 **	*8.81 ± 0.4*	*8.32 ± 0.55*	*14.03 ± 1.24 **	*22.63 ± 1.95 **
	Mean SCE/Chromosome ± SD	*0.18 ± 0.013*	*0.66 ± 0.04 **	*0.86 ± 0.06 **	*0.19 ± 0.08*	*0.18 ± 0.011*	*0.30 ± 0.03 **	*0.66 ± 0.04 **
	Mean PRI ± SD	*2.642 ± 0.16*	*2.18 ± 0.13 **	*2.23 ± 0.12 **	*2.612 ± 0.13*	*2.672 ± 0.11*	*2.696 ± 0.12*	*2.546 ± 0.16*
		Control	**ASA-A**	**ASA-B**	**ASA-C**	**POPA**	**3-AB**	**Olaparib**
UWB1.289	SCEs/Metaphase ± SEM	20.34 ± 1.68	73.78 ± 4.03 *	86.76 ± 4.97 *	24.53 ± 1.72	23.48 ± 1.58	31.91 ± 1.88 *	66.78 ± 4.45 *
	SCE/Chromosome ± SEM	0,27 ± 0.022	1.0 ± 0.054 *	1.17 ± 0.07 *	0.33 ± 0.02	0.32 ± 0.02	0.43 ± 0.025 *	0.90 ± 0.06 *
	PRI	2.18	1.88 *	1.77 *	2.15	2.22	2.14	1.98 *
UWB1.289 +BRCA1	SCEs/metaphase ± SEM	14.26 ± 1.14	63.86 ± 3.62 *	65.95 ± 3.47 *	15.32 ± 1.14	16.08 ± 1.26	14.86 ± 1.23	24.75 ± 1.86 *
	SCE/Chromosome ± SEM	0.19 ± 0.015	0.86 ± 0.05 *	0.89 ± 0.047 *	0.21 ± 0.02	0.22 ± 0.02	0.20 ± 0.017	0.33 ± 0.025 *
	PRI	2.09	1.85 *	1.88 *	2.08	2.11	2.16	2.07

4. Discussion

PARP-1, a multifunctional nucleus protein with a multidomain structure, has well documented roles in DNA damage repair, necrosis, apoptosis and DNA damage-dependent cancer progression. Recent studies also associate it with epithelial to mesenchymal transition (EMT), a biological phenomenon involved in the early progression of cancer metastasis, and the modulation of EMT regulators like Vimentin, Claudin-1 and other transcription factors, playing a dual role in EMT. In this sense, PARP-1 became a special target for cancer treatment, and several inhibitors of PARP-1 like Olaparib, Rucaparib and Veliparib are already in use [8].

Ovarian cancer is a disease with an extreme genetic complexity and a defective DNA repair via a homologous recombination (HR) pathway. Apart from BRCA1/2 mutations, additional mechanisms can also result in HR pathway alterations and, consequently, lead to a clinical benefit from PARP inhibitors. Thus, inhibitors of PARP have a significant therapeutic impact on the treatment of women with epithelial ovarian cancers, and in particular those with the most common histological subtype, high-grade serous cancer, because of the high rate of homologous recombination (HR) deficiency. The inhibition of PARP generates anticancer effects by disrupting DNA repair, thus causing genomic decay. PARP inhibitors are approved as a frontline maintenance treatment for patients with and without BRCA-associated cancers, and clinical data demonstrate the effectiveness of PARP inhibition in women with recurrent epithelial ovarian cancer harboring BRCA1/2 mutations and those with platinum-sensitive disease as a maintenance therapy or as a second-line treatment in recurrence [37,38].

It has currently been shown that the hybrid aza- (lactam) steroidal alkylating esters, which are chimeric steroid compounds, generate a dual inhibition on the RAS/PI3K/AKT and RAS/RAF/MEK/ERK signaling pathways [23]. In our case, the studied lactam steroidal alkylators ASA-A and ASA-B produced a significantly high anticancer activity in vitro, over against ASA-C, that was almost inactive. As previously reported [28,30,39,40] and demonstrated by our data, the existence of the hydrocarbon side chain on the C-17 position constrains the anticancer pharmacological effects and nearly inactivates the steroidal alkylating esters, probably due to the occurrence of some biochemical obstruction. In addition, the PARP-inhibitor-induced activation of the cytoprotective PI3K-Akt pathway likely contributes to the development of PARP inhibitor resistance. Recently, the molecular mechanism that explains how PARP inhibition induces Akt activation and may account for apoptosis resistance and mitochondrial protection in oxidative stress and in cancer has been discovered [41]. However, as previously reported [23], A- and D-lactam steroid alkylators (similar to the B-lactam-steroid alkylating agents tested in the present report) generate, apart from the suspension of PARP activation, a dual inhibition on the RAS/PI3K/AKT and RAS/RAF/MEK/ERK signal pathways, abrogating the undesirable PARP-inhibition-induced activation of the PI3K-Akt pathway, deploying a unique pharmacological molecular action. Furthermore, the inhibitory effects of the homo-aza-steroidal alkylating esters on the RAS/RAF/MEK/ERK signaling pathway induce a further indirect interruption of PARP-1 activation. As previously reported [42], PARP-1 is activated via a direct interaction with phosphorylated ERK2, and ERK2-induced PARP-1 activation dramatically amplifies ERK-signals, enhancing ERK-induced phosphorylation. Hence, PARP-1 activation mediates epigenetic mechanisms promoting growth, proliferation and differentiation regulated by the Raf-MEK-ERK phosphorylation cascade.

The B-lactam ASA-A and B,D-bilactam ASA-B steroid alkylators, unlike the nitrogen mustard alkylator POPA, presented excellent cytostatic and cytocidal activity against SKOV-3 and OVCAR-3 human ovarian cancer cells, which are insensitive to the anticancer activity of alkylators such as cisplatin and melphalan, as well as to the chemotherapeutic cytotoxic agent Adriamycin. Moreover, these active compounds show a selectivity of their cytostatic and cytotoxic effects against the ovarian cancer cells. These results further confirm that the hybrid aza-steroid alkylators are overcoming relevant cell drug resistance pharmacological mechanisms. Moreover, these chimeric compounds, that bear a modified steroid containing a lactamic group (-NH-C=O−) and carry an alkylating moiety, produce a higher antitumor activity than other alkylators conventionally used in cancer therapy [16,22,24,39,40,43,44].

Evidently, comparing the cell lines tested with the respective compounds, the anticancer outcome of ASA-A and ASA-B proved independent of the TP53 mutations borne by cancer cells. Moreover, cancer cells that have an active mutation on the PIK3CA gene and/or present a high microsatellite instability (MSI-high) seem to demonstrate a higher sensitivity to active lactam steroidal alkylators [39,40]. These are noteworthy observations because TP53 mutations are instrumental in the development of cancer cell resistance to chemotherapeutic drugs, such as classical cytotoxic agents including cisplatin, doxorubicin, 5-fluorouracil, temozolomide and paclitaxel, since the p53-based chemoresistance is highly associated with the chemical properties of the anticancer drug, the cellular drug target, the biological function being blocked by the chemotherapeutic agent, the genomic instability and alterations of the tumor, as well as its differentiation state [45]. It has been reported that the inactivation of TP53 mutations in the presence of BRCA mutations may be associated with cancer cell resistance in the treatment with PARP inhibitors. The synthetic lethality of PARP inhibition in these cases is reported as p53-dependent [45–47]. On the other hand, B-lactam-steroidal alkylating agents attain anticancer activity in both p53-mutated and non-mutated cancer cell lines, and this effect may be independent of their PARP inhibitory activity. The broad screening of A-, B- and D-lactam steroid alkylators for anticancer activity against several panels of human cancer cell lines (breast cancer, colon cancer, pancreatic cancer, small-cell-lung cancer cells) showed that these compounds produce their cytostatic and cytotoxic effects independently of the presence of different inactivating mutations of TP53 (data not shown).

It is also interesting to note that mutations of PIK3CA in ovarian and breast cancer have been associated with chemoresistance [48,49] and endocrine resistance to hormonal therapy [50,51]. The sensitivity of PIK3CA mutated cancer cells to the lactam steroidal alkylators is in accordance with previous findings that showed that these hybrid steroid compounds inhibit the PI3K/AKT signaling pathway [23].

As previously described [17,23,37,43] and indicated by our data, the expression of steroid receptors in ovarian cancer cells, contrary to breast cancer, is independent of, and does not affect, the cytostatic or cytotoxic antitumor effect of the homo-aza-steroidal alkylating esters. Earlier findings about the antitumor effects of homo-aza-steroidal alkylators against human breast cancer cells detected a weak to moderate dose-dependent correlation of their anticancer activity with the estrogen receptor expression [37,43].

As previously reported, A- and D-lactam steroidal alkylators are potent inhibitors of PARP activity by binding in PARP's catalytic site and blocking its activity [17]. Our data showed that the B,D-bilactam ASA-B and the B-lactam ASA-A steroid alkylators also exhibited inhibitory effects on PARP activity in a cell-free assay, with ASA-B prompting even stronger inhibitory effects than the well-established PARP inhibitor 3-AB. These inhibitory effects on PARP activity were significantly correlated to their cytostatic effects on UWB1.289, UWB1.289 + BRCA1, OVCAR-3 and SKOV-3 human ovarian cancer cells. Nevertheless, the well-established PARP inhibitor Olaparib demonstrated much higher inhibitory effects than the tested B-lactam steroid alkylators on PARP activity. However, Olaparib induced cytostatic effects only against the BRCA1-null UWB1.289 and less against OVCAR-3 cancer cells at the tested drug concentrations.

In order to repair the generated genotoxicity and DNA damage induced by the lactam steroidal alkylators ASA-A and ASA-B, the cells of all the examined human ovarian cancer cell lines responded with significantly elevated transcription levels of PARP1 and PARP2 mRNA. These increases of PARP1,2 mRNA were contingent on the diverse cell biology and sensitivity of the tested compounds and were correlated in proportion and respectively with the cytostatic and cytotoxic effects of the investigated agents. Hence, the lactam steroid alkylators ASA-B and ASA-A have equivocal anticancer properties by simultaneously inhibiting PARP activity and increasing the PARP 1 and 2 mRNA expression. On this aspect, it is verified by our experiments that POPA finally obtains a reduced anticancer "state" among the compounds studied; while it provokes alkylating DNA damage and thus an increase in PARP mRNA transcription, it cannot obstruct the specific cell repair mechanism, not being a PARP 1,2 inhibitor. This fact, in combination with its highly toxic nature, leads to its anti-tumor inferiority.

It has been shown as well [21] that the homo-aza-steroidal alkylating esters of nitrogen mustards exerted cytogenetic damage, which was enhanced synergistically in combinational treatment with the PARP inhibitor 3-amino-benzamide (3-AB). It has also been shown that treatment with the homo-aza-steroidal alkylators led to the depletion of cellular NAD, while the addition of 3-AB prevented the drop of NAD levels.

While it is argued that for minor DNA damage, cancer cells activate PARP-1 to help in the cell survival by repairing the mild DNA lesions, for high-level DNA damage PARP-1 is overactivated and uses NAD+ as a donor of ADP ribose for ADP-ribosylation, driving the cell into an imbalance of NAD+/ATP ratio, NAD depletion and eventually resulting in an energy-deprived necrosis [8,52,53]. Remarkably, each cell line seems to be "specialized" on PARP1 and/or PARP2 for the aforementioned repair mechanisms, and a ratio between them proportional to the degree of the DNA damage and the mutagenic effect of each alkylating compound is formed (Figures 5 and 7). This overactivation phenomenon, "emerging" both from PARP1 and PARP2 in all tested cell lines is also emphasized in our results in the wild-type cancer cells (without BRCAness), promoting the NAD depletion/energy cell "starvation" as an alternative procedure leading to cellular death, even for more enduring carcinomas, with a mechanism for these compounds similar to the one reported for lactam-steroidal alkylators [21].

The loss of DNA damage repair pathways is an early and frequent event in tumorigenesis, occurring in 40–50% of many cancer types. The basis of synthetic lethality in cancer therapy are the DNA damage-repair-deficient cancers dependent on backup DNA repair pathways. In cancer, the synthetic lethality mostly relates to pairs of genes, where the inactivation of one and the pharmacological inhibition of the other lead to the death of the cancer cells. The combination of PARP inhibitors and other DNA damage repair inhibitors, or the combination of multiple components of the same pathway, may have great potential for synthetic lethality efficiency. Both genetic and epigenetic alterations may serve as synthetic lethal therapeutic markers. PARP is a critical factor in the repair of single-stranded DNA damage via the base excision repair pathway, and PARP inhibitors have a substantial single-agent antitumor activity by inducing synthetic lethality, especially in tumors harboring deleterious germline or somatic BRCA mutations. PARP inhibition produces single-stranded DNA breaks, which may be repaired by homologous recombination, a process partially dependent on BRCA1 and BRCA2 [54–56].

Tumors harboring genetic defects in homologous recombination (HR), a DNA double-strand break repair pathway, are hypersensitive to PARP inhibitors. Unlike HR-defective cells, HR-proficient cells manifest very low cytotoxicity when exposed to PARP inhibitors, although they mount a DNA damage response. PARP1 deficiency is markedly increased when combined with other DNA damage response genetic deficiencies. Sister chromatid exchange (SCE) is a recombination event between identical sister chromatids following DNA replication, typically but not uniformly, and resulting in an equal exchange of genetic information. However, SCEs that occur at offset repeats can lead to the loss or gain of genetic information. Thus, increased SCEs are a measure of hyper-recombination leading to chromatid exchange, whether due to deficiencies in genes involved in the suppression of exchange or to unrepaired DNA damage encountered during DNA replication and the subsequent HR-mediated repair at stalled replication forks. An increase in SCEs correlates with the mutagenic potential and the induced loss of heterozygosity. Olaparib causes marked hypersensitivity in HR-deficient cells and increases SCE frequency in repair-proficient human cells. Less potent PARP inhibitors, which did not demonstrate synthetic lethality in BRCA-deficient tumor cells, have been shown to increase SCEs [57]. The B-lactam steroid alkylators ASA-A and ASA-B, although less potent PARP inhibitors compared to Olaparib, induced a significant increase of SCEs in either repair-proficient or repair-deficient normal and cancerous human cells. ASA-B was presented as the more potent genotoxic agent, including the BRCA1-null ovarian cancer UWB1.289 cells.

In HR-deficient tumors, the "synthetic lethality" strategy for PARP inhibition is dependent on unrepaired endogenous strand breaks that arise during DNA replication and normal cell metabolism. A current therapeutic strategy, investigated in ongoing clinical trials, is the combination of chemotherapeutics that directly induce DNA damage whilst also inhibiting single-strand breaks repair with PARP inhibition [58].

Our data revealed that the active lactam steroid alkylator ASA-B exhibited a significantly higher anticancer activity in the BRCA1-null ovarian cancer cell line UWB1.289 than in the respective BRCA1 wild-type UWB1.289 + BRCA1 ovarian cancer cell line. The BRCA1 mutated ovarian cell line UWB1.289 appeared to be considerably more prone to the cytostatic and cytocidal effects of the tested alkylating agents than the other ovarian cancer cell lines tested due to its BRCAness phenotype. The B,D-dilactam steroidal alkylator ASA-B presented a more clear synthetic lethality attribute in BRCA-1 mutated UWB1.289 ovarian cancer cells due to its corresponding cytostatic and cytotoxic effects in relation to its inhibitory activity on PARP enzymes. On the other hand, the B-lactam steroidal alkylator ASA-A induced the strongest cytostatic and cytotoxic antitumor results in all tested ovarian cancer cell lines, and while it proved a potent PARP inhibitor in a cell free assay, the synthetic lethality in BRCA-1 "null" UWB1.289 cells could not be as evident. Moreover, the B,D-bilactam ASA-B and the B-lactam ASA-A steroid alkylators instigated higher increases of PARP1 mRNA cellular content in the UWB1.289 + BRCA1 than in BRCA1 mutated UWB1.289 cancer cells. In addition, these compounds promoted a more appreciable elevation of the PARP2 mRNA transcription in the BRCA1-null UWB1.289 than BRCA1 wild-type, UWB1.289 + BRCA1 cancer cells.

The silencing of DNA damage repair genes by aberrant epigenetic changes provides new opportunities for synthetic lethal therapy in cancer [59]. These chimeric modified homo-aza-steroidal alkylators produce significant and characteristic anticancer features generated in the cancer cells: on the one hand, DNA damage, and on the other hand, disrupting DNA repair by inhibiting PARP activity and causing important cellular energy pool depletion through a high increase of PARP 1 and 2 mRNA transcription and expression. These hybrid lactam steroid alkylators act with multi-targeted pharmacological mechanisms of PARP "trapping", combined in one molecule, and can cause enhanced synthetic lethality in cancer cells.

Conclusively, the tested B-lactam steroid alkylators are non-typical inhibitors of PARP and are potentially relevant for clinical use due to their multiple targeting. However, the PARP inhibition by these compounds has only been demonstrated in a cell free assay, and it is currently unclear whether this is relevant or contributes to the effects observed in the cells. Nevertheless, despite their low (and unknown in cell conditions) potency in PARP inhibition compared to Olaparib, additional pharmacological mechanisms of action may contribute to B-aza-steroid alkylators' activity and potential to overcome the pharmacological resistance to current inhibitors.

5. Patents

This research is covered by the international patent WO2017001439.

Author Contributions: Conceptualization, D.T.T. and V.S.; methodology, D.T.T. and V.S.; software; validation, N.N., P.D. and A.P.; formal analysis, D.T.T., P.D. and N.N.; investigation, D.T.T., N.N., P.D., A.P., S.S., E.G.G., C.E.A., M.V.D. and V.S.; resources, D.T.T., V.S. and N.N.; data curation, P.D., N.N., D.T.T. and A.P.; writing—original draft preparation, N.N. and D.T.T.; writing—review and editing, D.T.T., N.N. and P.D.; visualization, D.T.T., N.N. and P.D.; supervision, D.T.T.; project administration, D.T.T.; funding acquisition, D.T.T. All authors have read and agreed to the published version of the manuscript.

Funding: This research was funded by GALENICA S.A. and ENERGON BIOTECH S.A., Greece, as well as, in part, by the Institute of Biomedical Sciences "ATHENA", Greece.

Institutional Review Board Statement: Not applicable.

Informed Consent Statement: Not applicable.

Data Availability Statement: Data are available on request from the authors.

Conflicts of Interest: The authors declare no conflict of interest.

References

1. Stewart, C.; Ralyea, C.; Lockwood, S. Ovarian Cancer: An Integrated Review. *Semin. Oncol. Nurs.* **2019**, *35*, 151–156. [CrossRef] [PubMed]
2. Markman, M. Optimal management of recurrent ovarian cancer. *Int. J. Gynecol. Cancer* **2009**, *19* (Suppl. 2), S40-3. [CrossRef] [PubMed]
3. Fraser, M.; Leung, B.; Jahani-Asl, A.; Yan, X.; Thompson, W.E.; Tsang, B.K. Chemoresistance in human ovarian cancer: The role of apoptotic regulators. *Reprod. Biol. Endocrinol.* **2003**, *1*, 66. [CrossRef]
4. Jelovac, D.; Armstrong, D.K. Recent progress in the diagnosis and treatment of ovarian cancer. *CA Cancer J. Clin.* **2011**, *61*, 183–203. [CrossRef]
5. Tutt, A.; Ashworth, A. The relationship between the roles of BRCA genes in DNA repair and cancer predisposition. *Trends Mol. Med.* **2002**, *8*, 571–576. [CrossRef]
6. Turner, N.; Tutt, A.; Ashworth, A. Hallmarks of "BRCAness" in sporadic cancers. *Nat. Rev. Cancer* **2004**, *4*, 814–819. [CrossRef]
7. Bryant, H.E.; Schultz, N.; Thomas, H.D.; Parker, K.M.; Flower, D.; Lopez, E.; Kyle, S.; Meuth, M.; Curtin, N.J.; Helleday, T. Specific killing of BRCA2-deficient tumours with inhibitors of poly(ADP-ribose) polymerase. *Nature* **2005**, *434*, 913–917. [CrossRef]
8. Kumar, M.; Jaiswal, R.K.; Yadava, P.K.; Singh, R.P. An assessment of poly (ADP-ribose) polymerase-1 role in normal and cancer cells. *Biofactors* **2020**, *46*, 894–905. [CrossRef] [PubMed]
9. Satoh, M.S.; Lindahl, T. Role of poly(ADP-ribose) formation in DNA repair. *Nature* **1992**, *356*, 356–358. [CrossRef]
10. Golstein, P.; Kroemer, G. Cell death by necrosis: Towards a molecular definition. *Trends Biochem. Sci.* **2007**, *32*, 37–43. [CrossRef] [PubMed]
11. Pacher, P.; Szabó, C. Role of poly(ADP-ribose) polymerase 1 (PARP-1) in cardiovascular diseases: The therapeutic potential of PARP inhibitors. *Cardiovasc. Drug Rev.* **2007**, *25*, 235–260. [CrossRef]
12. D'Amours, D.; Sallmann, F.R.; Dixit, V.M.; Poirier, G.G. Gain-of-function of poly(ADP-ribose) polymerase-1 upon cleavage by apoptotic proteases: Implications for apoptosis. *J. Cell Sci.* **2001**, *114*, 3771–3778. [CrossRef]
13. Pacher, P.; Szabo, C. Role of the peroxynitrite-poly(ADP-ribose) polymerase pathway in human disease. *Am. J. Pathol.* **2008**, *173*, 2–13. [CrossRef]
14. Evgenov, O.V.; Liaudet, L. Role of nitrosative stress and activation of poly(ADP-ribose) polymerase-1 in cardiovascular failure associated with septic and hemorrhagic shock. *Curr. Vasc. Pharmacol.* **2005**, *3*, 293–299. [CrossRef]
15. Jagtap, P.; Szabó, C. Poly(ADP-ribose) polymerase and the therapeutic effects of its inhibitors. *Nat. Rev. Drug Discov.* **2005**, *4*, 421–440. [CrossRef]
16. Curtin, N.J. PARP inhibitors for cancer therapy. *Expert Rev. Mol. Med.* **2005**, *7*, 1–20. [CrossRef]
17. Trafalis, D.T.; Polonifi, A.; Dalezis, P.; Nikoleousakos, N.; Katsamakas, S.; Sarli, V. Targeting on poly(ADP-ribose) polymerase activity with DNA-damaging hybrid lactam-steroid alkylators in wild-type and BRCA1-mutated ovarian cancer cells. *Chem. Biol. Drug Des.* **2017**, *90*, 854–866. [CrossRef] [PubMed]
18. Tselepi, M.R.; Zacharopoulou, A.; Demopoulos, N.A.; Catsoulacos, P. Induction of chromosomal aberrations in human lymphocytes by the antitumour alkylating agent homo-aza-steroidal ester of p-bis(2-chloroethyl)aminophenoxy acetic acid, in vitro. *Mutat. Res.* **1993**, *290*, 211–216. [CrossRef]
19. Stephanou, G.; Demopoulos, N.A.; Catsoulacos, P. Genotoxic effect of the antitumour agent homo-aza-steroidal ester of p-bis-(2-chloroethyl)aminophenoxy acetic acid (NSC 294859) tested in the Drosophila wing somatic mutation and recombination test. *Mutagenesis* **1991**, *6*, 325–328. [CrossRef] [PubMed]
20. Catsoulacos, P.; Camoutsis, C.; Papageorgiou, A.; Margariti, E.; Psaraki, K.; Demopoulos, N. Structure–anticancer and structure–genetic activity relationships of homo-aza-steroidal esters of N,N-bis(2-chloroethyl)aminocinnamic acid isomers. *J. Pharm. Sci.* **1993**, *82*, 204–208. [CrossRef] [PubMed]
21. Petrou, C.; Mourelatos, D.; Mioglou, E.; Dozi-Vassiliades, J.; Catsoulacos, P. Effects of alkylating antineoplastics alone or in combination with 3-aminobenzamide on genotoxicity, antitumor activity, and NAD levels in human lymphocytes in vitro and on Ehrlich ascites tumor cells in vivo. *Teratog. Carcinog. Mutagen.* **1990**, *10*, 321–331. [CrossRef] [PubMed]
22. Trafalis, D.; Geromichalou, E.; Dalezis, P.; Nikoleousakos, N.; Sarli, V. Synthesis and evaluation of new steroidal lactam conjugates with aniline mustards as potential antileukemic therapeutics. *Steroids* **2016**, *115*, 1–8. [CrossRef]
23. Dalezis, P.; Geromichalou, E.; Polonifi, A.; Sagredou, S.; Nikoleousakos, N.; Nikolaou, M.; Sarli, V.; Panayiotidis, M.I.; Trafalis, D.T. Azasteroid Alkylators as Dual Inhibitors of AKT and ERK Signaling for the Treatment of Ovarian Carcinoma. *Cancers* **2020**, *12*, 1263. [CrossRef]
24. Trafalis, D.; Dalezis, P.; Geromichalou, E.; Sagredou, S.; Sflakidou, E.; Voura, M.; Grammatikopoulou, M.; Gabriel, C.; Sarli, V. Discovery of steroidal lactam conjugates of POPAM-NH2 with potent anticancer activity. *Future Med. Chem.* **2020**, *12*, 19–35. [CrossRef] [PubMed]

25. Donà, F.; Chiodi, I.; Belgiovine, C.; Raineri, T.; Ricotti, R.; Mondello, C.; Scovassi, A.I. Poly(ADP-ribosylation) and neoplastic transformation: Effect of PARP inhibitors. *Curr. Pharm. Biotechnol.* **2013**, *14*, 524–536. [CrossRef]
26. DelloRusso, C.; Welcsh, P.L.; Wang, W.; Garcia, R.L.; King, M.-C.; Swisher, E.M. Functional characterization of a novel BRCA1-null ovarian cancer cell line in response to ionizing radiation. *Mol. Cancer Res.* **2007**, *5*, 35–45. [CrossRef] [PubMed]
27. Alley, M.C.; Scudiero, D.A.; Monks, A.; Hursey, M.L.; Czerwinski, M.J.; Fine, D.L.; Abbott, B.J.; Mayo, J.G.; Shoemaker, R.H.; Boyd, M.R. Feasibility of drug screening with panels of human tumor cell lines using a microculture tetrazolium assay. *Cancer Res.* **1988**, *48*, 589–601.
28. Trafalis, D.T.P.; Camoutsis, C.; Papageorgiou, A. Research on the anti-tumour effect of steroid lactam alkylator (NSC-294859) in comparison with conventional chemotherapeutics in malignant melanoma. *Melanoma Res.* **2005**, *15*, 273–281. [CrossRef] [PubMed]
29. Trafalis, D.T.; Camoutsis, C.; Dalezis, P.; Papageorgiou, A.; Kontos, M.; Karamanakos, P.; Giannakos, G.; Athanassiou, A.E. Antitumour effect of a- and d- lactam androgen nitrogen mustards on non-small cell lung carcinoma. *J. BUON* **2004**, *9*, 275–282.
30. Camoutsis, C.; Sambani, C.; Trafalis, D.T.; Peristeris, P. On the formation of steroidal amidoesters of 4-[N,N-bis(2-chloroethyl)amino]benzoic acid and their cytotoxic activity. *Eur. J. Med. Chem.* **1999**, *34*, 645–649. [CrossRef]
31. Jouan-Lanhouet, S.; Arshad, M.I.; Piquet-Pellorce, C.; Martin-Chouly, C.; Le Moigne-Muller, G.; Van Herreweghe, F.; Takahashi, N.; Sergent, O.; Lagadic-Gossmann, D.; Vandenabeele, P.; et al. TRAIL induces necroptosis involving RIPK1/RIPK3-dependent PARP-1 activation. *Cell Death Differ.* **2012**, *19*, 2003–2014. [CrossRef]
32. Meador, J.A.; Zhao, M.; Su, Y.; Narayan, G.; Geard, C.R.; Balajee, A.S. Histone H2AX is a critical factor for cellular protection against DNA alkylating agents. *Oncogene* **2008**, *27*, 5662–5671. [CrossRef]
33. Mathews, M.T.; Berk, B.C. PARP-1 inhibition prevents oxidative and nitrosative stress-induced endothelial cell death via transactivation of the VEGF receptor 2. *Arterioscler. Thromb. Vasc. Biol.* **2008**, *28*, 711–717. [CrossRef]
34. Denman, S.E.; McSweeney, C.S. Development of a real-time PCR assay for monitoring anaerobic fungal and cellulolytic bacterial populations within the rumen. *FEMS Microbiol. Ecol.* **2006**, *58*, 572–582. [CrossRef]
35. Arya, M.; Shergill, I.S.; Williamson, M.; Gommersall, L.; Arya, N.; Patel, H.R.H. Basic principles of real-time quantitative PCR. *Expert Rev. Mol. Diagn.* **2005**, *5*, 209–219. [CrossRef] [PubMed]
36. Trafalis, D.T.P.; Sambani, C.; Kapsimali, V.; Economidou, J.; Politis, G.; Catsoulacos, P. Effects of homo-aza-steroids on acute non-lymphocytic leukaemia cell proliferation in vitro. *Br. J. Haematol.* **1995**, *91*, 907–914. [CrossRef]
37. Kontos, M.; Nikolopoulou, M.; Trafalis, D.T.P.; Geromichalos, G.D.; Koukoulitsa, C.; Camoutsis, J.C.; Bastounis, E.; Karamanakos, P. The effect of an estrone D-lactam steroid ester derivative on breast cancer cells and its predicted binding interactions with the ligand binding domain of estrogen receptor-alpha. *Oncol. Res.* **2005**, *16*, 129–142. [CrossRef] [PubMed]
38. Cook, S.A.; Tinker, A.V. PARP Inhibitors and the Evolving Landscape of Ovarian Cancer Management: A Review. *BioDrugs* **2019**, *33*, 255–273. [CrossRef]
39. Trafalis, D.T.P.; Tsavdaridis, D.; Camoutsis, C.; Karayiani, V.; Mourelatos, D.; Dalezis, P.; Athanassiou, A.; Pangalis, G.A.; Papageorgiou, A. Preclinical studies on NSC290205 aza-steroid alkylator activity in combination with adriamycin against lymphoid leukaemia. *Br. J. Haematol.* **2005**, *128*, 343–350. [CrossRef] [PubMed]
40. Wampler, G.L.; Catsoulacos, P. Antileukemic effect of homo-aza-steroidal ester of [p-[bis(2-chloroethyl)amino]phenyl]acetic acid. *Cancer Treat. Rep.* **1977**, *61*, 37–41. [PubMed]
41. Gallyas, F., Jr.; Sumegi, B.; Szabo, C. Role of Akt Activation in PARP Inhibitor Resistance in Cancer. *Cancers* **2020**, *12*, 532. [CrossRef]
42. Cohen-Armon, M. PARP-1 activation in the ERK signaling pathway. *Trends Pharmacol. Sci.* **2007**, *28*, 556–560. [CrossRef]
43. Trafalis, D.T.P.; Geromichalos, G.D.; Koukoulitsa, C.; Papageorgiou, A.; Karamanakos, P.; Camoutsis, C. Lactandrate: A D-homo-aza-androsterone alkylator in the treatment of breast cancer. *Breast Cancer Res. Treat.* **2006**, *97*, 17–31. [CrossRef]
44. Trafalis, D.T.P. Hybrid aza-steroid alkylators in the treatment of colon cancer. *Cancer Lett.* **2006**, *243*, 202–210. [CrossRef] [PubMed]
45. Cao, X.; Hou, J.; An, Q.; Assaraf, Y.G.; Wang, X. Towards the overcoming of anticancer drug resistance mediated by p53 mutations. *Drug Resist. Updat.* **2020**, *49*, 100671. [CrossRef] [PubMed]
46. Collot, T.; Niogret, J.; Carnet, M.; Chevrier, S.; Humblin, E.; Favier, L.; Bengrine-Lefevre, L.; Desmoulins, I.; Arnould, L.; Boidot, R. PARP inhibitor resistance and TP53 mutations in patients treated with olaparib for BRCA-mutated cancer: Four case reports. *Mol. Med. Rep.* **2021**, *23*, 75. [CrossRef] [PubMed]
47. Sizemore, S.T.; Mohammad, R.; Sizemore, G.M.; Nowsheen, S.; Yu, H.; Ostrowski, M.C.; Chakravarti, A.; Xia, F. Synthetic Lethality of PARP Inhibition and Ionizing Radiation is p53-dependent. *Mol. Cancer Res.* **2018**, *16*, 1092–1102. [CrossRef] [PubMed]
48. Carden, C.P.; Stewart, A.; Thavasu, P.; Kipps, E.; Pope, L.; Crespo, M.; Miranda, S.; Attard, G.; Garrett, M.D.; Clarke, P.A.; et al. The association of PI3 kinase signaling and chemoresistance in advanced ovarian cancer. *Mol. Cancer Ther.* **2012**, *11*, 1609–1617. [CrossRef] [PubMed]
49. Santarpia, M.; Altavilla, G.; Margeli, M.; Cirauqui, B.; Mesiti, M.; Cavallari, V.; Ramirez, J.L.; Sanchez-Ronco, M.; Santarpia, L.; Taron, M.; et al. PIK3CA mutations and BRCA1 expression in breast cancer: Potential biomarkers for chemoresistance. *Cancer Investig.* **2008**, *26*, 1044–1051. [CrossRef]
50. Zhang, J.; Wang, Q.; Wang, Q.; Cao, J.; Sun, J.; Zhu, Z. Mechanisms of resistance to estrogen receptor modulators in ER+/HER2- advanced breast cancer. *Cell. Mol. Life Sci.* **2020**, *77*, 559–572. [CrossRef]

51. AlFakeeh, A.; Brezden-Masley, C. Overcoming endocrine resistance in hormone receptor-positive breast cancer. *Curr. Oncol.* **2018**, *25*, S18–S27. [CrossRef]
52. Bouchard, V.J.; Rouleau, M.; Poirier, G.G. PARP-1, a determinant of cell survival in response to DNA damage. *Exp. Hematol.* **2003**, *31*, 446–454. [CrossRef]
53. Ha, H.C.; Snyder, S.H. Poly(ADP-ribose) polymerase is a mediator of necrotic cell death by ATP depletion. *Proc. Natl. Acad. Sci. USA* **1999**, *96*, 13978–13982. [CrossRef]
54. Boussios, S.; Karihtala, P.; Moschetta, M.; Karathanasi, A.; Sadauskaite, A.; Rassy, E.; Pavlidis, N. Combined Strategies with Poly (ADP-Ribose) Polymerase (PARP) Inhibitors for the Treatment of Ovarian Cancer: A Literature Review. *Diagnostics* **2019**, *9*, 87. [CrossRef] [PubMed]
55. Cetin, B.; Wabl, C.A.; Gumusay, O. The DNA damaging revolution. *Crit. Rev. Oncol. Hematol.* **2020**, *156*, 103117. [CrossRef]
56. Zhou, P.; Wang, J.; Mishail, D.; Wang, C.-Y. Recent advancements in PARP inhibitors-based targeted cancer therapy. *Precis. Clin. Med.* **2020**, *3*, 187–201. [CrossRef] [PubMed]
57. Ito, S.; Murphy, C.G.; Doubrovina, E.; Jasin, M.; Moynahan, M.E. PARP Inhibitors in Clinical Use Induce Genomic Instability in Normal Human Cells. *PLoS ONE* **2016**, *11*, e0159341. [CrossRef] [PubMed]
58. O'Connor, M.J. Targeting the DNA damage response in cancer. *Mol. Cell* **2015**, *60*, 547–560. [CrossRef]
59. Gao, A.; Guo, M. Epigenetic based synthetic lethal strategies in human cancers. *Biomark. Res.* **2020**, *8*, 44. [CrossRef]

biomedicines

MDPI

Review

Intratumoural Cytochrome P450 Expression in Breast Cancer: Impact on Standard of Care Treatment and New Efforts to Develop Tumour-Selective Therapies

Smarakan Sneha [1], Simon C. Baker [2], Andrew Green [3], Sarah Storr [3], Radhika Aiyappa [3], Stewart Martin [3] and Klaus Pors [1,*]

[1] Institute of Cancer Therapeutics, School of Pharmacy and Medical Sciences, Faculty of Life Sciences, University of Bradford, Bradford BD7 1DP, UK; S.Swaroop@bradford.ac.uk
[2] Jack Birch Unit for Molecular Carcinogenesis, Department of Biology & York Biomedical Research Institute, University of York, Heslington, York YO10 5DD, UK; simon.baker@york.ac.uk
[3] Nottingham Breast Cancer Research Centre, School of Medicine, Biodiscovery Institute, University Park, Nottingham NG7 2RD, UK; andrew.green@nottingham.ac.uk (A.G.); sarah.storr@nottingham.ac.uk (S.S.); Radhika.Aiyappa@nottingham.ac.uk (R.A.); stewart.martin@nottingham.ac.uk (S.M.)
* Correspondence: k.pors1@bradford.ac.uk; Tel.: +44-(0)1274-236482 or +44-(0)1274-235866; Fax: +44-(0)1274-233234

Abstract: Despite significant advances in treatment strategies over the past decade, selective treatment of breast cancer with limited side-effects still remains a great challenge. The cytochrome P450 (CYP) family of enzymes contribute to cancer cell proliferation, cell signaling and drug metabolism with implications for treatment outcomes. A clearer understanding of CYP expression is important in the pathogenesis of breast cancer as several isoforms play critical roles in metabolising steroid hormones and xenobiotics that contribute to the genesis of breast cancer. The purpose of this review is to provide an update on how the presence of CYPs impacts on standard of care (SoC) drugs used to treat breast cancer as well as discuss opportunities to exploit CYP expression for therapeutic intervention. Finally, we provide our thoughts on future work in CYP research with the aim of supporting ongoing efforts to develop drugs with improved therapeutic index for patient benefit.

Keywords: Cytochrome P450; CYP1A1; CYP1B1; CYP2W1; breast cancer; prodrug; bioprecursor; duocarmycin; phortress; AQ4N

Citation: Sneha, S.; Baker, S.C.; Green, A.; Storr, S.; Aiyappa, R.; Martin, S.; Pors, K. Intratumoural Cytochrome P450 Expression in Breast Cancer: Impact on Standard of Care Treatment and New Efforts to Develop Tumour-Selective Therapies. *Biomedicines* **2021**, *9*, 290. https://doi.org/10.3390/biomedicines9030290

Academic Editor: Simon J. Allison

Received: 13 January 2021
Accepted: 7 March 2021
Published: 12 March 2021

Publisher's Note: MDPI stays neutral with regard to jurisdictional claims in published maps and institutional affiliations.

Copyright: © 2021 by the authors. Licensee MDPI, Basel, Switzerland. This article is an open access article distributed under the terms and conditions of the Creative Commons Attribution (CC BY) license (https://creativecommons.org/licenses/by/4.0/).

1. Introduction

As our understanding of cancer improves, it has become increasingly evident that targeted and patient-specific treatment strategies are essential for effective treatment. A one-size-fits-all approach has multiple flaws in the face of a wide range of cancer subtypes, with each patient harbouring specific genetic alterations and metabolic profiles. Immunohistochemical studies have classified breast cancer into four broad subtypes: oestrogen receptor-positive (ER+), progesterone receptor-positive (PR+), human epidermal growth factor receptor 2-positive (HER2+) and triple negative [1]. Patients harbouring tumours that do not express any ER, PR or HER2 receptors are classified as triple negative breast cancers (TNBC) [2]. Around 80% of the breast tumours are positive for the hormone receptors wherein binding of the requisite chemical messenger to their receptor causes changes in the cancer cell resulting in proliferation [3]. Although these receptors can be targeted by drugs that disrupt the signalling process, hormonal therapies are still associated with undesirable side effects including bone weakness and temporary or permanent menopausal symptoms [4].

TNBC is the least common of all subtypes, disproportionately affecting pre-menopausal African American women [5] and has been associated with poor clinical outcome. Treatment relies heavily on chemotherapy as endocrine and targeted (e.g., trastuzumab-based)

therapies are of no benefit in this disease [6]. As a consequence, the number of drug combinations that can be used are limited thereby increasing the risk of developing drug resistance [7]. Furthermore, there is far less understanding of how to treat TNBC compared to receptor-positive cases as there are currently no known biomarkers to classify TNBCs according to response. Further research is required to unravel the underlying causes of TNBC, with deficiencies in underlying DNA repair mechanisms often being implicated [8–10].

2. Cytochrome P450 Function

The cytochrome P450 (CYP) superfamily of enzymes consist of a group of 57 human isoforms that have the capacity to oxidise a vast variety of endogenous and xenobiotic compounds [11]. In mammals, these enzymes are primarily involved in the metabolism of drugs and fat-soluble vitamins, conversion of polyunsaturated fatty acids to biologically active compounds and in the synthesis of steroid hormones. Apart from the capacity to detoxify and inactivate many clinically used drugs, CYPs are also involved in conversion of prodrugs into cytotoxic metabolites [12]. The ability of these enzymes to metabolise both procarcinogens and anticancer drugs makes their study interesting in terms of cancer aetiology and therapy. Differential endogenous expression of certain CYP genes in tumour vs. normal tissue provides an opportunity for selective targeting with small molecules designed to capitalise on CYP catalysis [13,14].

3. CYPs in Breast Cancer

A clearer understanding of CYP expression is important in the pathogenesis of breast cancer as several isoforms play critical roles in metabolising steroid hormones. CYPs are also essential for the conversion of environmental chemicals such as polycyclic aromatic hydrocarbons (PAHs) to carcinogenic metabolites that contribute to the genesis of breast cancer [15]. Critically, certain drugs used to treat advanced stages of breast cancer such as paclitaxel display little or no cytotoxic activity after metabolism by CYP2C8 and CYP3A4 (Figure 1) [16,17]. Figure 2 outlines drugs used clinically to treat various stages of breast cancer and which either benefit or suffer from intratumoural CYP expression [6,18].

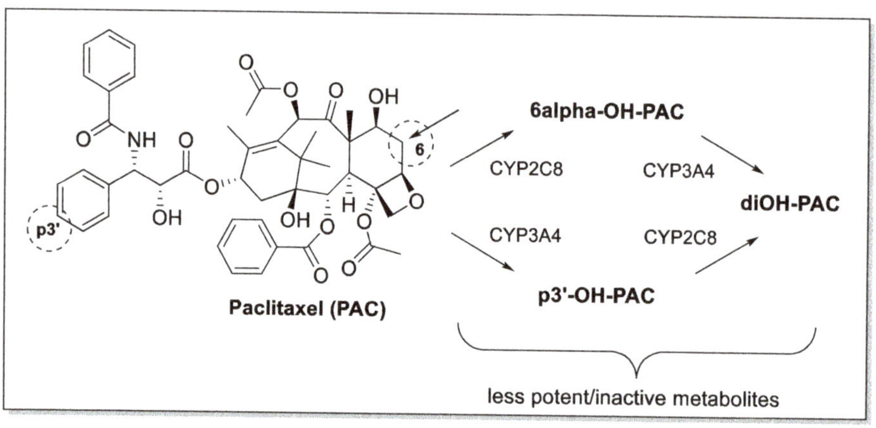

Figure 1. CYP metabolism contributes to paclitaxel treatment failure.

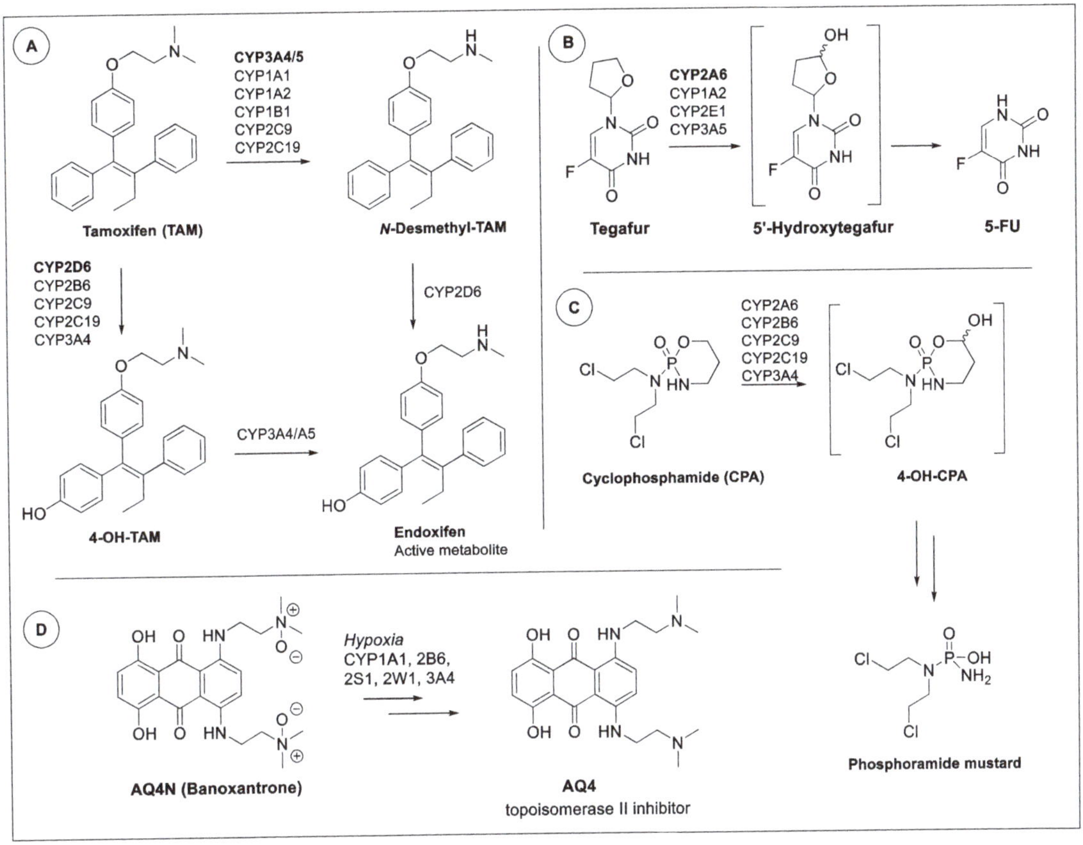

Figure 2. Non-selective bioactivation of prodrugs tamoxifen (**A**), tegafur (**B**), CPA (**C**) and tumour-selective hypoxia-activated prodrug AQ4N (**D**).

In breast cancer research, there has been a variability between laboratories in identifying a particular CYP expression profile at both mRNA and protein levels. In a study conducted by Murray et al., the expression of a panel of 21 CYPs was studied in a cohort of 170 breast tumours by immunostaining of tissue microarray [19]. CYP2S1, CYP2U1 and CYP4X1 exhibited the highest percentage of strong immunoreactivity in contrast to other isoforms such as CYP2J and CYP3A4, which displayed no reactivity in majority of the tumours. A positive correlation with increasing tumour grade has been observed with the expression of CYP4V2, CYP4X1 and CYP4Z1; while CYP1B1, CYP3A5 and CYP51 were significantly associated with the ER status of the tumour. Among the 21 CYPs examined, absence of expression of CYP2S1, CYP3A4, CYP4V2 and CYP26A1 correlated with better survival even though none of them turned out to be an independent prognostic factor [19]. On a genetic level, a comparable study in a set of 20 paired samples of tumour and adjacent normal breast tissues from patients with infiltrating ductal carcinoma identified the expression of *CYP1B1*, *CYP2B6*, *CYP2C*, *CYP2D6*, *CYP2E1*, *CYP4B1* and *CYP11A1* in both tumour and control tissues [15]. A very low level expression of *CYP2C9* and *CYP3A4* was observed by RT-PCR in a similar study conducted in a set of 40 breast cancer and adjacent normal tissues [20]. Nevertheless, *CYP1B1* was overexpressed in one third of the tumours with a mean value exceeding a 50-fold increase compared to *CYP2E1*. Significantly higher expression of *CYP2E1* was found to be correlated with an invasive lobular tumour type and advanced disease compared to the invasive ductal ones, suggesting its potential role as a prognosis marker [20]. Some of these CYPs have been extensively studied on account of their multifactorial roles in breast cancer aetiology and treatment. The purpose of this review is to provide an update on how the presence of CYPs impacts on standard of care (SoC) drugs used to treat breast cancer as well as discuss opportunities to exploit CYP regulation and/or function for therapeutic intervention.

3.1. *CYP1A1/A2*

Expression of CYP1A1 is induced through the binding of substrates to the cytosolic aryl hydrocarbon receptor (AhR) [21]. Elevated *CYP1A1* mRNA expression has been detected to varying degrees in many breast cancer subtypes which supports its potential as a drug target in breast cancer treatment [22]. Expression of CYP1A1 has been found to be high in breast tumour cells with a positive correlation to tumour grade and menopausal status in newly diagnosed patients with adenocarcinoma of the breast [23]. Additionally, it has been found to be overexpressed in breast cancers that are resistant to anti-oestrogen treatment [24]. Knockdown of *CYP1A1* has been shown to impair proliferation and survival in the breast cancer cell lines MCF-7 and MDA-MB-231 through activation of AMPK signaling and inhibition of the phosphorylation of AKT, ERK and P70S6K [22]. There is also some evidence that CYP1A1 is important for maintaining the integrity of breast cancer stem cells (BCSCs), possibly by acting through β-catenin and PTEN/AKT signaling [25]. Increased basal level expression of CYP1A1 has also been observed in spheroids cultured from MCF-7 breast cancer cells, compared to the monolayer, which was suppressed upon co-culture with $CD14^{+}$ cells in a macrophage-shaped environment [26]. This observation reinforces the suppressive effect of inflammatory conditions on the expression of CYP1A1 in mammary epithelial cells which is an important factor that has to be considered while designing immunomodulatory compounds and other chemotherapeutics that are bioactivated by CYP1A1.

Several studies have shown that polymorphisms in drug-metabolising genes not only pose as a risk factor for malignancy but also can serve as a predictive marker for drug efficacy and targeted therapies [27,28]. In metastatic breast cancer, patients with the *CYP1A1*2C* gene variant responded better to a treatment regimen involving docetaxel and capecitabine compared to those with the homozygous wildtype (*CYP1A1*1*) genotype (AA) who experienced a better efficacy with a combination of docetaxel with thiotepa [29]. In early breast cancer, *CYP1A2* rs762551 has been shown to be a potential predictive marker for patients treated with aromatase inhibitors paving the way towards individualised tailored assessment of therapy [30].

Prodrugs that are bioactivated by CYP1 enzymes can broadly be classified into four groups of agents and include flavonoids, benzothiazoles, stilbenes and alkylating agents [29]. Aminoflavone (NSC686288, Figure 3), a ligand of AhR, has been shown to selectively induce apoptosis in MCF-7 breast cancer cells by activating the transcription of *CYP1A1* and *CYP1B1* with functional activity confirmed using the ethoxyresorufin-O-deethylase (EROD) activity assay both in vitro and in xenograft models [31]. Specific activation of the lysine prodrug of aminoflavone (AFP464, NSC710464) and subsequent CYP1 metabolic activation has been observed to be particularly effective in ER-positive breast cancer cells compared with ER-negative counterparts. Epigenetic reactivation of ERα expression treatment with the histone deacetylase (HDAC) inhibitor vorinostat has been shown to sensitise resistant cells towards AFP464, through the transcriptional induction of CYP1A1 [32]. Furthermore, AFP464 also inhibits the growth of tamoxifen-resistant breast cancer cells exhibiting increased expression of α6-integrin linked to suppression of the α6-integrin-Src-Akt signaling pathway [32]. Tamoxifen is a selective oestrogen receptor modulator (SERM) used in the treatment of hormone-receptor-positive breast cancers. AF464 entered a phase 1 clinical trial for advanced solid tumours where a maximally tolerated dose (MTD) of 54 mg/m^2 on day 1, 8 and 15 of a 28-day cycle and dose-limiting pulmonary toxicity was observed [33]. Another prodrug, phortress, with selectivity to ER-positive cells, has also been shown to be an AhR ligand, effectively inducing CYP1A1 via the induction of the metabolite 5F203 to generate a number of metabolites including a presumed nitrenium ion intermediate capable of causing DNA damage (Figure 3) [34]. More recently, CYP2W1 has also been shown to contribute to the bioactivation of 5F203 [35]. In a phase 1 study, no MTD was determined for phortress while pulmonary and liver toxicities were observed [36]. The 50 patients that received phortress had advanced disease including bowel, lung, oesophageal and stomach cancer, and the study concluded that a "lack of efficacy" was observed, with no further clinical studies conducted. However, in preclinical studies, phortress was first and foremost being progressed as a prodrug with potential in ovarian and breast cancer but none of the 50 patients in the phase I trial harboured these cancers, meaning the true efficacy remains unknown [34,37]. Despite the discontinued clinical studies of AFP464 and phortress, AhR remains an interesting target for 5F 203 [38] and several promising preclinical strategies have been discussed in a recent review by McCluskey and co-workers and hence we refer to this for acquiring further knowledge [32].

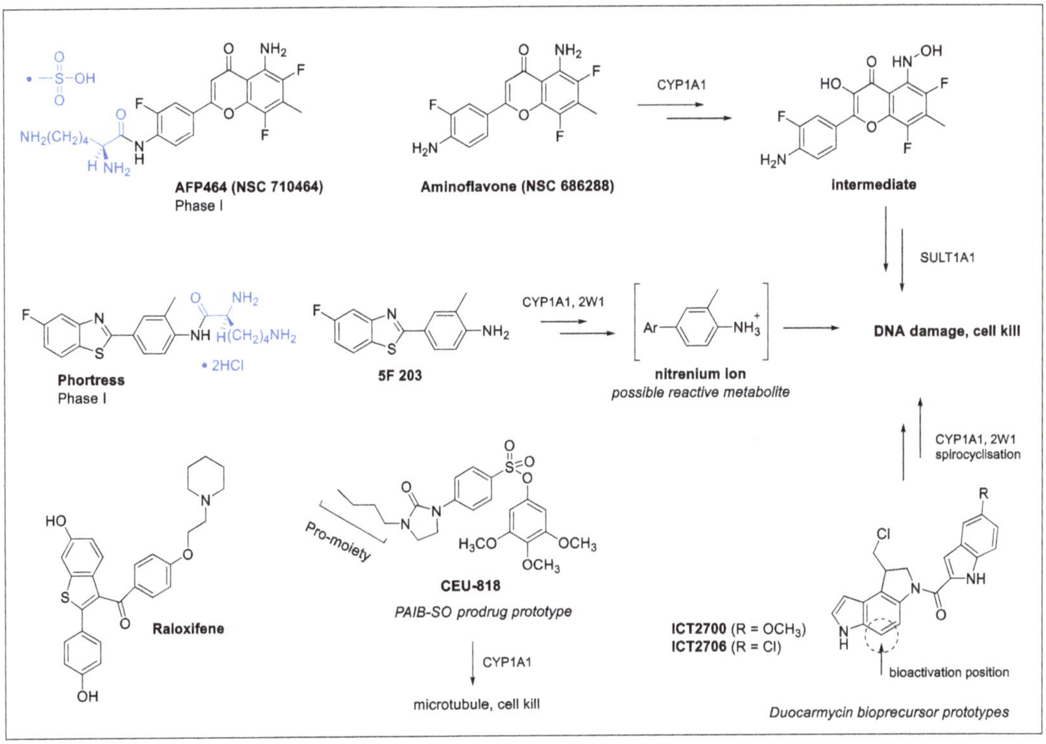

Figure 3. Prodrugs and bioprecursors targeting CYP1A1 or CYP1A1/2W1 for tumour-selective bioactivation.

Bioprecursor and prodrug approaches that rely on tumour-expressed CYP1A1 and not AhR induction for bioactivation includes our own work focused on reengineered duocarmycin molecules [39]. Duocarmycins are a class of exquisitely potent naturally occurring anti-tumour antibiotics that covalently bind the N3-position of adenine in the minor groove of DNA, subsequently resulting in cell death [40]. The clinical use of these compounds as anticancer agents is limited due to lack of tumour selectivity [41], however deactivation of the pharmacophore offers an opportunity to selectively target the oxidative capacity of CYP enzymes overexpressed in tumour tissue to restore the cell-killing ability of these compounds [42,43]. The xenograft proof-of-concept studies have been successful in bladder [44] and colorectal cancer [45] while preliminary data suggest the approach can be translated to breast cancer as well [39]. Patterson and co-workers have also shown that CYP1A1, as one of several CYPs, contributes in the conversion of the hypoxia-activated prodrug AQ4N to the topoisomerase II inhibitor AQ4 (Figure 2D), which has shown promise in a phase I trial that included six breast cancer patients [46,47]. Finally, CYP1A2 has been shown to contribute to bioactivation of the prodrug Tegafur to 5-FU [48], which is frequently used as a part of postoperative combination treatment that include tegafur–uracil (UFT) and tegafur–gimeracil–oteracil potassium (TS-1).

Another class of agent, the phenyl 4-(2-oxo-3-alkylimidazolidin-1-yl) benzenesulfonates (PAIB-SOs), has recently been reported as CYP1A1-targeting prodrugs of anti-microtubule-binding cytotoxins. The PAIB-SOs have been shown to display potent cytotoxic activity in both CYP1A1-expressing hormone-dependent and chemo-resistant breast cancers cells both in vitro and in vivo via *N*-dealkylation and prodrug activation (Figure 3) [49].

3.2. CYP1B1

Extrahepatic enzyme CYP1B1 has approximately 40% homology with CYP1A1 or CYP1A2 and plays a vital role in the treatment of receptor-positive breast cancers due to its ability to metabolise steroid hormones [12,50]. Expression of CYP1B1 is induced through the AhR receptor by ligands such as polycyclic hydrocarbons and dioxins akin to CYP1As and is mainly observed in tissues dependent on hormones such as mammary gland, ovary and uterus. *CYP1B1* is the most frequently expressed gene in breast cancer among the CYP1 family members [51]. However, there have been conflicting reports regarding the expression of CYP1B1 in tumours vs. the adjacent non tumour mammary gland tissues [20]. Although one reason for this disparity could be the increased sensitivity of transcript detection compared to protein, the lower expression of *CYP1B1* in cancerous tissue relative to adjacent could also be attributed to the possible downregulation of the gene by the AhR repressor in the tumour or by methylation of the promoter region, resulting in epigenetic silencing [52]. Nonetheless, expression of CYP1B1 in hormone-mediated malignancies is acknowledged to be important in the regulation of progression, metabolism, treatment and resistance of breast tumours [53]. Along with inactivating clinically relevant anticancer agents such as docetaxel and tamoxifen, CYP1B1 has also been shown to metabolise tamoxifen, 17 β-estradiol and bioactivate the natural product resveratrol to piceatannol, a metabolite with antiproliferative activity [54]. Single nucleotide polymorphisms (SNPs) have been found in the *CYP1B1* gene, which accounts for the variation of drug efficacy and toxicity between individual patients. A recent study reported that the expression of *CYP1B1* and SNPs 142 C > G, 4326 C > G were observed in TNBC patients with a higher rate of metastasis and recurrence following administration with taxane, adriamycin and cyclophosphamide (TAC) chemotherapeutic regimen [52]. Moreover, a higher expression of CYP1B1 was observed to be associated with increased drug resistance in breast cancer cells corroborating its role as a predictor of drug resistance and a prognostic indicator of treatment [55]. An overexpression of CYP1B1 and AhR has also been observed in inflammatory breast cancer (IBC) which correlated significantly with tumour grade, lymphovascular invasion, metastatic lymph nodes and the expression of Ki67 indicating cell proliferation [56]. Inhibiting the expression of AhR using CRISPR-Cas9 resulted in the decreased expression of CYP1B1, Wnt5a/b, and β-catenin in the IBC cell line SUM 149 [56]. Flow cytometry analysis also revealed that the subset of cells expressing a CSC phenotype (CD44+ CD24-/low) was positively correlated with the expression of AhR/CYP1B1, Wnt5a/b and β-catenin in the carcinoma tissues of IBC patients. This reinstated the role of AhR signaling in maintaining BCSCs subpopulations [25,56,57]. Whilst this study did not provide a clear understanding of the potential role of other AhR-induced CYP1 isoforms in breast CSCs, it partially substantiates the involvement of the receptor in regulating the stem cell compartment within the tumour microenvironment.

3.3. CYP2A6/CYP2A7

CYP2A6 is primarily involved in the metabolism of nicotine and hence is associated with tobacco-related disorders including lung cancer. It is also one of the multiple CYPs involved in the metabolism of tamoxifen (Figure 2A). In a study by Bieche et al., *CYP2A6* was identified as one among seven genes coding for major xenobiotic-metabolising enzymes involved in breast tumourigenesis and which displayed a higher expression in ER-α-positive breast tumours compared with ER-α-negative tumours and normal mammary tissues [58]. A 25% upregulation in *CYP2A6* expression has been observed among ER-α-positive post-menopausal breast cancer patients, which is contradictory to a study conducted by Iscan and colleagues that failed to detect any mRNA expression of *CYP2A6* in healthy or cancerous breast tissues [15]. A possible explanation for these contradictory results may be due to the fact that the latter study was conducted only in infiltrating ductal carcinoma subtype of breast cancer without considering the hormone receptor status of the patients and also had a limited sample size. Clearly, a more exact understanding of CYP2A6 expression is significant for those patients who are being administered the prodrugs Tegafur and cyclophosphamide (CPA) [14,59] that in part rely on CYP2A6 bioactivation (Figure 2B,C).

3.4. CYP2A13/2B6

In addition to CYP2A6, CYP2A13 also plays a significant role in smoking-induced lung cancer due to its involvement in metabolising tobacco-specific nitrosamines [60]. In breast cancer, mRNA expression of *CYP2B6* has been detected in both healthy and cancerous breast tissues [15]. In pre-menopausal women with ER- and/or PR-positive breast cancer, SNPs in *CYP2B6* such as *CYP2B6* rs4802101 (T/T) and *CYP2B6* rs3211371 (T/C) have been associated with adverse prognosis [61]. The variability in the expression and function of this isoform between individuals stem from its inducibility and frequent polymorphism, respectively. CYP2B6 also plays major role in activating CPA in the liver by oxidising the prodrug to a 4-hydroxycyclophosphamide (4-OH-CPA) which subsequently produces a cytotoxic alkylating agent that damages DNA and eradicates tumour cells [14] (Figure 2C). SNPs in *CYB2B6* can have considerable impact on the expression, function and pharmacokinetics of CPA. Variation in the alleles of *CYP2B6* such as *CYP2B6* *2, *4, *8 and *9 have been correlated with poor outcome while *CYP2B6* *5 has been associated with longer progression-free survival and higher dose delay incidence [62]. In metastatic breast cancer patients receiving a combination of the chemotherapeutic drugs docetaxel and thiotepa, it was observed that altered *CYP2B6* genotypes such as AG genotype of rs2279343 and a CT/TT genotype of rs8192719 were beneficial and that the former was associated with a decreased death risk (47%) and an overall survival increase of 6 months among patients with liver metastasis [63]. Moreover, an in vitro study has shown 17 β-estradiol to upregulate the expression of CYP2B6 exclusively in T-47D but not in MCF-7 breast cancer cells by increasing the recruitment of ER-α and nuclear receptor coactivator 3 (NCoA3) to the 5′ flanking region of the enzyme [64]. ER binding sites were also observed in an upstream 5′ regulatory region of the *CYP2B6* gene in several breast cancer cell lines.

3.5. CYP2C 8/9/18/19

Among all enzymes belonging to the CYP2C family, which contribute around 20% of the total CYP transcript expression in human liver, *CYP2C9* is the most abundant. CYP2C enzymes are known to contribute to the activation of prodrugs such as the oxazaphosphorines [65], including CPA shown in Figure 2C. Both RNA and protein-level expression of CYP2C9 and gene-level expression of *CYP2C8* have been reported in a small cohort of 10 breast tumour samples [66]. No expression of *CYP2C18* and *CYP2C19* was observed in these samples, which could be attributed to the downregulation of these CYPs by IL-6 cytokines in an autocrine manner, similar to observations in liver tissues [66]. Among genetic variants of these genes, *CYP2C9**2 polymorphism was found to be significantly correlated with inferior therapeutic benefit from neoadjuvant chemotherapy [67]. Similarly, the efficacy of tamoxifen treatment in ER-positive breast cancer patients might be influenced by an rs4917623 SNP in the *CYP2C19* gene [68].

3.6. CYP2D6

Bioactivation of tamoxifen to a more potent, pharmacologically active and abundant anti-oestrogenic metabolite "endoxifen" (4-hydroxy-*N*-desmethyltamoxifen) is a two-stage process that involves *N*-desmethylation and 4-hydroxylation steps, which depends mainly on the enzymatic activity of the rate-limiting enzyme CYP2D6 (Figure 2A). As a result, the clinical efficacy of tamoxifen in each patient is dependent on the concentration of endoxifen which further relies on the activity of the highly polymorphic enzyme CYP2D6 [69,70]. This makes polymorphisms a key predictor of tamoxifen outcome in breast cancer patients [71]. Genetic variants of CYP2D6 can be classified based on the activity of the enzyme as extensive, intermediate and poor metabolisers and these genotypes have a gene-dose effect on the endoxifen concentrations in patients treated with tamoxifen [72]. In a study conducted by Blancas et al. in 2018, a significantly lower disease-free survival was observed in a set of 87 patients taking adjuvant tamoxifen treatment with slow metabolisers compared to the rapid counterparts [73]; we refer the reader to in-depth reviews [74,75] of how the variant alleles of *CYP2D6* can be used to predict the outcome in tamoxifen-treated breast cancer patients. A recent study conducted with data from Swedish breast cancer cohorts demonstrated a worse prognosis for both rapid and slow metabolisers (compared to normal), which could be due to the occurrence of adverse effects of the treatment [76].

3.7. CYP2E1

Apart from being a phase 1 drug-metabolising enzyme, CYP2E1 is also associated with several diseases like obesity, type-2 diabetes and liver diseases related to alcohol consumption and cancer [77]. Increased expression of CYP2E1 in breast tumours correlated with more serious and invasive lobular types of tumours [20]. Ectopic expression of the enzyme in breast cancer cells has been shown to induce oxidative and ER stress, autophagy while an inhibitory effect on metastasis was modulated by p53 as a possible upstream regulator [77]. Furthermore, evidence also suggests that CYP2E1 regulates cellular energy metabolism in a manner dependent on cell type and potentially on the clinical staging of the disease and suggest a possible biomarker role [78]. The genetic polymorphic variant CYP2E1 *6 *Dra1 (D/C)* has also been shown to correlate with higher risk of developing breast cancer and warrants further investigation [79].

3.8. CYP2S1 and CYP2W1

Known as orphan CYPs, the biological functions of dioxin-inducible CYP2S1 and CYP2W1 have not been elucidated [80]. Both isoforms appear to be differentially expressed in cancerous tissues and their orphan status makes it interesting to explore their potential as therapeutic targets. In a similar manner to CYP1 family members, AhR and ARNT regulate the induction of CYP2S1 and the enzyme has been linked to metabolism of toxic carcinogenic compounds via the peroxidase shunt [80,81]. The expression of CYP2S1 is distributed among a variety of extrahepatic epithelial tissues that are susceptible to the exposure of carcinogens and xenobiotics and also in several epithelial-derived cancers including breast [19,80]. Contrarily, *CYP2W1* transcripts are mainly restricted to foetal and tumour tissues with minimal extrahepatic expression [82,83]. Apart from metabolising procarcinogens such as PAHs, CYP2W1 is also involved in the oxidation of indoline substrates, lysophospholipids, free fatty acids and 17 β-estradiol [83]. Hlavac et al. in 2014 assessed transcript levels of 10 CYP enzymes in 68 post treatment tumours of breast carcinoma patients and reported significantly higher intratumoural CYP2W1 in responders of neoadjuvant chemotherapy compared to non- responders [84]. In breast cancer cell lines MCF-7 and MDA-MB-231, constitutive expression of both CYP2S1 and CYP2W1were detected, with greater abundance in MDA-MB-231 cells [82]. Moreover, treating the cells with methoxy stilbenes or resveratrol increased the expression of *CYP2S1* and *CYP2W1* in

MCF-7 but not in MDA-MB-231 cells at the genetic level. The benzothiazole 5F203 (and analogue GW-610) was found to selectively induce the expression of CYP2S1 and CYP2W1 in isogenic breast cancer cells depleted for the isoforms [85]. Gene knockdown studies in breast cancer cells identified the role of CYP2S1 in metabolically inactivating benzothiazoles while induction of CYP1A1 remained crucial for their activation and anti-tumour properties [85]. In the molecular taxonomy of breast cancer international consortium (METABRIC) cohort, high *CYP2S1* mRNA was significantly associated with basal like breast cancers compared to *CYP2W1* which was predominantly expressed in luminal A tumours (SM and RA unpublished data).

3.9. CYP3A4/A5

CYP3A4 is abundantly expressed and plays major role in metabolising 60% of the marketed drugs including chemotherapy used to treat breast cancer patients such as CPA, docetaxel, paclitaxel, etoposide, irinotecan, imatinib, lapatinib, trastuzumab emtansine and tamoxifen [18,86]. CYP3A5 is commonly expressed in HER-2-positive breast cancers [84] and is reported to be involved in the bioactivation of lapatinib [87]. Lapatinib, used in the treatment of HER-2-positive metastatic breast cancer is metabolised by CYP3A4/5 in the liver. The metabolites have been known to cause idiosyncratic hepatocellular toxicity [88]. In addition, the nitroso metabolites of lapatinib can bind CYP3A4 forming a metabolite–inhibitor complex that can inactivate CYP3A4 [89], however the clinical consequences of the latter is unknown. Tumours negative for the expression of CYP3A4 have been shown to have a better response rate to the antineoplastic drug docetaxel than those expressing the enzyme [12]. Furthermore, higher expression of CYP3A4 has been shown to be related to tumour progression in ER+ breast cancer cells, in part via the synthesis of epoxyeicosatrienoic acids [90]. In 2001, Schmidt and colleagues identified the presence of CYP3A4 in breast cancer microsomes which aided the metabolism and activation of the CPA prodrug analogue ifosfamide [91]. DMI—a component of the trastuzumab–DMI conjugate used in the treatment of high-risk early stage and metastatic HER-2-positive breast cancer [92], is metabolised by CYP3A4 and to a lesser extent by CYP3A5. Administering trastuzumab–DMI with other agents that can act as CYP3A4 inhibitors (e.g., atazanavir, clarithromycin, ketoconazole and indinavir) should be avoided to minimise potential drug-related toxicity.

In postmenopausal tamoxifen-treated breast cancer patients, a genetic polymorphism in *CYP3A5* (*CYP3A5*3*) was found to be correlated with a positive response [93]. This finding is contrary to that of Spitman et al. who suggested that there is no added benefit of *CYP3A* genotype additionally to *CYP2D6* variants on tamoxifen metabolism [94]. As the authors claim, this discrepancy could possibly be due to the use of CYP3A4/5 inhibitors during the latter study, which were not properly evaluated owing to the minimal data availability. However, the mechanisms leading to tamoxifen resistance are less likely to be caused by a single gene polymorphism since the metabolism of the drug is complex involving several CYP enzymes.

3.10. CYP4Z1/Z2P

Orphan cytochrome CYP4Z1 is also known as a breast cancer-associated cytochrome due to its putative role in breast cancer through the formation of a signaling molecule 20-hydroxyeicosatetraenoic acid (20-HETE) [95]. *CYP4Z1* and its transcribed pseudogene *CYP4Z2P* (96% identity) were initially cloned from SKBr3 breast cancer cells. Both these enzymes are overexpressed in breast cancer, with CYP4Z2P having a similar expression pattern to CYP4Z1 but at a generally lower level which is likely due to the degraded non-functional protein product [96]. Expression of these enzymes is restricted to the mammary glands while they have been shown to be overexpressed in breast carcinoma tissues, which correlated with increasing tumour grade and poor prognosis [97]. Stable expression of CYP4Z1 in breast cancer cells has been shown to enhance angiogenesis and tumour growth in vivo by activating PI3K/Akt and ERK1/2 signaling pathways [97]. Intriguingly, in the breast cancer line MCF-7, CYP4Z1 was detected on the outside of the plasma membrane; however, no isotype-matched antibodies were included as negative controls and so the results must be treated with caution [98]. The study also detected autoantibodies against the enzyme in the sera of patients with breast cancer in comparison to the control samples suggesting a prospective role as a tumour-associated antigen [98]. The miRNA-mediated competing endogenous RNA network (ceRNET) between *CYP4Z1* and *CYP4Z2P* has been suggested to inhibit apoptosis in breast tumours by serving as a sub-ceRNET to *TERT* and also promotes angiogenesis and CDK3 expression [99].

3.11. CYP17A1 and CYP19A1

Endocrine therapy includes the use of SERMs such as tamoxifen and raloxifene and the aromatase inhibitors: anastrozole, letrozole and examestane. The latter target CYP aromatase (CYP19A1, oestrogen synthase) by intervening in the synthesis of aromatic C18 oestrogens (oestrone and oestradiol) from C19 androgens (androstenedione and testosterone). Mechanistic insights into CYP19A1 function, polymorphism and implication have recently been reported and we refer to a recent review for further reading [100].

Although the steroidogenic biosynthetic pathway is complex and multifactorial, it does offer targets for therapeutic intervention [101]. In men, CYP17A1 is a successful target for abiraterone treatment of patients diagnosed with castration-resistant prostate cancer. Abiraterone has also been explored in the treatment of patients with ER+ metastatic breast cancer, as a single agent or in combination with exemestane but with no improvement in progression-free survival [102]. Nonetheless, it is possible that a CYP17A1-targeting agent such as abiraterone could be used with benefit in a subset of TNBC patients harbouring tumours no longer responsive to standard treatment [103].

3.12. CYP24

The mitochondrial enzyme CYP24A1 (24-hydroxylase) manifests its key role in inactivating 1 α, 25-dihydroxyvitamin D_3 (1,25-$(OH)_2D_3$), the biologically active form of the secosteroid hormone vitamin D and is a candidate oncogene [104]. 1,25-$(OH)_2D_3$ binds to the vitamin D receptor (VDR) which is a ligand-activated nuclear receptor, which activates a network of genes that regulate calcium homeostasis, cell growth, differentiation and functional activities of different cell types [105]. Elevated expression of *CYP24A1*, an immediate target gene of VDR, has been observed in breast cancer cells with aberrant amplification of the chromosomal loci encoding the gene [106,107]. In the absence of ligand, the basal mRNA expression of *CYP24A1* correlated inversely with VDR protein levels in several breast cancer cell lines and the unliganded VDR exerts a repressive action on CYP24A1 [108]. Knockdown of *CYP24A1* using shRNA in mammary tumour lines resulted in the suppression of the metabolism of vitamin D and abrogated tumour growth in vivo, supporting the hypothesis that deficiency of vitamin D could be related to breast cancer incidence [109]. There are several inhibitors for CYP24A1 including the naturally occurring soy-derived isoflavane genestein which inhibits both CYP24A1 and CYP27B1 in a time- and dose-dependent manner; the latter is an enzyme that also catalyses the conversion of 25-hydroxyvitamin D_3 to 1,25-D_3 [106,110].

3.13. CYP26A1/B1

CYP26 family generally functions by metabolising and clearing all-*trans*-retinoic acid (ATRA) which is an active metabolite of vitamin A or retinol, involved in regulating cellular differentiation, proliferation, migration and apoptosis [111]. The major role of the highly conserved members of this family CYP26A1, CYP26B1 and CYP26C1 appears to be in the conversion of retinoic acid (RA) to its metabolites, primarily 4-OH-RA. Upregulation or overexpression of CYP26A1 triggers the signals for cell survival and anti-apoptosis by inducing oncogenes and downregulating tumour suppressor genes along with deregulating the cell cycle and DNA repair genes, perhaps indicating properties akin to an oncogene [112,113]. CYP26A1 has been shown to be expressed at a constitutively higher level in breast adenocarcinomas which can be selectively inhibited, e.g., by use of a RA metabolism blocking agent R116010, demonstrating an anti-tumour effect in ER+ mouse tumours [106]. Osanai et al. have demonstrated a strong cytoplasmic expression of CYP26A1 in 42% of the primary breast carcinomas samples by immunohistochemistry with significantly more intense staining observed in metastatic tumours. Breast cancer cells overexpressing *CYP26A1* displayed oncogenic properties, improved cell survival, higher resistance to anoikis and enhanced tumorigenicity and metastatic potential [113]. RA metabolism blocking agents (RAMBAs) together with ATRA have been shown to exhibit a synergistic effect in regressing ER+ MCF-7 and MCF-7Ca xenografts in nude mice [114]. CYP26A1 also regulates the expression of the actin-bundling protein fascin by modulating the levels of RA thereby affecting the malignant behaviour of CYP26A1 expressing breast cancer cells [115].

3.14. CYP27A1

As a part of regulating intracellular cholesterol homeostasis, CYP27A1 (27-hydroxylase) catalyses the hydroxylation of cholesterol to an oxysterol known as 27-hydroxycholesterol (27HC), which is a SERM that can act as a partial agonist of the ER in breast cancers subsequently stimulating growth and metastasis [116]. Differential expression of CYP27A1 has been observed in breast tumours correlating with the phenotype and prognosis of the disease [117]. The role of CYP27A1 in metastasis is mediated by the action of circulating metabolite 27HC on the function of myeloid cells located at distant metastatic niche inducing a suppressed immune microenvironment [118,119]. Treatment with statins, drugs used to manage hypercholesterolemia, have been shown to reduce the concentration of 27HC in the serum together with deregulation of the expression of this enzyme in breast tumours [117]. Inhibition of CYP27A1 with aromatase inhibitors, such anastrozole and fradrozole, has also proved to be beneficial in treating breast tumours dependent on ER signaling [120].

4. CYP Isoform Expression in the Distinct Breast Cancer Subtypes and Implication in Patient Survival

The impact of pharmacogenomics on cancer treatment together with our increasing understanding of the role and presence of drug-metabolising enzymes supports the choice of SoC drugs as well as in the discovery of new therapies. In breast cancer, a more routine molecular profiling will allow the assessment of CYP expression patterns in support of patient treatment selection, e.g., tamoxifen is mostly effective in patients expressing CYP2D6 while in contrast paclitaxel is largely ineffective in patients expressing high levels of CYP2C8 and CYP3A4. Design of targeted approaches exploiting the presence of selective CYPs is key for tumour-selective therapy with minimised side effects. This is relevant in neoadjuvant or adjuvant settings where conventional chemotherapeutic regimens are used to reduce tumour bulk and/or growth and metastasis. However, certain factors should be taken into consideration while utilising pharmacogenetics in personalising treatment. Data from several studies suggests that multiple drug-metabolising genes are involved in the metabolism of anticancer agents and signaling of oestrogen receptors. In this regard, profiling of CYPs prior to treatment will support clinicians in predicting drug efficacy and treatment outcomes. Moreover, there are possibilities of other accompanying medications which can inhibit CYP enzymes thereby altering the drug metabolism and outcome [121]. Discontinuation of a therapy is another variable which is pivotal as a poor metaboliser may benefit more from a continuous treatment than an extensive metaboliser who discontinues the medication due to its side effects. Despite these shortcomings, tailoring patient-specific CYP-directed therapeutic regimens in combination with routine chemotherapeutic agents is a promising way to manage breast cancer. Interestingly, despite the vast number of papers involving CYP expression in breast cancer, sparse information is available on CYP isoform expression in the distinct breast cancer subtypes. Principal component analysis using The Cancer Genome Atlas Breast Invasive Carcinoma (TCGA)-BRCA mRNAseq transcriptomic counts data (n = 935) with PAM50 subtype classifications ("Basal-like",

"Her2-enriched", "Luminal A", "Luminal B", "Normal-like") [122] revealed the CYP4 members *CYP4X1* and *CYP4Z1* to be highly expressed but not variable in a way that is not associated with the PAM50 classifications (Figure 4 and Figure S1). *CYP2A6*, *CYP2B6*, *CYP2D6*, *CYP3A4* and *CYP3A5* involved in drug metabolism of the breast cancer prodrugs CPA, tegafur and tamoxifen (Figure 3) were expressed at various levels, but no significant differences between the PAM50 groups were observed (data not shown). The divergence of CYPs expression variance from the PAM50 classifications is concerning because it means they will require separate analysis to form a distinct CYP profile. Currently we have only a limited basis for the interpretation of a molecular CYP profile; however, we hope the collection of these data will lead to enhanced therapeutic efficacy. Care must be taken with interpretation of any molecular CYP profile in light of the transient nature of CYP expression, their rapid response to any ongoing treatment will make rigorous collection of metadata critical to objective assessment.

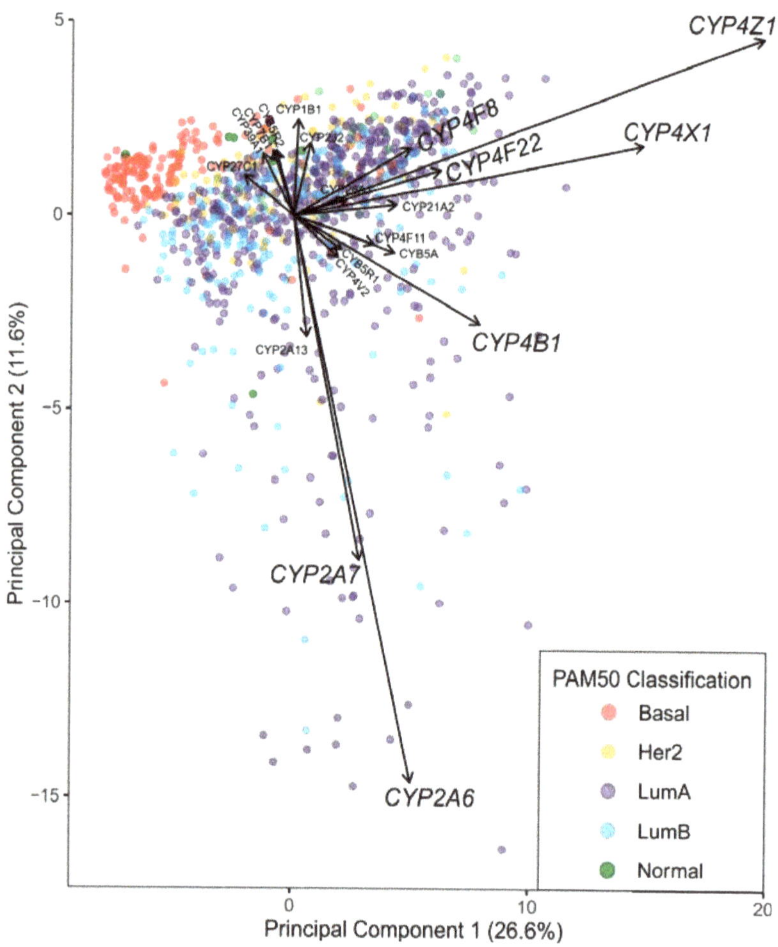

Figure 4. Variance in cytochrome P450 (*CYP*) gene expression (and their essential reductases) was analysed by principal component analysis (PCA). In the PCA biplot, each point represents an individual tumour from The Cancer Genome Atlas (TCGA) breast cancer cohort (n = 935) coloured according to the PAM50 classification [122]. The size and direction of arrows assigned to the top 20 most variant genes in the cohort represent the PCA loadings and reflect how strongly each gene influences principal components 1 and 2.

CYP1 analysis of TCGA-BRCA RNAseq data revealed *CYP1A2* expression to be very low (max TPM = 1.3 with a median TPM = 0), while expression of *CYP1A1* was high in a subset of tumours (max TPM = 78.3 but median TPM = 0.04) and *CYP1B1* was generally more abundant (max TPM = 1611.7 and median TPM = 31.9). Both *CYP1A1* and *CYP1B1* showed significant ($p < 0.0001$) correlation with *AhR* expression (Spearman Rho = 0.14 and 0.41, respectively). Interestingly, the analysis indicates that *CYP1B1* is likely regulated by two different mechanisms because basal *CYP1B1* expression is retained even in *AhR*-low tumours, a feature that is not evident for *CYP1A1* (Figure 5). CYP1B1 is implicated in drug sensitivity so this observation is important for treatment-naïve patients. Future studies are needed to evaluate transcript data from patients who have received AI or chemotherapy treatment and it is possible much higher expression of CYPs induced by AhR, CAR and PXR would be noticeable and an important factor to consider in patients with relapsed tumours.

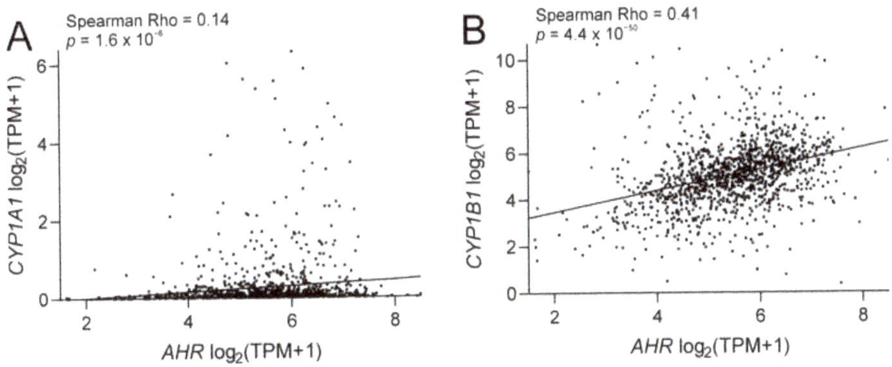

Figure 5. Scatter plots showing the relationship between aryl hydrocarbon receptor (*AHR*) and *CYP1A1* (**A**) or *CYP1B1* (**B**) gene expression, found both *CYPs* were significantly correlated with *AHR* in TCGA breast tumours ($n = 1222$).

The METABRIC analysis of CYP genes involved in the metabolism of anti-tumour agents in breast cancer unravelled the association of some of these isoforms with patient survival. In total, mRNA expression of 24 CYP genes was analysed using the METABRIC cohort, of which expression of 9 genes were significant and expression of 15 were not significant with respect to patient survival.

Kaplan Meier survival analysis shows that low *CYP2A6* ($p = 0.001$), *CYP2D6* ($p = 0.024$), *CYP2C8* ($p = 0.013$), *CYP2E1* ($p = 0.011$), *CYP26A1* ($p < 0.0001$) and high *CYP2C9* ($p = 0.034$), *CYP2C18* ($p = 0.003$), *CYP26B1* ($p = 0.026$) and *CYP17A1* ($p = 0.049$) mRNA expression, in the total patient cohort, was significantly associated with adverse breast cancer-specific survival (Figure 6). Overall, 434/1603 (27%) patients with low *CYP2A6* expression died compared to 70/353 (19%) patients with high expression. A similar trend was observed for *CYP2D6* expression, wherein 68/220 (30%) patients with low expression had worse survival compared to 437/1738 (25%) patients with high expression. A proportion of 305/1125 (27%) patients with low *CYP2C8* expression had adverse survival compared to 200/834 (24%) with high expression. Similarly, low *CYP2E1* (85/281, 30%) and *CYP26A1* (87/251, 35%) expressing patients had adverse survival compared to those with high expression (420/1678, 25% and 418/1708, 24%, respectively).

An opposite trend was observed for *CYP2C9* with higher expression resulting in adverse survival, with 355/1297 (27%) events compared to 150/662 (22%) in patients with low *CYP2C9*. Similarly, patients with high *CYP2C18* (459/1698, 27%), *CYP26B1* (134/475, 28%) and *CYP17A1* (190/670, 28%) had worse survival compared to those with low expression (i.e., 46/261, 18%, 371/1484, 25% and 315/1289, 24%, respectively).

mRNA expression of *CYP1A1* (p = 0.123), *CYP1B1* (p = 0.109), *CYP2A13* (p = 0.195), CYP2C19 (p = 0.130), *CYP2S1* (p = 0.279), *CYP2W1* (p = 0.121), *CYP3A4* (p= 0.194), *CYP3A5* (p = 0.082), *CYP4Z1* (p = 0.113), *CYP4Z2P* (p = 0.092), *CYP19A1* (p = 0.054), *CYP24A1* (p = 0.285), *CYP26C1* (p = 0.055), *CYP27A1* (p = 0.074) and *CYP27B1* (p = 0.126), did not show any significant association with survival (data not shown). Survival analysis could not be conducted for *CYP1A2* or *CYP2B6* due to lack of gene expression information in the METABRIC dataset.

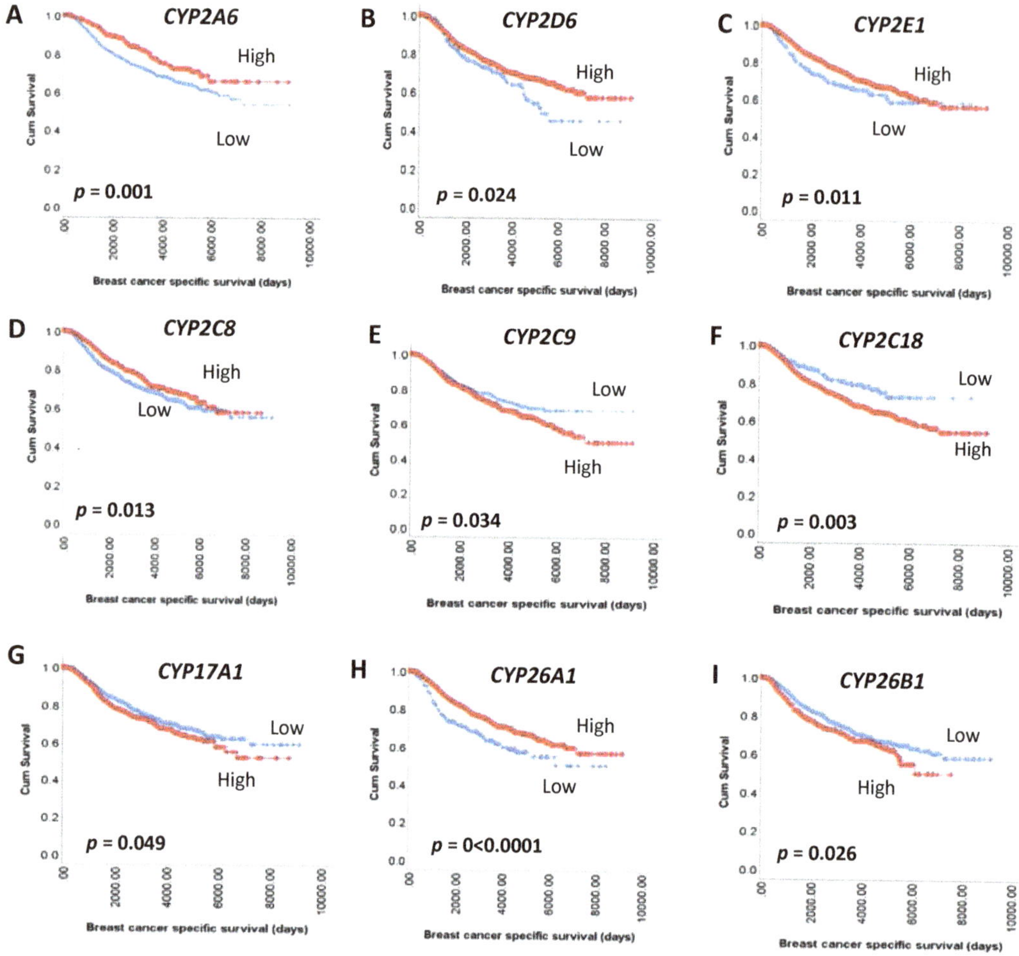

Figure 6. Kaplan–Meier survival analysis of CYP mRNA on breast cancer-specific survival. Survival curves showing the impact of low (blue line) and high (red line) mRNA expression: (**A**) *CYP2A6* (**B**) *CYP2D6* (**C**) *CYP2E1* (**D**) *CYP2C8* (**E**) *CYP2C9* (**F**) *CYP2C18* (**G**) *CYP17A1* (**H**) *CYP26A1* (**I**) *CYP26B1*. Significance was determined using the log-rank test. Significant p-values (<0.05) are indicated in bold.

Lower expression of CYP2A6 and CYP2D6, involved in the metabolism of tamoxifen, tegafur and CPA [123,124] may result in inefficient metabolism/activation of these agents which may affect survival of breast cancer patients as demonstrated in the current study. Higher expression of CYP2C9 has been associated with cancer progression and aggressiveness with certain polymorphisms (*CYP2C9*2* and *CYP2C9*3*) encoding enzymes with reduced activity [37]. Such polymorphic alleles may reduce responsiveness of patients to agents such as tamoxifen and CPA [125].

5. Discussion and Future Directions

In spite of the significant advances in treatment strategies over the past decade, selective treatment of breast cancer with limited off-target effects still remains a great challenge. CYP17A1 and CYP19A1 have been successfully pursued as targets for therapeutic intervention and small molecule-based drugs such as abiraterone and the AIs anastrozole, letrozole and exemastane are now established treatment options for prostate and breast cancer, respectively. Although not discovered through rationalised drug discovery, tamoxifen, tegafur and CPA are examples of CYP-activated prodrugs used to treat breast cancer. As these prodrugs are not bioactivated in a tumour-selective manner, administration of these agents is associated with side effects. Several prodrug strategies focused on targeting the AhR and CYP1A1 and CYP1B1 expression have not yet translated into drug approval for clinical use, but promising research in this area has been reported.

Considering the heterogeneity of breast tumours, it is well established that hypoxia plays a key part in both chemo- and radio-resistance. Hypoxia occurs in approximately 40% of breast cancers and is more commonly seen in locally advanced primary breast cancer [126]. Hypoxic regions have been found to occur in all breast cancer stages, from in situ carcinomas to both lobular and ductal invasive cancers [127]. Studies have reported that the mean pO_2 in breast cancer ranges between 2.5 and 28mmHg (0.3–3.5% O_2), with a median value of 10 mmHg (1.3% O_2), in comparison to 65 mmHg (8% O_2) in normal breast [128]. The presence of hypoxia in breast cancer decreases the effectiveness of hormonal therapy due to loss of ERα [129]. In patients with ductal carcinoma in situ (DCIS), the presence of hypoxia or necrosis is associated with a shorter time to recurrence [130]. Furthermore, the presence of hypoxia in primary breast cancers has been associated with hypoxia markers in their lymph node metastases, indicating that the hypoxia phenotype is maintained even in metastatic lesions [131]. Accordingly, the presence of hypoxia can be exploited for patient benefit via use of hypoxia-activated prodrug (HAP) therapy [132]. A rationalised drug design and understanding of the bioreductive capacity of CYPs [133] in solid tumours affected by hypoxic stress led to the development of AQ4N (Banoxantrone), which has reached clinical evaluation [47] but not yet market approval. AQ4N has been shown to be bioactivated under low oxygen tension by at least five CYPs (see Figure 2D) to generate AQ4, a potent topoisomerase II poison. Given that patients harbouring breast tumours with reduced oxygen tension have been linked to increased risk of mortality and metastasis [134], there is a real opportunity for use of HAPs as an adjuvant in combination with SoC treatment. Sadly, despite much promise, no HAP has yet been approved for clinical use [135] but this field warrants further investigation in defining predictive biomarkers for HAPs and overcoming the technical challenges of assaying them in clinical settings, as argued elegantly in a recent review by Wilson and co-workers [136].

Regardless, of inhibitor or prodrug approach relying on intratumoural CYP expression for therapeutic intervention, it is important to advance the understanding of CYP expression, function and regulation in the tumour microenvironment. This would include careful investigations of 2D and 3D breast cancer models including co-culture systems, spheroids, organoids, patient-derived xenograft models and clinical specimens. Analysis of data from TGCA database did not reveal clear expression differences associated with the PAM50 classification system as we have reported here, however more work is required to fully understand regulation and expression in treatment-naïve and differently treated patient cohorts. Furthermore, as alluded to in this review, it is also essential to take into consideration the presence of CYP polymorphic variants and their function to better understand how these impact on treatment outcomes.

In conclusion, acquiring in-depth knowledge on CYP expression and function is important for improving current SoC treatment as well as guiding new and ongoing efforts to develop drugs with improved therapeutic index for patient benefit.

6. Methods

6.1. TCGA Analysis

TCGA-BRCA mRNAseq transcriptomic counts data were downloaded from the National Cancer Institute (NCI) Genomic Data Commons dbGaP accession phs000178. The results published here are based upon data generated by The Cancer Genome Atlas (TCGA) managed by the NCI and National Human Genome Research Institute. Information about TCGA can be found at http://cancergenome.nih.gov (accessed 18 January 2019). PAM50 subtype classifications ("Basal-like", "Her2-enriched", "Luminal A", "Luminal B", "Normal-like") were taken from Ciriello et al. [122]. TCGA-BRCA mRNAseq data were converted to transcripts per million (TPM) and log2 (TPM+1) values were used in principal component analysis (PCA). Values for protein-coding CYP genes were extracted and combined with those for the two essential reductases (P450 oxidoreductase: POR and cytochrome b5: CYB5A). Gene expression was analysed in an unscaled PCA using prcomp in R (version 3.6.1). Graphs were drawn by fviz_pca (version 1.0.3) in R. Plots display contributions of the top 20 genes for clarity. All other graphs and statistical analyses were generated in Prism (v.8.3.0; Graphpad).

6.2. METABRIC Analysis

Patient cohort

The molecular taxonomy of breast cancer international consortium (METABRIC) dataset, containing sequencing data of 1980 primary breast cancer patients [137], was used to assess the significance of CYP mRNA associations with patient survival. Information about the METABRIC data set is published in Curtis et al., 2012. Patient tumours were collected from 5 centres in the UK and Canada between 1977 and 2005. Median follow-up was 141 months. Patients who were ER-positive and/or lymph node-negative did not receive adjuvant chemotherapy, whereas all ER-negative and/or lymph node-positive patients were administered adjuvant therapy. mRNA was isolated from primary patient tumours in the METABRIC cohort and assayed using an Illumina HT-12 v3 platform [137].

6.3. Statistical Analysis

METABRIC cohort CYP data were stratified based on breast cancer-specific survival (BCSS) using X-Tile software version 3.6.1 [138]. To assess the relationship between mRNA expression and survival outcomes, survival curves were plotted using the Kaplan–Meier method with significance determined using the Log-rank test. A p-value of <0.05 was considered as statistically significant.

Supplementary Materials: The following are available online at https://www.mdpi.com/2227-9059/9/3/290/s1, Figure S1: (A) CYP4Z1 and (B) CYP4X1 expression analysis using PAM50 classification.

Author Contributions: S.S. (Smarakan Sneha) and K.P. wrote the first draft. S.C.B. performed TCGA analysis, A.G., S.S. (Sarah Storr), R.A. performed METABRIC analysis and interpretation of data. S.C.B., S.M. and K.P. critically revised manuscript. All authors have read and agreed to the published version of the manuscript.

Funding: Our work focused on exploring CYPs as targets for prodrug development is funded by Breast Cancer Now and Secondary1st (Grant: 2017NovPhD1056) and Yorkshire Cancer Research (Grant No: B381PA). S. (Sneha) is funded by the Schlumberger Foundation Faculty for the Future program.

Institutional Review Board Statement: Not applicable.

Informed Consent Statement: Not applicable.

Data Availability Statement: Not applicable.

Acknowledgments: The authors would like to thank Andrew Mason (University of York) for downloading TCGA-BRCA counts data and performing the TPM conversion.

Conflicts of Interest: The authors declare no conflict of interest.

References

1. Harbeck, N.; Penault-Llorca, F.; Cortes, J.; Gnant, M.; Houssami, N.; Poortmans, P.; Ruddy, K.; Tsang, J.; Cardoso, F. Breast cancer. *Nat. Rev. Dis. Primers* **2019**, *5*, 66. [CrossRef]
2. Foulkes, W.D.; Smith, I.E.; Reis, J.S. Triple-Negative Breast Cancer. *N. Engl. J. Med.* **2010**, *363*, 1938–1948. [CrossRef]
3. Williams, C.; Lin, C.Y. Oestrogen receptors in breast cancer: Basic mechanisms and clinical implications. *Ecancermedicalscience* **2013**, *7*, 370. [CrossRef] [PubMed]
4. Cusack, L.; Brennan, M.; Baber, R.; Boyle, F. Menopausal symptoms in breast cancer survivors: Management update. *Br. J. Gen. Pract.* **2013**, *63*, 51–52. [CrossRef] [PubMed]
5. Anders, C.; Carey, L.A. Understanding and treating triple-negative breast cancer. *Oncology* **2008**, *22*, 1233–1239; discussion 1239–1240, 1243. [PubMed]
6. Twelves, C.; Jove, M.; Gombos, A.; Awada, A. Cytotoxic chemotherapy: Still the mainstay of clinical practice for all subtypes metastatic breast cancer. *Crit. Rev. Oncol. Hematol.* **2016**, *100*, 74–87. [CrossRef]
7. Nedeljkovic, M.; Damjanovic, A. Mechanisms of Chemotherapy Resistance in Triple-Negative Breast Cancer-How We Can Rise to the Challenge. *Cells* **2019**, *8*, 957. [CrossRef] [PubMed]
8. Staaf, J.; Glodzik, D.; Bosch, A.; Vallon-Christersson, J.; Reutersward, C.; Hakkinen, J.; Degasperi, A.; Amarante, T.D.; Saal, L.H.; Hegardt, C.; et al. Whole-genome sequencing of triple-negative breast cancers in a population-based clinical study. *Nat. Med.* **2019**, *25*, 1526–1533. [CrossRef]
9. Turner, N.C. Signatures of DNA-Repair Deficiencies in Breast Cancer. *N. Engl. J. Med.* **2017**, *377*, 2490–2492. [CrossRef]
10. Chopra, N.; Tovey, H.; Pearson, A.; Cutts, R.; Toms, C.; Proszek, P.; Hubank, M.; Dowsett, M.; Dodson, A.; Daley, F.; et al. Homologous recombination DNA repair deficiency and PARP inhibition activity in primary triple negative breast cancer. *Nat. Commun.* **2020**, *11*, 2662. [CrossRef] [PubMed]
11. Estabrook, R.W. A passion for P450s (rememberances of the early history of research on cytochrome P450). *Drug Metab. Dispos.* **2003**, *31*, 1461–1473. [CrossRef]
12. Rodriguez-Antona, C.; Ingelman-Sundberg, M. Cytochrome P450 pharmacogenetics and cancer. *Oncogene* **2006**, *25*, 1679–1691. [CrossRef] [PubMed]
13. Riddick, D.S.; Lee, C.; Ramji, S.; Chinje, E.C.; Cowen, R.L.; Williams, K.J.; Patterson, A.V.; Stratford, I.J.; Morrow, C.S.; Townsend, A.J.; et al. Cancer chemotherapy and drug metabolism. *Drug Metab. Dispos.* **2005**, *33*, 1083–1096. [CrossRef] [PubMed]
14. Ortiz de Montellano, P.R. Cytochrome P450-activated prodrugs. *Future Med. Chem.* **2013**, *5*, 213–228. [CrossRef] [PubMed]
15. Iscan, M.; Klaavuniemi, T.; Coban, T.; Kapucuoglu, N.; Pelkonen, O.; Raunio, H. The expression of cytochrome P450 enzymes in human breast tumours and normal breast tissue. *Breast Cancer Res. Treat.* **2001**, *70*, 47–54. [CrossRef] [PubMed]
16. Harris, J.W.; Rahman, A.; Kim, B.R.; Guengerich, F.P.; Collins, J.M. Metabolism of taxol by human hepatic microsomes and liver slices: Participation of cytochrome P450 3A4 and an unknown P450 enzyme. *Cancer Res.* **1994**, *54*, 4026–4035.
17. Rahman, A.; Korzekwa, K.R.; Grogan, J.; Gonzalez, F.J.; Harris, J.W. Selective biotransformation of taxol to 6 alpha-hydroxytaxol by human cytochrome P450 2C8. *Cancer Res.* **1994**, *54*, 5543–5546.
18. van Eijk, M.; Boosman, R.J.; Schinkel, A.H.; Huitema, A.D.R.; Beijnen, J.H. Cytochrome P450 3A4, 3A5, and 2C8 expression in breast, prostate, lung, endometrial, and ovarian tumors: Relevance for resistance to taxanes. *Cancer Chemother. Pharmacol.* **2019**, *84*, 487–499. [CrossRef]
19. Murray, G.I.; Patimalla, S.; Stewart, K.N.; Miller, I.D.; Heys, S.D. Profiling the expression of cytochrome P450 in breast cancer. *Histopathology* **2010**, *57*, 202–211. [CrossRef]
20. Vaclavikova, R.; Hubackova, M.; Stribrna-Sarmanova, J.; Kodet, R.; Mrhalova, M.; Novotny, J.; Gut, I.; Soucek, P. RNA expression of cytochrome P450 in breast cancer patients. *Anticancer Res.* **2007**, *27*, 4443–4450.
21. Androutsopoulos, V.P.; Tsatsakis, A.M.; Spandidos, D.A. Cytochrome P450 CYP1A1: Wider roles in cancer progression and prevention. *BMC Cancer* **2009**, *9*, 187. [CrossRef]
22. Rodriguez, M.; Potter, D.A. CYP1A1 regulates breast cancer proliferation and survival. *Mol. Cancer Res.* **2013**, *11*, 780–792. [CrossRef] [PubMed]
23. Vinothini, G.; Nagini, S. Correlation of xenobiotic-metabolizing enzymes, oxidative stress and NFkappaB signaling with histological grade and menopausal status in patients with adenocarcinoma of the breast. *Clin. Chim. Acta* **2010**, *411*, 368–374. [CrossRef] [PubMed]
24. Brockdorff, B.L.; Skouv, J.; Reiter, B.E.; Lykkesfeldt, A.E. Increased expression of cytochrome p450 1A1 and 1B1 genes in anti-estrogen-resistant human breast cancer cell lines. *Int. J. Cancer* **2000**, *88*, 902–906. [CrossRef]
25. Al-Dhfyan, A.; Alhoshani, A.; Korashy, H.M. Aryl hydrocarbon receptor/cytochrome P450 1A1 pathway mediates breast cancer stem cells expansion through PTEN inhibition and beta-Catenin and Akt activation. *Mol. Cancer* **2017**, *16*, 14. [CrossRef] [PubMed]
26. Winslow, S.; Scholz, A.; Rappl, P.; Brauss, T.F.; Mertens, C.; Jung, M.; Weigert, A.; Brune, B.; Schmid, T. Macrophages attenuate the transcription of CYP1A1 in breast tumor cells and enhance their proliferation. *PLoS ONE* **2019**, *14*, e0209694. [CrossRef]
27. Sacco, K.; Grech, G. Actionable pharmacogenetic markers for prediction and prognosis in breast cancer. *EPMA J.* **2015**, *6*, 15. [CrossRef]
28. Bertholee, D.; Maring, J.G.; van Kuilenburg, A.B. Genotypes Affecting the Pharmacokinetics of Anticancer Drugs. *Clin. Pharmacokinet.* **2017**, *56*, 317–337. [CrossRef] [PubMed]

75. Bezerra, L.S.; Santos-Veloso, M.A.O.; Bezerra Junior, N.D.S.; Fonseca, L.C.D.; Sales, W.L.A. Impacts of Cytochrome P450 2D6 (CYP2D6) Genetic Polymorphism in Tamoxifen Therapy for Breast Cancer. *Rev. Bras. Ginecol. Obstet.* **2018**, *40*, 794–799. [CrossRef] [PubMed]
76. He, W.; Grassmann, F.; Eriksson, M.; Eliasson, E.; Margolin, S.; Thoren, L.; Hall, P.; Czene, K. CYP2D6 Genotype Predicts Tamoxifen Discontinuation and Prognosis in Patients With Breast Cancer. *J. Clin. Oncol.* **2020**, *38*, 548–557. [CrossRef] [PubMed]
77. Leung, T.; Rajendran, R.; Singh, S.; Garva, R.; Krstic-Demonacos, M.; Demonacos, C. Cytochrome P450 2E1 (CYP2E1) regulates the response to oxidative stress and migration of breast cancer cells. *Breast Cancer Res.* **2013**, *15*, R107. [CrossRef] [PubMed]
78. Singh, S.; Rajendran, R.; Kuroda, K.; Isogai, E.; Krstic-Demonacos, M.; Demonacos, C. Oxidative stress and breast cancer biomarkers: The case of the cytochrome P450 2E1. *J. Cancer Metastasis Treat.* **2016**, *2*, 268–276. [CrossRef]
79. Lu, Y.; Zhu, X.; Zhang, C.; Jiang, K.; Huang, C.; Qin, X. Role of CYP2E1 polymorphisms in breast cancer: A systematic review and meta-analysis. *Cancer Cell Int.* **2017**, *17*, 11. [CrossRef] [PubMed]
80. Saarikoski, S.T.; Rivera, S.P.; Hankinson, O.; Husgafvel-Pursiainen, K. CYP2S1: A short review. *Toxicol. Appl. Pharmacol.* **2005**, *207*, 62–69. [CrossRef]
81. Bui, P.; Imaizumi, S.; Beedanagari, S.R.; Reddy, S.T.; Hankinson, O. Human CYP2S1 metabolizes cyclooxygenase- and lipoxygenase-derived eicosanoids. *Drug Metab. Dispos.* **2011**, *39*, 180–190. [CrossRef] [PubMed]
82. Szaefer, H.; Licznerska, B.; Cykowiak, M.; Baer-Dubowska, W. Expression of CYP2S1 and CYP2W1 in breast cancer epithelial cells and modulation of their expression by synthetic methoxy stilbenes. *Pharmacol. Rep.* **2019**, *71*, 1001–1005. [CrossRef] [PubMed]
83. Zhao, Y.; Wan, D.; Yang, J.; Hammock, B.D.; Ortiz de Montellano, P.R. Catalytic Activities of Tumor-Specific Human Cytochrome P450 CYP2W1 Toward Endogenous Substrates. *Drug Metab. Dispos.* **2016**, *44*, 771–780. [CrossRef]
84. Hlavac, V.; Brynychova, V.; Vaclavikova, R.; Ehrlichova, M.; Vrana, D.; Pecha, V.; Trnkova, M.; Kodet, R.; Mrhalova, M.; Kubackova, K.; et al. The role of cytochromes p450 and aldo-keto reductases in prognosis of breast carcinoma patients. *Medicine* **2014**, *93*, e255. [CrossRef]
85. Tan, B.S.; Tiong, K.H.; Muruhadas, A.; Randhawa, N.; Choo, H.L.; Bradshaw, T.D.; Stevens, M.F.; Leong, C.O. CYP2S1 and CYP2W1 mediate 2-(3,4-dimethoxyphenyl)-5-fluorobenzothiazole (GW-610, NSC 721648) sensitivity in breast and colorectal cancer cells. *Mol. Cancer Ther.* **2011**, *10*, 1982–1992. [CrossRef]
86. van Schaik, R.H. Cancer treatment and pharmacogenetics of cytochrome P450 enzymes. *Invest. New Drugs* **2005**, *23*, 513–522. [CrossRef]
87. Towles, J.K.; Clark, R.N.; Wahlin, M.D.; Uttamsingh, V.; Rettie, A.E.; Jackson, K.D. Cytochrome P450 3A4 and CYP3A5-Catalyzed Bioactivation of Lapatinib. *Drug Metab. Dispos.* **2016**, *44*, 1584–1597. [CrossRef]
88. Gomez, H.L.; Doval, D.C.; Chavez, M.A.; Ang, P.C.; Aziz, Z.; Nag, S.; Ng, C.; Franco, S.X.; Chow, L.W.; Arbushites, M.C.; et al. Efficacy and safety of lapatinib as first-line therapy for ErbB2-amplified locally advanced or metastatic breast cancer. *J. Clin. Oncol.* **2008**, *26*, 2999–3005. [CrossRef] [PubMed]
89. Takakusa, H.; Wahlin, M.D.; Zhao, C.; Hanson, K.L.; New, L.S.; Chan, E.C.; Nelson, S.D. Metabolic intermediate complex formation of human cytochrome P450 3A4 by lapatinib. *Drug Metab. Dispos.* **2011**, *39*, 1022–1030. [CrossRef] [PubMed]
90. Mitra, R.; Guo, Z.; Milani, M.; Mesaros, C.; Rodriguez, M.; Nguyen, J.; Luo, X.; Clarke, D.; Lamba, J.; Schuetz, E.; et al. CYP3A4 mediates growth of estrogen receptor-positive breast cancer cells in part by inducing nuclear translocation of phospho-Stat3 through biosynthesis of (+/-)-14,15-epoxyeicosatrienoic acid (EET). *J. Biol. Chem.* **2011**, *286*, 17543–17559. [CrossRef] [PubMed]
91. Schmidt, R.; Baumann, F.; Hanschmann, H.; Geissler, F.; Preiss, R. Gender difference in ifosfamide metabolism by human liver microsomes. *Eur. J. Drug Metab. Pharm.* **2001**, *26*, 193–200. [CrossRef] [PubMed]
92. Amiri-Kordestani, L.; Blumenthal, G.M.; Xu, Q.C.; Zhang, L.; Tang, S.W.; Ha, L.; Weinberg, W.C.; Chi, B.; Candau-Chacon, R.; Hughes, P.; et al. FDA approval: Ado-trastuzumab emtansine for the treatment of patients with HER2-positive metastatic breast cancer. *Clin. Cancer Res.* **2014**, *20*, 4436–4441. [CrossRef]
93. Wegman, P.; Elingarami, S.; Carstensen, J.; Stal, O.; Nordenskjold, B.; Wingren, S. Genetic variants of CYP3A5, CYP2D6, SULT1A1, UGT2B15 and tamoxifen response in postmenopausal patients with breast cancer. *Breast Cancer Res.* **2007**, *9*, R7. [CrossRef] [PubMed]
94. Sanchez Spitman, A.B.; Moes, D.; Gelderblom, H.; Dezentje, V.O.; Swen, J.J.; Guchelaar, H.J. Effect of CYP3A4*22, CYP3A5*3, and CYP3A combined genotypes on tamoxifen metabolism. *Eur. J. Clin. Pharmacol.* **2017**, *73*, 1589–1598. [CrossRef] [PubMed]
95. McDonald, M.G.; Ray, S.; Amorosi, C.J.; Sitko, K.A.; Kowalski, J.P.; Paco, L.; Nath, A.; Gallis, B.; Totah, R.A.; Dunham, M.J.; et al. Expression and Functional Characterization of Breast Cancer-Associated Cytochrome P450 4Z1 in Saccharomyces cerevisiae. *Drug Metab. Dispos.* **2017**, *45*, 1364–1371. [CrossRef] [PubMed]
96. Rieger, M.A.; Ebner, R.; Bell, D.R.; Kiessling, A.; Rohayem, J.; Schmitz, M.; Temme, A.; Rieber, E.P.; Weigle, B. Identification of a novel mammary-restricted cytochrome P450, CYP4Z1, with overexpression in breast carcinoma. *Cancer Res.* **2004**, *64*, 2357–2364. [CrossRef] [PubMed]
97. Yu, W.; Chai, H.; Li, Y.; Zhao, H.; Xie, X.; Zheng, H.; Wang, C.; Wang, X.; Yang, G.; Cai, X.; et al. Increased expression of CYP4Z1 promotes tumor angiogenesis and growth in human breast cancer. *Toxicol. Appl. Pharmacol.* **2012**, *264*, 73–83. [CrossRef] [PubMed]
98. Nunna, V.; Jalal, N.; Bureik, M. Anti-CYP4Z1 autoantibodies detected in breast cancer patients. *Cell Mol. Immunol.* **2017**, *14*, 572–574. [CrossRef] [PubMed]
99. Li, C.; Zheng, L.; Xin, Y.; Tan, Z.; Zhang, Y.; Meng, X.; Wang, Z.; Xi, T. The competing endogenous RNA network of CYP4Z1 and pseudogene CYP4Z2P exerts an anti-apoptotic function in breast cancer. *FEBS Lett.* **2017**, *591*, 991–1000. [CrossRef] [PubMed]

100. Hamadeh, I.S.; Patel, J.N.; Rusin, S.; Tan, A.R. Personalizing aromatase inhibitor therapy in patients with breast cancer. *Cancer Treat. Rev.* **2018**, *70*, 47–55. [CrossRef] [PubMed]
101. Pippione, A.C.; Boschi, D.; Pors, K.; Oliaro-Bosso, S.; Lolli, M.L. Androgen-AR axis in primary and metastatic prostate cancer: Chasing steroidogenic enzymes for therapeutic intervention. *J. Cancer Metastasis Treat.* **2017**, *3*, 328–361. [CrossRef]
102. O'Shaughnessy, J.; Campone, M.; Brain, E.; Neven, P.; Hayes, D.; Bondarenko, I.; Griffin, T.W.; Martin, J.; De Porre, P.; Kheoh, T.; et al. Abiraterone acetate, exemestane or the combination in postmenopausal patients with estrogen receptor-positive metastatic breast cancer. *Ann. Oncol.* **2016**, *27*, 106–113. [CrossRef]
103. Gerratana, L.; Basile, D.; Buono, G.; De Placido, S.; Giuliano, M.; Minichillo, S.; Coinu, A.; Martorana, F.; De Santo, I.; Del Mastro, L.; et al. Androgen receptor in triple negative breast cancer: A potential target for the targetless subtype. *Cancer Treat. Rev.* **2018**, *68*, 102–110. [CrossRef] [PubMed]
104. Horvath, H.C.; Lakatos, P.; Kosa, J.P.; Bacsi, K.; Borka, K.; Bises, G.; Nittke, T.; Hershberger, P.A.; Speer, G.; Kallay, E. The candidate oncogene CYP24A1: A potential biomarker for colorectal tumorigenesis. *J. HistoChem. CytoChem.* **2010**, *58*, 277–285. [CrossRef]
105. Bortman, P.; Folgueira, M.A.; Katayama, M.L.; Snitcovsky, I.M.; Brentani, M.M. Antiproliferative effects of 1,25-dihydroxyvitamin D3 on breast cells: A mini review. *Braz. J. Med. Biol. Res.* **2002**, *35*, 1–9. [CrossRef] [PubMed]
106. McFadyen, M.C.; Melvin, W.T.; Murray, G.I. Cytochrome P450 enzymes: Novel options for cancer therapeutics. *Mol. Cancer Ther.* **2004**, *3*, 363–371.
107. Luo, W.; Hershberger, P.A.; Trump, D.L.; Johnson, C.S. 24-Hydroxylase in cancer: Impact on vitamin D-based anticancer therapeutics. *J. Steroid. BioChem. Mol. Biol.* **2013**, *136*, 252–257. [CrossRef] [PubMed]
108. Alimirah, F.; Vaishnav, A.; McCormick, M.; Echchgadda, I.; Chatterjee, B.; Mehta, R.G.; Peng, X. Functionality of unliganded VDR in breast cancer cells: Repressive action on CYP24 basal transcription. *Mol. Cell BioChem.* **2010**, *342*, 143–150. [CrossRef]
109. Osanai, M.; Lee, G.H. CYP24A1-induced vitamin D insufficiency promotes breast cancer growth. *Oncol. Rep.* **2016**, *36*, 2755–2762. [CrossRef]
110. Bruno, R.D.; Njar, V.C. Targeting cytochrome P450 enzymes: A new approach in anti-cancer drug development. *Bioorg. Med. Chem.* **2007**, *15*, 5047–5060. [CrossRef] [PubMed]
111. Stevison, F.; Jing, J.; Tripathy, S.; Isoherranen, N. Role of Retinoic Acid-Metabolizing Cytochrome P450s, CYP26, in Inflammation and Cancer. *Adv. Pharmacol.* **2015**, *74*, 373–412. [CrossRef]
112. Chang, C.L.; Hong, E.; Lao-Sirieix, P.; Fitzgerald, R.C. A novel role for the retinoic acid-catabolizing enzyme CYP26A1 in Barrett's associated adenocarcinoma. *Oncogene* **2008**, *27*, 2951–2960. [CrossRef]
113. Osanai, M.; Sawada, N.; Lee, G.H. Oncogenic and cell survival properties of the retinoic acid metabolizing enzyme, CYP26A1. *Oncogene* **2010**, *29*, 1135–1144. [CrossRef] [PubMed]
114. Patel, J.B.; Mehta, J.; Belosay, A.; Sabnis, G.; Khandelwal, A.; Brodie, A.M.; Soprano, D.R.; Njar, V.C. Novel retinoic acid metabolism blocking agents have potent inhibitory activities on human breast cancer cells and tumour growth. *Br. J. Cancer* **2007**, *96*, 1204–1215. [CrossRef]
115. Osanai, M.; Lee, G.H. The retinoic acid-metabolizing enzyme CYP26A1 upregulates fascin and promotes the malignant behavior of breast carcinoma cells. *Oncol. Rep.* **2015**, *34*, 850–858. [CrossRef] [PubMed]
116. Nelson, E.R.; Wardell, S.E.; Jasper, J.S.; Park, S.; Suchindran, S.; Howe, M.K.; Carver, N.J.; Pillai, R.V.; Sullivan, P.M.; Sondhi, V.; et al. 27-Hydroxycholesterol links hypercholesterolemia and breast cancer pathophysiology. *Science* **2013**, *342*, 1094–1098. [CrossRef]
117. Kimbung, S.; Chang, C.Y.; Bendahl, P.O.; Dubois, L.; Thompson, J.W.; McDonnell, D.P.; Borgquist, S. Impact of 27-hydroxylase (CYP27A1) and 27-hydroxycholesterol in breast cancer. *Endocr. Relat. Cancer* **2017**, *24*, 339–349. [CrossRef] [PubMed]
118. Baek, A.E.; Yu, Y.A.; He, S.; Wardell, S.E.; Chang, C.Y.; Kwon, S.; Pillai, R.V.; McDowell, H.B.; Thompson, J.W.; Dubois, L.G.; et al. The cholesterol metabolite 27 hydroxycholesterol facilitates breast cancer metastasis through its actions on immune cells. *Nat. Commun.* **2017**, *8*, 864. [CrossRef] [PubMed]
119. Ma, L.; Wang, L.; Nelson, A.T.; Han, C.; He, S.; Henn, M.A.; Menon, K.; Chen, J.J.; Baek, A.E.; Vardanyan, A.; et al. 27-Hydroxycholesterol acts on myeloid immune cells to induce T cell dysfunction, promoting breast cancer progression. *Cancer Lett.* **2020**. [CrossRef] [PubMed]
120. Mast, N.; Lin, J.B.; Pikuleva, I.A. Marketed Drugs Can Inhibit Cytochrome P450 27A1, a Potential New Target for Breast Cancer Adjuvant Therapy. *Mol. Pharmacol.* **2015**, *88*, 428–436. [CrossRef] [PubMed]
121. Wegman, P.; Vainikka, L.; Stal, O.; Nordenskjold, B.; Skoog, L.; Rutqvist, L.E.; Wingren, S. Genotype of metabolic enzymes and the benefit of tamoxifen in postmenopausal breast cancer patients. *Breast Cancer Res.* **2005**, *7*, R284–R290. [CrossRef]
122. Ciriello, G.; Gatza, M.L.; Beck, A.H.; Wilkerson, M.D.; Rhie, S.K.; Pastore, A.; Zhang, H.; McLellan, M.; Yau, C.; Kandoth, C.; et al. Comprehensive Molecular Portraits of Invasive Lobular Breast Cancer. *Cell* **2015**, *163*, 506–519. [CrossRef] [PubMed]
123. Desta, Z.; Ward, B.A.; Soukhova, N.V.; Flockhart, D.A. Comprehensive evaluation of tamoxifen sequential biotransformation by the human cytochrome P450 system in vitro: Prominent roles for CYP3A and CYP2D6. *J. Pharmacol. Exp. Ther.* **2004**, *310*, 1062–1075. [CrossRef] [PubMed]
124. Wang, H.; Bian, T.; Liu, D.; Jin, T.; Chen, Y.; Lin, A.; Chen, C. Association analysis of CYP2A6 genotypes and haplotypes with 5-fluorouracil formation from tegafur in human liver microsomes. *Pharmacogenomics* **2011**, *12*, 481–492. [CrossRef] [PubMed]
125. Jernstrom, H.; Bageman, E.; Rose, C.; Jonsson, P.E.; Ingvar, C. CYP2C8 and CYP2C9 polymorphisms in relation to tumour characteristics and early breast cancer related events among 652 breast cancer patients. *Br. J. Cancer* **2009**, *101*, 1817–1823. [CrossRef]

126. Vaupel, P.; Mayer, A.; Briest, S.; Hockel, M. Hypoxia in breast cancer: Role of blood flow, oxygen diffusion distances, and anemia in the development of oxygen depletion. *Adv. Exp. Med. Biol.* **2005**, *566*, 333–342. [CrossRef] [PubMed]
127. Hohenberger, P.; Felgner, C.; Haensch, W.; Schlag, P.M. Tumor oxygenation correlates with molecular growth determinants in breast cancer. *Breast Cancer Res. Treat.* **1998**, *48*, 97–106. [CrossRef] [PubMed]
128. Vaupel, P.; Schlenger, K.; Knoop, C.; Hockel, M. Oxygenation of human tumors: Evaluation of tissue oxygen distribution in breast cancers by computerized O2 tension measurements. *Cancer Res.* **1991**, *51*, 3316–3322. [PubMed]
129. Kurebayashi, J.; Otsuki, T.; Moriya, T.; Sonoo, H. Hypoxia reduces hormone responsiveness of human breast cancer cells. *Jpn. J. Cancer Res.* **2001**, *92*, 1093–1101. [CrossRef]
130. Cornfield, D.B.; Palazzo, J.P.; Schwartz, G.F.; Goonewardene, S.A.; Kovatich, A.J.; Chervoneva, I.; Hyslop, T.; Schwarting, R. The prognostic significance of multiple morphologic features and biologic markers in ductal carcinoma in situ of the breast: A study of a large cohort of patients treated with surgery alone. *Cancer* **2004**, *100*, 2317–2327. [CrossRef] [PubMed]
131. Van den Eynden, G.G.; Van der Auwera, I.; Van Laere, S.J.; Colpaert, C.G.; Turley, H.; Harris, A.L.; van Dam, P.; Dirix, L.Y.; Vermeulen, P.B.; Van Marck, E.A. Angiogenesis and hypoxia in lymph node metastases is predicted by the angiogenesis and hypoxia in the primary tumour in patients with breast cancer. *Br. J. Cancer* **2005**, *93*, 1128–1136. [CrossRef]
132. Wilson, W.R.; Hay, M.P. Targeting hypoxia in cancer therapy. *Nat. Rev. Cancer* **2011**, *11*, 393–410. [CrossRef] [PubMed]
133. Patterson, L.H. Rationale for the use of aliphatic N-oxides of cytotoxic anthraquinones as prodrug DNA binding agents: A new class of bioreductive agent. *Cancer Metastasis Rev.* **1993**, *12*, 119–134. [CrossRef]
134. Vaupel, P.; Mayer, A.; Hockel, M. Tumor hypoxia and malignant progression. *Methods EnzyMol.* **2004**, *381*, 335–354. [CrossRef] [PubMed]
135. McKenna, D.J.; Errington, R.; Pors, K. Current challenges and opportunities in treating hypoxic prostate tumors. *J. Cancer Metastasis Treat.* **2018**, *4*, 11. [CrossRef]
136. Hunter, F.W.; Wouters, B.G.; Wilson, W.R. Hypoxia-activated prodrugs: Paths forward in the era of personalised medicine. *Br. J. Cancer* **2016**, *114*, 1071–1077. [CrossRef] [PubMed]
137. Curtis, C.; Shah, S.P.; Chin, S.F.; Turashvili, G.; Rueda, O.M.; Dunning, M.J.; Speed, D.; Lynch, A.G.; Samarajiwa, S.; Yuan, Y.; et al. The genomic and transcriptomic architecture of 2,000 breast tumours reveals novel subgroups. *Nature* **2012**, *486*, 346–352. [CrossRef] [PubMed]
138. Camp, R.L.; Dolled-Filhart, M.; Rimm, D.L. X-tile: A new bio-informatics tool for biomarker assessment and outcome-based cut-point optimization. *Clin. Cancer Res.* **2004**, *10*, 7252–7259. [CrossRef] [PubMed]

biomedicines

Review

Anti-*c-myc* RNAi-Based Onconanotherapeutics

Saffiya Habib, Mario Ariatti and Moganavelli Singh *

Nano-Gene and Drug Delivery Group, Discipline of Biochemistry, University of KwaZulu-Natal, Private Bag, Durban X54001, South Africa; saffiya.habib@gmail.com (S.H.); ariattim@ukzn.ac.za (M.A.)
* Correspondence: singhm1@ukzn.ac.za; Tel.: +27-31-2607170

Received: 17 November 2020; Accepted: 5 December 2020; Published: 15 December 2020

Abstract: Overexpression of the *c-myc* proto-oncogene features prominently in most human cancers. Early studies established that inhibiting the expression of oncogenic *c-myc*, produced potent anti-cancer effects. This gave rise to the notion that an appropriate *c-myc* silencing agent might provide a broadly applicable and more effective form of cancer treatment than is currently available. The endogenous mechanism of RNA interference (RNAi), through which small RNA molecules induce gene silencing by binding to complementary mRNA transcripts, represents an attractive avenue for *c-myc* inhibition. However, the development of a clinically viable, anti-*c-myc* RNAi-based platform is largely dependent upon the design of an appropriate carrier of the effector nucleic acids. To date, organic and inorganic nanoparticles were assessed both in vitro and in vivo, as carriers of small interfering RNA (siRNA), DICER-substrate siRNA (DsiRNA), and short hairpin RNA (shRNA) expression plasmids, directed against the *c-myc* oncogene. We review here the various anti-*c-myc* RNAi-based nanosystems that have come to the fore, especially between 2005 and 2020.

Keywords: *c-myc*; RNA interference; siRNA; oncogene; gene silencing; expression; nanosystems

1. Introduction

The *c-myc* gene encodes a nuclear phosphoprotein that is widely recognized for its role as a transcription factor. The c-Myc protein is believed to participate in the regulation of 10–15% of all genes [1]. These include genes involved in cell cycle progression [2,3], metabolism [4], cell growth [5–7], differentiation [8], adhesion [9], and apoptosis [10].

Due to its function in regulating essential cellular functions, expression of the *c-myc* gene and activity of the c-Myc protein is, under normal circumstances, tightly controlled. Control points include the transcriptional regulation of the *c-myc* gene itself [11,12], the activity of translation initiation factor eIF4E, which ensures that only faithful *c-myc* mRNA transcripts are exported to the cytoplasm [13]; the short half-life of *c-myc* mRNA [14], post-translational modifications such as phosphorylation, acetylation, and ubiquitinylation [15]; and proteins which either directly interact with c-Myc [16] or influence dimerization with its obligate partner protein, Max [17]. As shown in Figure 1, *c-myc* functions in response to signals from several ligand membrane receptor complexes, which cause either positive or negative regulation. When associated with Max, c-Myc binds to DNA E-boxes and this, in turn, regulates the transcription of its target genes [18].

Abnormal *c-myc* expression can occur due to genetic events that include translocations [19], rearrangements [20], and amplification [21], as well as flaws in the pathways implicated in the regulation of this gene or the protein that it encodes [22]. Research carried out in the 1980s showed an association between the deregulated expression of *c-myc* and tumorigenesis [23,24]. Further work showed that abnormal *c-myc* expression causes neoplastic changes, by eliminating check-points in the cell cycle [25,26], prompting genomic instability [27], and through association with other oncogenes [28,29]. In fact, tumor cells often rely on *c-myc* expression for the maintenance of the cancerous state. This phenomenon, known as the oncogene addiction, was emphasized by studies that showed that *c-myc* inactivation caused tumor regression in transgenic mice [30], by inhibiting the cellular proliferation and inducing senescence or apoptosis and differentiation [31]. Moreover, the effects of systemic *c-myc* inhibition were found to be mild in normal tissues, and were well tolerated over time [32]. These findings, together with an estimation that *c-myc* is deregulated in up to 70% of human cancers [18], making it the most frequently altered oncogene, motivate strongly for the therapeutic value of inhibiting oncogenic *c-myc*.

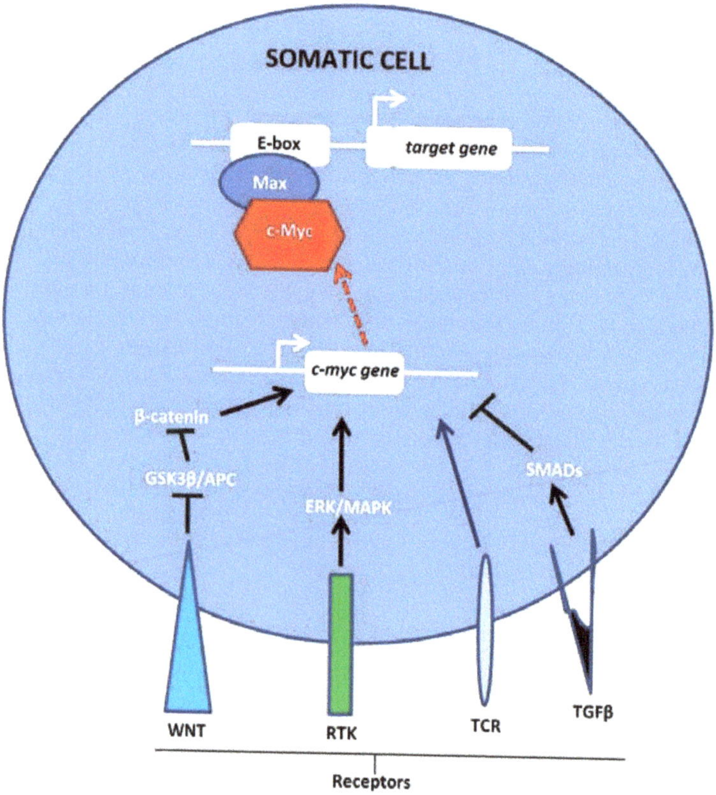

Figure 1. Transcription of the *c-myc* gene in normal cells occurs in response to signals from membrane-anchored receptors. White arrows show the direction of transcription (adapted from [18]).

In theory, the oncogenic activity of *c-myc* can be eliminated by inhibiting the expression of the activated gene, inhibiting inter-protein associations that are critical for c-Myc function, or by disrupting pathways that support *c-myc* deregulation in cancer cells. This provided a basis for the design and evaluation of several potential anti-cancer strategies. The antisense oligonucleotides were featured in some of the earliest reports of *c-myc* inhibition [33–35]. The application of antisense technology to c-Myc inhibition expanded with nucleotide modifications designed to confer greater stability and specificity [36,37]. However, Nobel Prize-winning work that described an endogenous gene silencing mechanism, known as RNA interference (RNAi) [38], presented further possibilities.

Short RNA duplexes of 19–21 base pairs with 2 nucleotide 3′ overhangs, known as small interfering RNA (siRNA), are the key mediators of this pathway. siRNA associates with a network of cytoplasmic proteins to form the RNA-induced silencing complex (RISC), through which it guides the degradation of mRNA, bearing a complementary sequence [39]. Short dsRNA molecules, lacking the dinucleotide overhangs that typify siRNA, termed DICER-substrate siRNA (DsiRNA), can also induce RNAi, with a reportedly higher efficiency. These are processed by the enzyme DICER into siRNA molecules that associate with the RNAi machinery [40]. DNA-directed RNAi, a strategy that generates specific siRNA molecules in vivo, is a useful gene-silencing tool [41]. This involves the construction of a RNA pol-driven plasmid expression vector, into which an antigene sequence of at least 19 nucleotides is inserted, together with appropriate termination signals. When introduced into cells, the antigene sequence is transcribed in the nucleus as a stem-loop structure, which is essentially 2 complementary sequences, 19–22 ribonucleotides in length, linked by a short loop of 4–11 ribonucleotides. This is known as short hairpin RNA (shRNA). The shRNA is exported to the cytoplasm where it is processed into siRNA by DICER [42]. In theory, effective silencing of *c-myc*, or any oncogene, might be achieved using endogenous cellular machinery, provided that the appropriately designed effector nucleic acid is successfully introduced. However, several factors militate against the success of naked nucleic acids in vivo. Naked nucleic acids are highly susceptible to serum nucleases [43] and are rapidly cleared by the kidneys [44]. However, studies have reported that chemical modifications such 2′-*O*-methylation of the guide strand [45], and the use of a passenger 3′ 19ntDNA/siRNA construct (previously referred to as "crook siRNA"), endowed the nucleic acid with nuclease resistance [46].

Furthermore, the size and net negative charge prevent passage across biological membranes [47]. Much effort was focused on the design of delivery agents that would mask its negative charge, protect its integrity, prevent its early removal from the body and facilitate cellular entry. In this regard, nano delivery systems received considerable attention, many of which are based on the principle that nucleic acids can electrostatically associate with positively charged agents [48]. This review discusses the potential application of anti-*c-myc* RNAi nanosystems in cancer treatment.

2. Anti-*c-myc*-siRNA

2.1. Lipid-Based Nanosystems

The liposome is arguably the least complicated lipid-based delivery agent—the simplest of which is a self-assembled phospholipid bilayer that encircles an aqueous core in which a variety of molecules might be entrapped [49]. It is this carrying capability that was exploited for the delivery of several therapeutically important molecules, including siRNA. A neutral liposome composed of dioleoylphosphatidylcholine (DOPC), cholesterol (Chol), and distearoylphosphatidylethanolamine-poly(ethylene glycol) (DSPE-PEG) was used to encapsulate and deliver anti-*c-myc* siRNA in vivo [50]. Pegylation, the introduction of the PEG polymer, served to create a hydration shell around the liposome that sterically inhibits adverse interparticle associations that reduce nanoparticle longevity in the body [51]. Systemic administration of the DOPC/Chol/DSPE-PEG/siRNA complex reduced the growth of ovarian cancer xenograft tumors and did not inhibit the growth of cells with low *c-myc* expression [41]. Anti-*c-myc* siRNA delivered via pegylated DOPC liposomes also showed promise in the treatment of cisplatin-resistant ovarian tumors [52]. Figure 2 provides a representation of some lipid-based delivery systems for anti-*c-myc*-siRNA.

Felgner et al. [53] first reported that the hydration of a mixture containing a synthetic cationic lipid and zwitterionic phospholipid create vesicles that bear a net positive charge, and paved the way for the use of cationic liposomes in nucleic acid delivery. Unlike neutral liposomes in which the siRNA must be encapsulated, cationic liposomes electrostatically associate with siRNA to form nanostructures, known as lipoplexes [54]. Early experiments involved the use of a commercially available cationic liposomal reagent, LipofectamineTM 2000, to demonstrate the therapeutic value of siRNA-mediated c-myc inhibition in human colon cancer [55]. Later, liposomes prepared from equimolar quantities of the cationic lipid *N*,*N*-dimethylaminopropylamidosuccinyl- cholesterylformylhydrazide (MS09), and Chol proved to be simple, but effective anti-*c-myc* agents, which elicited apoptotic cancer cell death and loss of migratory potential in colorectal and breast carcinoma cell lines [56].

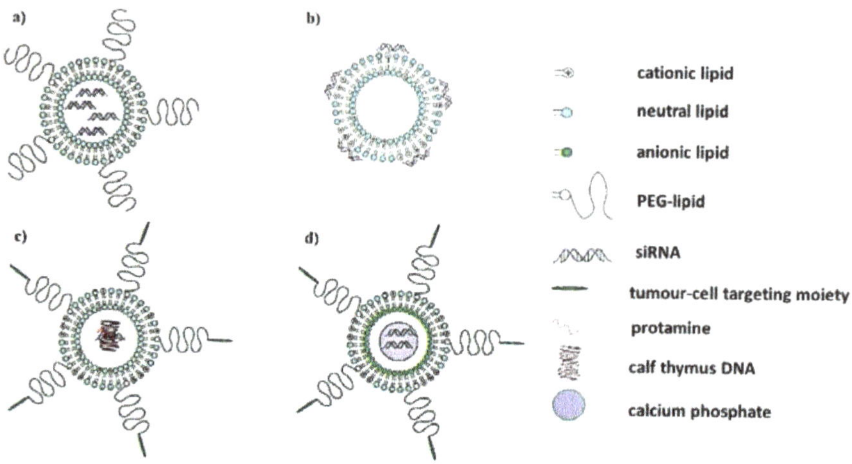

Figure 2. Lipid-based delivery agents for anti-*c-myc* siRNA (**a**) pegylated, neutral liposome, (**b**) cationic liposome, (**c**) liposome-polycation-DNA (LPD) nanoparticle, and (**d**) lipid calcium phosphate (LCP) nanoparticle. Images were created using DesignSpark Mechanical 2.0 software.

Besides being limited to use in cationic liposome formulations, cationic lipids contributed to the development of more elaborate lipid nanoparticles. For example, Chen et al. [57] used a traditional cationic liposome made up of 1,2-dioleoyl-3-trimethylammonium-propane (DOTAP) and Chol, to envelope a core of protamine-bound anti-*c-myc* siRNA and calf thymus DNA (Figure 2). This is known as a liposome-polycation-DNA (LPD) nanoparticle. Surface modifications included post-inserted PEG chains for steric stabilization and a peptide directed to aminopeptidase N, that is overexpressed by cancer cells. Effective siRNA delivery, *c-myc* inhibition, and tumor cell apoptosis were noted after these nanoparticles were intravenously administered in a xenograft model. Co-formulation of the anti-cancer drug doxorubicin with siRNA in targeted LPD nanoparticles further improved treatment efficacy [57]. Following the concept of stabilized core/shell lipid nano-assemblies, Zhang et al. [58] used a DOTAP/Chol/PEG formulation as the outer coating of a calcium phosphate core containing anti-*c-myc* siRNA (Figure 2). The resulting lipid calcium phosphate (LCP) nanoparticle was directed to sigma receptor-positive tumor cells by attachment of anisamide to the distal ends of PEG chains. Similar to the findings of Chen et al. [57], co-encapsulation of anti-*c-myc* siRNA and a chemotherapeutic agent, in this case, gemcitabine, gave a more pronounced anti-cancer effect.

Physical agents might prove useful in promoting the deposition of systemically introduced liposomal anti-*c-myc* siRNA nanoparticles in tumors. A tumor-targeted formulation of 3β[*N*-(*N*′, *N*′-dimethylaminoethane)-carbamoyl] cholesterol (DC-Chol), Chol, and DSPE-PEG with a photolabile-caged cell-penetrating peptide was used to deliver anti-*c-myc* siRNA. The application of near-infrared light at the tumor site, activated the cell-penetrating ability of the peptide [59]. Liposomes were also used as ultrasound cavitation agents for site-specific release of anti-*c-myc* siRNA conjugated to a cell-penetrating peptide [60]. In both instances, treatment delayed tumor progression in fibrosarcoma xenograft models.

Anti-*c-myc* siRNA was included in multi-targeted anti-cancer strategies, which involve the combined delivery of siRNAs against several genes implicated in cancer. A mixture of siRNAs against *c-myc*, *MDM2*, and *VEGF* was shown to inhibit tumor growth more effectively than the individual siRNAs [61]. Li et al. [62] co-encapsulated siRNA molecules against the same targets in a pegylated LPD nanocarrier, for systemic administration in a murine model of metastatic lung cancer. This treatment simultaneously silenced all three genes in cancerous tissue, reduced metastasis by approximately 80%, and extended survival time, with minimal toxicity. Similar results were obtained when siRNAs against the aforementioned oncogenes were pooled in pegylated LCP nanoparticles [63]. Later, a mechanistic study showed that this system impaired the growth of tumors in mice, by simultaneously inhibiting cell proliferation and angiogenesis [64].

Besides delivery via synthetic lipid vesicles, it is worth mentioning that siRNA can also be loaded in exosomes. Exosomes are vesicles that are naturally released by cells for the purposes of intercellular communication and represent an emerging nanocarrier system for a variety of medically relevant molecules [65]. The potential for exosome-mediated anti-*c-myc* siRNA delivery was demonstrated by Lunavat et al. [66] in vitro.

2.2. Miscellaneous Organic Nanosystems

Other organic anti-*c-myc* nano delivery systems reported are often complex polymer- and peptide-based nanocomposites. Folate-targeted, pegylated chitosan nanoparticles were used to encapsulate anti-*c-myc* siRNA associated with packaging RNA, to give a dual-targeting anti-tumor system that improved cellular uptake, gene silencing, and cancer cell death [67]. As a further example, Raichur et al. [68] used a layer-by-layer approach to associate anti-*c-myc* siRNA with poly(lactic-co-glycolic acid) hollow nanoparticles. In vitro experiments showed that the nanoparticles were taken up by aggressive cancer cells and reduced *c-myc* expression with loss of cell viability. More recently, Misra et al. [69] achieved approximately 90% growth inhibition in human melanoma with a nano assembly of palmitoyl-bioconjugated acetyl coenzyme-A termed "siNozyme", which co-delivered anti-*c-myc* siRNA and the chemotherapeutic agent, amonafide.

Anti-*c-myc* siRNA is associated with cell-penetrating peptides (CPP), the simplest of which is an epidermal growth factor receptor-targeted fusion peptide, SPACE–EGF, for topical application to skin cancers [70]. A more complex peptide assembly that contained cationic peptides for siRNA-binding, pH-sensitive peptides for endosomal escape, and a tumor-targeting motif was used for the simultaneous delivery of siRNA against *c-myc* and *Stat3*. This system markedly reduced anchorage-independent growth in recalcitrant breast cancer cells [71]. An elaborate system consisting of a disulfide linked anti-*c-myc* siRNA-CPP encapsulated by a thermosensitive liposome, decorated with a tumor-targeting peptide motif showed effective *c-myc* silencing and antitumor activity in a fibrosarcoma xenograft model [72]. In a related study, an anti-*c-myc* siRNA-CPP conjugate contained within thermal and magnetic dual-responsive liposomes gave encouraging results in a murine breast cancer model [73].

2.3. Inorganic Nanosystems

The use of inorganic nanoparticles in siRNA delivery was explored in recent years. These are often modified with organic components to improve surface properties and reduce toxicity [74–78]. Attachment of siRNA involves either covalent conjugation or electrostatic association with positively charged groups introduced on the surface of the nanoparticle [79,80]. Anti-*c-myc* siRNA carried by PEG- [81], poly(ethylene imine)- [82], and chitosan- [83] functionalized gold nanoparticles was shown to reduce *c-myc* expression in human cervical, liver, and breast cancer cell lines, respectively. In separate in vivo experiments, gold nanoparticles modified with cationic [84] and anionic polymer shells [85], glucose residues [86], and an RGD tumor-targeting peptide [87], delivered anti-*c-myc* siRNA and suppressed the growth of lung tumors.

Nanoparticles based on selenium [77,88] and graphene oxide [82] was also introduced as potential carriers of anti-*c-myc* siRNA. Huang et al. [88] modified a doxorubicin selenium core with RGD-linked polyamidoamine for cancer cell-specific combination therapy, using anti-*c-myc* siRNA. The resultant nanostructure was serum-stable, successfully penetrated the blood–brain barrier and inhibited the growth of glioblastoma spheroids in vitro. In the same year, Imani et al. [89] showed that nano-graphene oxide with PEG and octaarginine conjugation effectively delivered anti-*c-myc* siRNA to human breast cancer cell lines, due to its superior stability and cell-penetrating ability. Some proof of principle studies using nanoparticles such as gold [78], selenium [77], and hydrotalcites [90], showed the potential for the delivery of anti-*Luc*-siRNA, paving the way for the delivery of other therapeutic siRNA molecules, including anti-*c-myc* siRNA.

3. DsiRNA

Like conventional siRNA, DsiRNA requires a vehicle for successful entry. Of significance to this discussion is the fact that pharmaceutical company, Dicerna, reported on a DsiRNA specific for the c-*myc* oncogene, DCR-MYC, delivered using a proprietary EnCore™ lipid nanoparticle. DCR-MYC in this delivery platform is the first, and only anti-*c-myc* RNAi system, to date, to have reached clinical trials [91]. Although the outcome of the initial trial was encouraging, a subsequent trial showed an unsatisfactory knockdown efficiency and its development was discontinued [92].

4. Anti-*c-myc*-shRNA

Most experiments with anti-*c-myc* shRNA plasmids involved their introduction into cells in culture, with the aid of commercial cationic lipid transfection reagents. In one such study, plasmid-driven anti-*c-myc* shRNA silenced *c-myc* expression by as much as 80%, reduced the colony-forming ability, and promoted apoptosis in MCF-7 breast cancer cells [93]. A similar plasmid system impaired proliferation, invasion, and motility in the hepatocellular carcinoma cell line, HepG2 [94]. The transfection of colon cancer cells with anti-*c-myc* shRNA plasmids not only reduced *c-myc* expression, but also that of the human telomerase reverse transcriptase gene (*hTERT*), which is under the transcriptional regulation of *c-myc*, and also contributes towards carcinogenesis, when abnormally expressed [95].

As with siRNA, the effect of multigene silencing using shRNA expression plasmids was also explored. A single plasmid was engineered to direct the transcription of shRNAs against *c-myc*, *VEGF*, *hTERT*, and *BIRC5*, which encodes *Survivin*. This produced a more effective anti-cancer effect than shRNA plasmids targeting individual oncogenes [96]. Similarly, Tai et al. [97] observed a synergistic anti-cancer effect, in colon cancer cells, when the cells were co-transfected with two shRNA plasmids, each separately targeting *c-myc* and *VEGF*. The field of anti-*c-myc* RNAi also benefitted from advances in the design of shRNA-encoding vectors. Recently Cheng et al. [98] used branched PCR technology to introduce a multisite-targeting c-Myc shRNA array into DNA nanovectors that reduced cellular *c-myc* mRNA levels by approximately 98%.

Thus far, only one in vivo experiment with anti-*c-myc* shRNA was reported. In this study, a poly(ethylene imine)-grafted polyglycidal methacrylate nanoparticle was used as a carrier of the shRNA expression vector. Anti-*c-myc* shRNA delivered in this manner suppressed tumor growth in murine models of breast and colon cancer [99].

Table 1 provides a summary of the anti-*c-myc* RNAi-based systems developed to date. The systems discussed were selectively used for the delivery of anti-*c-myc* RNAi molecules. Nano-delivery systems such as mesoporous silica nanoparticles [100,101] and magnetic nanoparticles [102,103] are not discussed in this review, but showed potential in drug delivery, which can be considered for as future carriers of siRNA, shRNA, or DsiRNA.

Table 1. Summary of anti-*c-myc* RNAi-based nanosystems developed to date. Symbols ✔ = present/yes; ✘ = absent/no.

Nucleic Acid	Carrier	Polymer Modification (✔/✘)	Ligand-Targeting Modification (✔/✘)	Disease State	Test System	Advantages	Disadvantages	Clinical Trial (✔/✘)	Reference
siRNA	neutral liposome	✔	✘	Ovarian cancer	Xenograft tumors	High encapsulation efficiency Good biocompatibility Low toxicity Weak immunogenicity	Low transfection efficiency High production cost	✘	[50,52]
	cationic liposome	✘	✘	Colon cancer Breast cancer	HT-29 cells MCF-7 cells	High encapsulation efficiency Good biocompatibility Low toxicity Weak immunogenicity	Low transfection efficiency High production cost	✘	[55,56]
	liposome-polycation-DNA nanoparticle	✔	✔	Fibrosarcoma	HT-1080 cells Xenograft tumors	High transfection efficiency Good biocompatibility	Complex structure and synthesis	✘	[57]
	lipid calcium phosphate nanoparticle	✔	✔	Non-small cell lung cancer	Xenograft tumors	High transfection efficiency Good biocompatibility Low toxicity	Complex structure and synthesis	✘	[58]
	exosomes	✘	✘	-	Mouse λ820 cells	High encapsulation efficiency Natural carriers Good biocompatibility Steady release profile	Lack of standardized techniques for isolation and purification	✘	[66]
	chitosan nanoparticles	✔	✔	Breast cancer	MCF-7 cells Xenograft tumors	Small particle size Good biocompatibility	Low transfection efficiency	✘	[67]
	poly(lactic-co-glycolic acid) nanocapsule	✘	✘	Neuroblastoma	-	High stability Biodegradability FDA-approved material Sustained release	Low transfection efficiency	✘	[68]
	siNozyme	✘	✘	melanoma	-	Biocompatibility Good bioavailability	Complex structure and synthesis	✘	[69]
	cell-penetrating peptide	✘	✔	melanoma	B16 cells Xenograft tumors	High transfection efficiency Low toxicity	The possible need for covalent conjugation Low cell specificity	✘	[70]

Table 1. *Cont.*

Nucleic Acid	Carrier	Polymer Modification (✔/✘)	Ligand-Targeting Modification (✔/✘)	Disease State	Test System	Advantages	Disadvantages	Clinical Trial (✔/✘)	Reference
siRNA	multi-peptide complex	✘	✔	Breast cancer	MDA-MB-231	High transfection efficiency Low toxicity	Complex structure and synthesis	✘	[71]
	gold nanoparticles	✔	✔	Cervical cancer Breast cancer Lung cancer	HeLa cells MCF-7 cells Xenograft tumors A549 cells CMT/167 cells	Large surface area-high loading capacity Amenable to chemical manipulation	Toxicity	✘	[81–87]
	selenium nanoparticles	✔	✔	Glioblastoma	U251 tumor spheroids	Large surface area-high loading capacity Amenable to chemical manipulation	Toxicity	✘	[88]
	nano-graphene oxide	✔	✘	Breast cancer	MCF-7 cells MDA-MB-231 cells	High surface area to volume ratio Flexibility for cargo loading Amenable to functionalization	Adverse interactions with proteins Toxicity Immunogenicity	✘	[89]
DsiRNA	EnCore™ lipid nanoparticle	✘	✘	Advanced solid tumors Multiple myeloma Lymphoma	-Patients	High carrying capacity Good biocompatibility	Poor tumor penetration Unsatisfactory knockdown efficiency Expensive Labor intensive	✔	[91]
shRNA expression plasmid	cationic liposome	✘	✘	Breast cancer Liver cancer Colon cancer Nasopharyngeal cancer	MCF-7 cells Xenograft tumors HepG2 cells Colo320 cells CNE-2Z	High encapsulation efficiency Good biocompatibility Low toxicity Weak immunogenicity	Low transfection efficiency High production cost	✘	[93–96]
	polyglycidal methacrylate nanoparticle	✔	✘	Breast cancer Colorectal cancer	-	Low toxicity at high concentrations	Low transfection efficiency Complex structure and synthesis	✘	[99]

5. Conclusions

Anti-*c-myc* RNAi-based nanosystems have, in many instances, induced potent anti-cancer effects in vitro and in vivo. To date, only DsiRNA was evaluated as an alternative cancer treatment in clinical trials but did not progress further. Several groups and companies are pursuing the idea of inhibiting *c-myc* at the level of translation as a means of designing a clinically viable anti-*c-myc* agent. Hence, RNAi-based strategies are currently significant [93]. Although longer lasting oncogene inhibition can be achieved with DNA-directed RNAi [104], mature siRNA molecules are easily synthesized and pose fewer delivery concerns, as they are of lower molecular weight and do not require genome integration [48,105]. Hence, siRNA is considered more suitable for therapeutic use.

While gene expression might be interrupted by other means such as the restriction enzyme-based system, CRISPR/Cas9, RNAi is most likely the better strategy for *c-myc* inhibition. Given that the RNAi apparatus is present in all mammalian somatic cells, no prior genetic manipulation of the diseased cell line is needed [106]. This is a massive advantage because simple, transient transfection with anti-*c-myc* siRNA is sufficient to achieve anticancer activity [106–108]. Moreover, since RNAi occurs in the cytoplasm, there are no issues with chromatin accessibility, which can perturb gene-editing attempts with CRISPR/Cas9 technology. It is worth mentioning at this point that RNAi is not without its drawbacks, notably the occurrence of off-target effects. However, these are relatively easily attenuated by careful optimization of the design and dose of anti-*c-myc* siRNA molecules [106].

Research to date has emphasized that the development of a suitable anti-c-myc agent is largely dependent upon the design of an appropriate nano delivery system. Great strides were made since the first anti-*c-myc* oligomers were introduced in nanoparticle form [109–111]. siRNA, which functions catalytically and non-stoichiometrically, has surpassed the potency of the antisense oligomers. In recent years, nanosystems were used to deliver small molecule inhibitors of c-Myc-Max dimerization in prodrug form. Initially plagued by issues such as poor bioavailability, low solubility, rapid metabolism, and low potency; their incorporation into nanoparticles showed promise [112–114]. However, anti-*c-myc* RNAi nanotechnology is, at present, a more developed field, and presents a large body of knowledge upon which to improve.

Of all nucleic acid carriers explored thus far, the most significant development were made in the field of lipid-based delivery. Inorganic nanoplatforms, an emerging field, served to solidify the notion that anti-*c-myc* RNAi is a potent anti-cancer instrument. However, with systemic administration, RNAi nanoparticles still have significant challenges to overcome. Their entry into clinical trials might highlight difficulties that include poor retention time in the body and target-site penetration [115]. To this end, polymer and ligand-targeting modifications are common features of several anti-*c-myc* RNAi nanosystems. Such advances in nanoparticle design and expanding test systems for the newly developed nanoparticles represent additional avenues of research in this field. Hence, the advent of a clinically viable anti-*c-myc* RNAi-based anti-neoplastic agent is eagerly awaited.

Author Contributions: Conceptualization, S.H.; M.A. and M.S.; software, S.H.; resources, M.S.; writing—original draft preparation, S.H.; writing—review and editing, M.S. and M.A.; supervision, M.S. and M.A.; project administration, M.S.; funding acquisition, M.S. All authors have read and agreed to the published version of the manuscript.

Funding: This review received no external funding, but research in this area was funded by the National Research Foundation of South Africa, grant numbers 113850 and 120455.

Acknowledgments: The authors acknowledge members of the Nano-Gene and Drug Delivery group for their advice and technical support.

Conflicts of Interest: The authors declare no conflict of interest.

References

1. Zeller, K.I.; Zhao, X.; Lee, C.W.H.; Chiu, K.P.; Yao, F.; Yustein, J.T.; Ooi, H.S.; Orlov, Y.L.; Shahab, A.; Yong, H.C.; et al. Global mapping of c-Myc binding sites and target gene networks in human B cells. *Proc. Natl. Acad. Sci. USA* **2006**, *103*, 17834–17839. [CrossRef] [PubMed]
2. Berns, K.; Hijmans, E.M.; Bernards, R. Repression of c-Myc responsive genes in cycling cells causes G1 arrest through reduction of cyclin E/CDK2 kinase activity. *Oncogene* **1997**, *15*, 1347–1356. [CrossRef] [PubMed]
3. Hermeking, H.; Rago, C.; Schuhmacher, M.; Li, Q.; Barrett, J.F.; Obaya, A.; O'Connell, B.C.; Mateyak, M.K.; Tam, W.; Kohlhuber, F.; et al. Identification of CDK4 as a target of c-MYC. *Proc. Natl. Acad. Sci. USA* **2000**, *97*, 2229–2234. [CrossRef] [PubMed]
4. Kim, J.-W.; Zeller, K.I.; Wang, Y.; Jegga, A.G.; Aronow, B.J.; O'Donnell, K.A.; Dang, C.V. Evaluation of Myc E-Box Phylogenetic Footprints in Glycolytic Genes by Chromatin Immunoprecipitation Assays. *Mol. Cell. Biol.* **2004**, *24*, 5923–5936. [CrossRef] [PubMed]
5. Liu, Y.-C.; Li, F.; Handler, J.; Huang, C.R.L.; Xiang, Y.; Neretti, N.; Sedivy, J.M.; I Zeller, K.; Dang, C.V. Global Regulation of Nucleotide Biosynthetic Genes by c-Myc. *PLoS ONE* **2008**, *3*, e2722. [CrossRef] [PubMed]
6. Schuhmacher, M.; Staege, M.S.; Pajic, A.; Polack, A.; Weidle, U.H.; Bornkamm, G.W.; Eick, D.; Kohlhuber, F. Control of cell growth by c-Myc in the absence of cell division. *Curr. Biol.* **1999**, *9*, 1255–1258. [CrossRef]
7. Van Riggelen, J.; Alper, S.L.; Felsher, D.W. MYC as a regulator of ribosome biogenesis and protein synthesis. *Nat. Rev. Cancer* **2010**, *10*, 301–309. [CrossRef] [PubMed]
8. Wilson, A.; Murphy, M.J.; Oskarsson, T.; Kaloulis, K.; Bettess, M.D.; Oser, G.M.; Pasche, A.-C.; Knabenhans, C.; Macdonald, H.R.; Trumpp, A. c-Myc controls the balance between hematopoietic stem cell self-renewal and differentiation. *Genes Dev.* **2004**, *18*, 2747–2763. [CrossRef]
9. Gebhardt, A.; Frye, M.; Herold, S.; Benitah, S.A.; Braun, K.; Samans, B.; Watt, F.M.; Elsässer, H.-P.; Eilers, M. Myc regulates keratinocyte adhesion and differentiation via complex formation with Miz1. *J. Cell Biol.* **2006**, *172*, 139–149. [CrossRef] [PubMed]
10. Morrish, F.; Giedt, C.; Hockenbery, D. c-MYC apoptotic function is mediated by NRF-1 target genes. *Genes Dev.* **2003**, *17*, 240–255. [CrossRef]
11. Eick, D.; Bornkamm, G.W. Transcriptional arrest within the first exon is a fast control mechanism in c-myc gene expression. *Nucleic Acids Res.* **1986**, *14*, 8331–8346. [CrossRef] [PubMed]
12. Bentley, D.L.; Groudine, M. Novel promoter upstream of the human c-myc gene and regulation of c-myc expression in B-cell lymphomas. *Mol. Cell. Biol.* **1986**, *6*, 3481–3489. [CrossRef] [PubMed]
13. Culjkovic, B.; Topisirovic, I.; Skrabanek, L.; Ruiz-Gutierrez, M.; Borden, K.L.B. eIF4E is a central node of an RNA regulon that governs cellular proliferation. *J. Cell Biol.* **2006**, *175*, 415–426. [CrossRef] [PubMed]
14. Dani, C.; Blanchard, J.M.; Piechaczyk, M.; El Sabouty, S.; Marty, L.; Jeanteur, P. Extreme instability of myc mRNA in normal and transformed human cells. *Proc. Natl. Acad. Sci. USA* **1984**, *81*, 7046–7050. [CrossRef] [PubMed]
15. Vervoorts, J.; Lüscher-Firzlaff, J.; Lüscher, B. The Ins and Outs of MYC Regulation by Posttranslational Mechanisms. *J. Biol. Chem.* **2006**, *281*, 34725–34729. [CrossRef] [PubMed]
16. Dai, M.-S.; Arnold, H.; Sun, X.-X.; Sears, R.; Lü, H. Inhibition of c-Myc activity by ribosomal protein L11. *EMBO J.* **2009**, *28*, 993. [CrossRef]
17. Grandori, C.; Cowley, S.M.; James, L.P.; Eisenman, R.N. The Myc/Max/Mad Network and the Transcriptional Control of Cell Behavior. *Annu. Rev. Cell Dev. Biol.* **2000**, *16*, 653–699. [CrossRef]
18. Dang, C.V. MYC on the Path to Cancer. *Cell* **2012**, *149*, 22–35. [CrossRef]
19. Battey, J.; Moulding, C.; Taub, R.; Murphy, W.; Stewart, T.; Potter, H.; Lenoir, G.; Leder, P. The human c-myc oncogene: Structural consequences of translocation into the igh locus in Burkitt lymphoma. *Cell* **1983**, *34*, 779–787. [CrossRef]
20. Dalla-Favera, R.; Martinotti, S.; Gallo, R.C.; Erikson, J.; Croce, C.M. Translocation and rearrangements of the c-myc oncogene locus in human undifferentiated B-cell lymphomas. *Science* **1983**, *219*, 963–967. [CrossRef]
21. Treszl, A.; Ádány, R.; Rákosy, Z.; Kardos, L.; Bégány, Á.; Gilde, K.; Balazs, M. Extra copies of c-mycare more pronounced in nodular melanomas than in superficial spreading melanomas as revealed by fluorescence in situ hybridisation. *Cytom. Part B Clin. Cytom.* **2004**, *60*, 37–46. [CrossRef] [PubMed]

22. Smith, D.R.; Myint, T.; Goh, H.S. Over-expression of the c-myc proto-oncogene in colorectal carcinoma. *Br. J. Cancer* **1993**, *68*, 407–413. [CrossRef] [PubMed]
23. Adams, J.M.; Harris, A.W.; Pinkert, C.A.; Corcoran, L.M.; Alexander, W.S.; Cory, S.; Palmiter, R.D.; Brinster, R.L. The c-myc oncogene driven by immunoglobulin enhancers induces lymphoid malignancy in transgenic mice. *Nat. Cell Biol.* **1985**, *318*, 533–538. [CrossRef]
24. Leder, A.; Pattengale, P.K.; Kuo, A.; Stewart, T.A.; Leder, P. Consequences of widespread deregulation of the c-myc gene in transgenic mice: Multiple neoplasms and normal development. *Cell* **1986**, *45*, 485–495. [CrossRef]
25. Li, Q.; Dang, C.V. c-Myc Overexpression Uncouples DNA Replication from Mitosis. *Mol. Cell. Biol.* **1999**, *19*, 5339–5351. [CrossRef]
26. Gil, J.J.; Kerai, P.P.; Lleonart, M.M.; Bernard, D.; Cigudosa, J.C.J.; Peters, G.G.; Carnero, A.; Beach, D.D. Immortalization of Primary Human Prostate Epithelial Cells by c-Myc. *Cancer Res.* **2005**, *65*, 2179–2185. [CrossRef]
27. Kuzyk, A.; Mai, S. c-MYC-Induced Genomic Instability. *Cold Spring Harb. Perspect. Med.* **2014**, *4*, a014373. [CrossRef] [PubMed]
28. Vaux, D.L.; Cory, S.; Adams, J.M. Bcl-2 gene promotes haemopoietic cell survival and cooperates with c-myc to immortalize pre-B cells. *Nat. Cell Biol.* **1988**, *335*, 440–442. [CrossRef]
29. Wang, C.; Lisanti, M.P.; Liao, D.J. Reviewing once more the c-myc and Ras collaboration. *Cell Cycle* **2011**, *10*, 57–67. [CrossRef]
30. Arvanitis, C.; Felsher, D.W. Conditional transgenic models define how MYC initiates and maintains tumorigenesis. *Semin. Cancer Biol.* **2006**, *16*, 313–317. [CrossRef]
31. Felsher, D.W. MYC Inactivation Elicits Oncogene Addiction through Both Tumor Cell-Intrinsic and Host-Dependent Mechanisms. *Genes Cancer* **2010**, *1*, 597–604. [CrossRef]
32. Soucek, L.; Whitfield, J.R.; Martins, C.P.; Finch, A.J.; Murphy, D.J.; Sodir, N.M.; Karnezis, A.N.; Swigart, L.B.; Nasi, S.; Evan, G.I. Modelling Myc inhibition as a cancer therapy. *Nat. Cell Biol.* **2008**, *455*, 679–683. [CrossRef] [PubMed]
33. Holt, J.T.; Redner, R.L.; Nienhuis, A.W. An oligomer complementary to c-myc mRNA inhibits proliferation of HL-60 promyelocytic cells and induces differentiation. *Mol. Cell. Biol.* **1988**, *8*, 963–973. [CrossRef] [PubMed]
34. Loke, S.L.; Stein, C.; Zhang, X.; Avigan, M.; Cohen, J.; Neckers, L.M. Delivery of c-myc Antisense Phosphorothioate Oligodeoxynucleotides to Hematopoietic Cells in Culture by Liposome Fusion: Specific Reduction in c-myc Protein Expression Correlates with Inhibition of Cell Growth and DNA Synthesis. *Endogenous ADP-Ribosylation* **1988**, *141*, 282–289. [CrossRef]
35. Wickstrom, E.L.; Bacon, T.A.; Gonzalez, A.; Freeman, D.L.; Lyman, G.H.; Wickstrom, E. Human promyelocytic leukemia HL-60 cell proliferation and c-myc protein expression are inhibited by an antisense pentadecadeoxynucleotide targeted against c-myc mRNA. *Proc. Natl. Acad. Sci. USA* **1988**, *85*, 1028–1032. [CrossRef] [PubMed]
36. Cutrona, G.; Carpaneto, E.M.; Ponzanelli, A.; Ulivi, M.; Millo, E.; Scarfi, S.; Ferrarini, M. Inhibition of the translocated c-myc in Burkitt's lymphoma by a PNA Complementary to the Eμ Enhancer. *Cancer Res.* **2003**, *63*, 6144–6148. [PubMed]
37. Hudziak, R.M.; Summerton, J.; Weller, D.D.; Iversen, P.L. Antiproliferative Effects of Steric Blocking Phosphorodiamidate Morpholino Antisense Agents Directed against c-myc. *Antisense Nucleic Acid Drug Dev.* **2000**, *10*, 163–176. [CrossRef]
38. Fire, A.; Xu, S.; Montgomery, M.K.; Kostas, S.A.; Driver, S.E.; Mello, C.C. Potent and specific genetic interference by double-stranded RNA in Caenorhabditis elegans. *Nature* **1998**, *391*, 806–811. [CrossRef]
39. Elbashir, S.M.; Lendeckel, W.; Tuschl, T. RNA interference is mediated by 21- and 22-nucleotide RNAs. *Genes Dev.* **2001**, *15*, 188–200. [CrossRef]
40. Kim, D.-H.; Behlke, M.A.; Rose, S.D.; Chang, M.-S.; Choi, S.; Rossi, J.J. Synthetic dsRNA Dicer substrates enhance RNAi potency and efficacy. *Nat. Biotechnol.* **2005**, *23*, 222–226. [CrossRef]
41. Ambesajir, A.; Kaushik, A.; Kaushik, J.J.; Petros, S.T. RNA interference: A futuristic tool and its therapeutic applications. *Saudi J. Biol. Sci.* **2012**, *19*, 395–403. [CrossRef] [PubMed]
42. Taxman, D.J.; Moore, C.B.; Guthrie, E.H.; Huang, M.T.-H. Short Hairpin RNA (shRNA): Design, Delivery, and Assessment of Gene Knockdown. In *Structural Genomics and Drug Discovery*; Springer Science and Business Media LLC: New York, NY, USA, 2010; Volume 629, pp. 139–156.

43. Cao, B.; Ji, A.-M. Construction of small interfering RNA targeting mouse vascular endothelial growth factor receptor-2: Its serum stability and gene silencing efficiency in vitro. *J. South. Med. Univ.* **2009**, *29*, 864–867.
44. Huang, Y.; Hong, J.; Zheng, S.; Ding, Y.; Guo, S.; Zhang, H.; Zhang, X.; Du, Q.; Liang, Z. Elimination Pathways of Systemically Delivered siRNA. *Mol. Ther.* **2011**, *19*, 381–385. [CrossRef] [PubMed]
45. Hoerter, J.A.H.; Walter, N.G. Chemical modification resolves the asymmetry of siRNA strand degradation in human blood serum. *RNA* **2007**, *13*, 1887–1893. [CrossRef] [PubMed]
46. Allison, S.J.; Milner, J. RNA Interference by Single- and Double-stranded siRNA with a DNA Extension Containing a 3′ Nuclease-resistant Mini-hairpin Structure. *Mol. Ther. Nucl. Acids* **2014**, *2*, e141. [CrossRef]
47. Akhtar, S.; Benter, I. Toxicogenomics of non-viral drug delivery systems for RNAi: Potential impact on siRNA-mediated gene silencing activity and specificity. *Adv. Drug Deliv. Rev.* **2007**, *59*, 164–182. [CrossRef]
48. Wang, J.; Lu, Z.; Wientjes, M.G.; Au, J.L.-S. Delivery of siRNA Therapeutics: Barriers and Carriers. *AAPS J.* **2010**, *12*, 492–503. [CrossRef]
49. Batzri, S.; Korn, E.D. Single bilayer liposomes prepared without sonication. *Biochim. Biophys. Acta (BBA) Biomembr.* **1973**, *298*, 1015–1019. [CrossRef]
50. Reyes-González, J.M.; Armaiz-Peña, G.N.; Mangala, L.S.; Valiyeva, F.; Ivan, C.; Pradeep, S.; Echevarría-Vargas, I.M.; Rivera-Reyes, A.; Sood, A.K.; Vivas-Mejía, P.E. Targeting c-MYC in Platinum-Resistant Ovarian Cancer. *Mol. Cancer Ther.* **2015**, *14*, 2260–2269. [CrossRef]
51. Suk, J.S.; Xu, Q.; Kim, N.; Hanes, J.; Ensign, L.M. PEGylation as a strategy for improving nanoparticle-based drug and gene delivery. *Adv. Drug Deliv. Rev.* **2016**, *99*, 28–51. [CrossRef]
52. Vivas-Mejia, P.E.; Gonzalez, J.M.R.; Sood, A.K. Nanoliposomal c-MYC-siRNA Inhibits In Vivo Tumor Growth of Cisplatin-Resistant Ovarian Cancer. U.S. Patent No. 10,548,842, 4 February 2020.
53. Felgner, P.L.; Gadek, T.R.; Holm, M.; Roman, R.; Chan, H.W.; Wenz, M.; Northrop, J.P.; Ringold, G.M.; Danielsen, M. Lipofection: A highly efficient, lipid-mediated DNA-transfection procedure. *Proc. Natl. Acad. Sci. USA* **1987**, *84*, 7413–7417. [CrossRef] [PubMed]
54. Khatri, N.; Baradia, D.; Vhora, I.; Rathi, M.; Misra, A. Development and Characterization of siRNA Lipoplexes: Effect of Different Lipids, In Vitro Evaluation in Cancerous Cell Lines and In Vivo Toxicity Study. *AAPS PharmSciTech* **2014**, *15*, 1630–1643. [CrossRef] [PubMed]
55. Zhang, X.; Ge, Y.; Tian, R.-H. The knockdown of c-myc expression by RNAi inhibits cell proliferation in human colon cancer HT-29 cells in vitro and in vivo. *Cell. Mol. Biol. Lett.* **2009**, *14*, 305–318. [CrossRef] [PubMed]
56. Habib, S.; Daniels, A.; Ariatti, M.; Singh, M. Anti-c-myc cholesterol based lipoplexes as onco-nanotherapeutic agents in vitro. *F1000Research* **2020**, *9*, 770. [CrossRef]
57. Chen, Y.; Wu, J.J.; Huang, L. Nanoparticles Targeted With NGR Motif Deliver c-myc siRNA and Doxorubicin for Anticancer Therapy. *Mol. Ther.* **2010**, *18*, 828–834. [CrossRef] [PubMed]
58. Zhang, Y.; Peng, L.; Mumper, R.J.; Huang, L. Combinational delivery of c-myc siRNA and nucleoside analogs in a single, synthetic nanocarrier for targeted cancer therapy. *Biomaterials* **2013**, *34*, 8459–8468. [CrossRef]
59. Yang, Y.; Yang, Y.; Xie, X.; Wang, Z.; Gong, W.; Zhang, H.; Li, Y.; Yu, F.; Li, Z.; Mei, X. Dual-modified liposomes with a two-photon-sensitive cell penetrating peptide and NGR ligand for siRNA targeting delivery. *Biomaterials* **2015**, *48*, 84–96. [CrossRef]
60. Xie, X.; Lin, W.; Li, M.; Yang, Y.; Deng, J.; Liu, H.; Chen, Y.; Fu, X.; Liu, H.; Yang, Y. Efficient siRNA Delivery Using Novel Cell-Penetrating Peptide-siRNA Conjugate-Loaded Nanobubbles and Ultrasound. *Ultrasound Med. Biol.* **2016**, *42*, 1362–1374. [CrossRef]
61. Song, E.; Zhu, P.; Lee, S.-K.; Chowdhury, D.; Kussman, S.; Dykxhoorn, D.M.; Feng, Y.; Palliser, D.; Weiner, D.B.; Shankar, P.; et al. Antibody mediated in vivo delivery of small interfering RNAs via cell-surface receptors. *Nat. Biotechnol.* **2005**, *23*, 709–717. [CrossRef]
62. Li, S.-D.; Chono, S.; Huang, L. Efficient Oncogene Silencing and Metastasis Inhibition via Systemic Delivery of siRNA. *Mol. Ther.* **2008**, *16*, 942–946. [CrossRef]
63. Yang, Y.; Li, J.; Liu, F.; Huang, L. Systemic Delivery of siRNA via LCP Nanoparticle Efficiently Inhibits Lung Metastasis. *Mol. Ther.* **2012**, *20*, 609–615. [CrossRef] [PubMed]
64. Yang, Y.; Hu, Y.; Wang, Y.; Li, J.; Liu, F.; Huang, L. Nanoparticle Delivery of Pooled siRNA for Effective Treatment of Non-Small Cell Lung Caner. *Mol. Pharm.* **2012**, *9*, 2280–2289. [CrossRef] [PubMed]

65. Ha, D.; Yang, N.; Nadithe, V. Exosomes as therapeutic drug carriers and delivery vehicles across biological membranes: Current perspectives and future challenges. *Acta Pharm. Sin. B* **2016**, *6*, 287–296. [CrossRef] [PubMed]
66. Lunavat, T.R.; Jang, S.C.; Nilsson, L.; Park, H.T.; Repiska, G.; Lässer, C.; Nilsson, J.A.; Gho, Y.S.; Lötvall, J. RNAi delivery by exosome-mimetic nanovesicles–Implications for targeting c-Myc in cancer. *Biomaterials* **2016**, *102*, 231–238. [CrossRef] [PubMed]
67. Li, L.; Hu, X.; Zhang, M.; Ma, S.; Yu, F.; Zhao, S.; Liu, N.; Wang, Z.; Wang, Y.; Guan, H.; et al. Dual Tumor-Targeting Nanocarrier System for siRNA Delivery Based on pRNA and Modified Chitosan. *Mol. Ther. Nucleic Acids* **2017**, *8*, 169–183. [CrossRef] [PubMed]
68. Raichur, A.; Nakajima, Y.; Nagaoka, Y.; Matsumoto, K.; Mizuki, T.; Kato, K.; Maekawa, T.; Kumar, D.S. Strategist PLGA Nano-capsules to Deliver siRNA for Inhibition of Carcinoma and Neuroblastoma Cell Lines by Knockdown of MYC Proto-oncogene using CPPs and PNA. *NanoWorld J.* **2015**, *1*, 32–45. [CrossRef]
69. Misra, S.K.; Kampert, T.L.; Pan, D. Nano-Assembly of Pamitoyl-Bioconjugated Coenzyme-A for Combinatorial Chemo-Biologics in Transcriptional Therapy. *Bioconjugate Chem.* **2018**, *29*, 1419–1427. [CrossRef]
70. Ruan, R.; Chen, M.; Sun, S.; Wei, P.; Zou, L.; Liu, J.; Gao, D.; Wen, L.; Ding, W. Topical and Targeted Delivery of siRNAs to Melanoma Cells Using a Fusion Peptide Carrier. *Sci. Rep.* **2016**, *6*, 29159. [CrossRef]
71. Bjorge, J.D.; Pang, A.; Fujita, D.J. Delivery of gene targeting siRNAs to breast cancer cells using a multifunctional peptide complex that promotes both targeted delivery and endosomal release. *PLoS ONE* **2017**, *12*, e0180578. [CrossRef]
72. Yang, Y.; Yang, Y.; Xie, X.; Xu, X.; Xia, X.; Wang, H.; Li, L.; Dong, W.; Ma, P.; Liu, Y. Dual stimulus of hyperthermia and intracellular redox environment triggered release of siRNA for tumor-specific therapy. *Int. J. Pharm.* **2016**, *506*, 158–173. [CrossRef] [PubMed]
73. Yang, Y.; Xie, X.; Xu, X.; Xia, X.; Wang, H.; Li, L.; Dong, W.; Ma, P.; Yang, Y.; Liu, Y.; et al. Thermal and magnetic dual-responsive liposomes with a cell-penetrating peptide-siRNA conjugate for enhanced and targeted cancer therapy. *Colloids Surf. B Biointerfaces* **2016**, *146*, 607–615. [CrossRef] [PubMed]
74. Chaudhary, S.; Umar, A.; Mehta, S.K. Surface functionalized selenium nanoparticles for biomedical applications. *J. Biomed. Nanotechnol.* **2014**, *10*, 3004–3042. [CrossRef] [PubMed]
75. Fraga, S.; Brandão, A.; Soares, M.E.; Morais, T.; Duarte, J.A.; Pereira, L.; Soares, L.; Neves, C.S.; Pereira, E.; Bastos, M.D.L.; et al. Short- and long-term distribution and toxicity of gold nanoparticles in the rat after a single-dose intravenous administration. *Nanomed. Nanotechnol. Biol. Med.* **2014**, *10*, 1757–1766. [CrossRef] [PubMed]
76. Zhou, J.; Ralston, J.; Sedev, R.; Beattie, D.A. Functionalized gold nanoparticles: Synthesis, structure and colloid stability. *J. Colloid Interface Sci.* **2009**, *331*, 251–262. [CrossRef] [PubMed]
77. Maiyo, F.; Singh, M. Polymerized Selenium Nanoparticles for Folate-Receptor-Targeted Delivery of Anti-Luc-siRNA: Potential for Gene Silencing. *Biomaterials* **2020**, *8*, 76. [CrossRef]
78. Mbatha, L.S.; Maiyo, F.C.; Singh, M. Dendrimer functionalized folate-targeted gold nanoparticles for luciferase gene silencing in vitro: A proof of principle study. *Acta Pharm.* **2019**, *69*, 49–61. [CrossRef]
79. Conde, J.; Ambrosone, A.; Sanz, V.; Hernandez, Y.; Marchesano, V.; Tian, F.; Child, B.H.; Berry, C.C.; Ibarra, M.R.; Baptista, P.V.; et al. Design of Multifunctional Gold Nanoparticles for In Vitro and In Vivo Gene Silencing. *ACS Nano* **2012**, *6*, 8316–8324. [CrossRef]
80. Xia, Y.; Guo, M.; Xu, T.; Li, Y.; Wang, C.; Lin, Z.; Zhao, M.; Zhu, B. siRNA-loaded selenium nanoparticle modified with hyaluronic acid for enhanced hepatocellular carcinoma therapy. *Int. J. Nanomed.* **2018**, *13*, 1539–1552. [CrossRef]
81. McCully, M.; Hernandez, Y.; Conde, J.; Baptista, P.V.; De La Fuente, J.M.; Hursthouse, A.; Stirling, D.; Berry, C.C. Significance of the balance between intracellular glutathione and polyethylene glycol for successful release of small interfering RNA from gold nanoparticles. *Nano Res.* **2015**, *8*, 3281–3292. [CrossRef]
82. Shaat, H.; Mostafa, A.; Moustafa, M.; Gamal-Eldeen, A.; Emam, A.; El-Hussieny, E.; Elhefnawi, M. Modified gold nanoparticles for intracellular delivery of anti-liver cancer siRNA. *Int. J. Pharm.* **2016**, *504*, 125–133. [CrossRef] [PubMed]
83. Daniels, A.N.; Singh, M. Sterically stabilized siRNA:gold nanocomplexes enhance c-MYC silencing in a breast cancer cell model. *Nanomedicine* **2019**, *14*, 1387–1401. [CrossRef] [PubMed]
84. Kim, H.S.; Son, Y.J.; Mao, W.; Leong, K.W.; Yoo, H.S. Atom Transfer Radical Polymerization of Multishelled Cationic Corona for the Systemic Delivery of siRNA. *Nano Lett.* **2017**, *18*, 314–325. [CrossRef] [PubMed]

85. Mao, W.; Lee, S.; Shin, J.U.; Yoo, H.S. Surface-Initiated Atom Transfer Polymerized Anionic Corona on Gold Nanoparticles for Anti-Cancer Therapy. *Pharmaceutics* **2020**, *12*, 261. [CrossRef] [PubMed]
86. Conde, J.; Tian, F.; Hernandez, Y.; Bao, C.; Baptista, P.V.; Cui, D.; Stoeger, T.; De La Fuente, J.M. RNAi-based glyconanoparticles trigger apoptotic pathways for in vitro and in vivo enhanced cancer-cell killing. *Nanoscale* **2015**, *7*, 9083–9091. [CrossRef] [PubMed]
87. Conde, J.; Tian, F.; Hernández, Y.; Bao, C.; Cui, D.; Janssen, K.-P.; Ibarra, M.R.; Baptista, P.V.; Stoeger, T.; De La Fuente, J.M. In vivo tumor targeting via nanoparticle-mediated therapeutic siRNA coupled to inflammatory response in lung cancer mouse models. *Biomaterials* **2013**, *34*, 7744–7753. [CrossRef]
88. Huang, W.; Liang, Y.; Sang, C.; Mei, C.; Li, X.; Chen, T. Therapeutic nanosystems co-deliver anticancer drugs and oncogene SiRNA to achieve synergetic precise cancer chemo-gene therapy. *J. Mater. Chem. B* **2018**, *6*, 3013–3022. [CrossRef] [PubMed]
89. Imani, R.; Prakash, S.; Vali, H.; Faghihi, S. Polyethylene glycol and octa-arginine dual-functionalized nanographene oxide: An optimization for efficient nucleic acid delivery. *Biomater. Sci.* **2018**, *6*, 1636–1650. [CrossRef]
90. Nundkumar, N.; Singh, S.; Singh, M. Amino Acid Functionalized Hydrotalcites for Gene Silencing. *J. Nanosci. Nanotechnol.* **2020**, *20*, 3387–3397. [CrossRef] [PubMed]
91. Tolcher, A.W.; Papadopoulos, K.P.; Patnaik, A.; Rasco, D.W.; Martinez, D.; Wood, D.L.; Fielman, B.; Sharma, M.; Janisch, L.A.; Brown, B.D.; et al. Safety and activity of DCR-MYC, a first-in-class Dicer-substrate small interfering RNA (DsiRNA) targeting MYC, in a phase I study in patients with advanced solid tumors. *J. Clin. Oncol.* **2015**, *33*, 11006. [CrossRef]
92. Whitfield, J.R.; Beaulieu, M.-E.; Soucek, L. Strategies to Inhibit Myc and Their Clinical Applicability. *Front. Cell Dev. Biol.* **2017**, *5*, 10. [CrossRef]
93. Wang, Y.; Liu, S.; Zhang, G.; Zhou, C.-Q.; Zhu, H.-X.; Zhou, X.-B.; Quan, L.-P.; Bai, J.-F.; Xu, N. Knockdown of c-Myc expression by RNAi inhibits MCF-7 breast tumor cells growth in vitro and in vivo. *Breast Cancer Res.* **2004**, *7*, R220–R228. [CrossRef] [PubMed]
94. Zhao, Y.; Wang, J.; Gao, W.; Zheng, Y.-X.; Wang, Y.-K.; Zhou, Z.-Q.; Zhang, H.; Wang, C.-J. RNAi silencing of c-Myc inhibits cell migration, invasion, and proliferation in HepG2 human hepatocellular carcinoma cell line: C-Myc silencing in hepatocellular carcinoma cell. *Cancer Cell Int.* **2013**, *13*, 23. [CrossRef] [PubMed]
95. Huang, H.; Nancai, Y.; Lei, F.; Xiong, W.; Wen, S.; Guofu, H.; Yanxia, W.; Hanju, H.; Qian, L.; Hong, X. siRNA directed against c-Myc inhibits proliferation and downregulates human telomerase reverse transcriptase in human colon cancer Colo 320 cells. *J. Exp. Clin. Cancer Res.* **2008**, *27*, 27. [CrossRef]
96. Song, Y.; Dong, M.-M.; Yang, H.-F. Effects of RNA interference targeting four different genes on the growth and proliferation of nasopharyngeal carcinoma CNE-2Z cells. *Cancer Gene Ther.* **2011**, *18*, 297–304. [CrossRef] [PubMed]
97. Tai, J.; Wang, G.; Liu, T.; Wang, L.; Lin, C.; Li, F. Effects of siRNA Targeting c-Myc and VEGF on Human Colorectal Cancer Volo Cells. *J. Biochem. Mol. Toxicol.* **2012**, *26*, 499–505. [CrossRef] [PubMed]
98. Cheng, L.; Ma, D.; Lu, L.; Ouyang, D.; Xi, Z. Building Customizable Multisite-Targeting c-Myc shRNA Array into Branch-PCR-Constructed DNA Nanovectors for Enhanced Tumor Cell Suppression. *ChemistrySelect* **2020**, *5*, 10250–10255. [CrossRef]
99. Tangudu, N.K.; Verma, V.K.; Clemons, T.D.; Beevi, S.S.; Hay, T.; Mahidhara, G.; Raja, M.; Nair, R.A.; Alexander, L.E.; Patel, A.B.; et al. RNA Interference Using c-Myc-Conjugated Nanoparticles Suppresses Breast and Colorectal Cancer Models. *Mol. Cancer Ther.* **2015**, *14*, 1259–1269. [CrossRef]
100. Moodley, T.; Singh, M. Polymeric Mesoporous Silica Nanoparticles for Enhanced Delivery of 5-Fluorouracil In Vitro. *Pharmaceutics* **2019**, *11*, 288. [CrossRef]
101. Moodley, T.; Singh, M. Sterically Stabilised Polymeric Mesoporous Silica Nanoparticles Improve Doxorubicin Efficiency: Tailored Cancer Therapy. *Molecules* **2020**, *25*, 742. [CrossRef] [PubMed]
102. Mngadi, S.; Mokhosi, S.; Singh, M.; Mdlalose, W. Chitosan-Functionalized Mg0.5Co0.5Fe2O4 Magnetic Nanoparticles Enhance Delivery of 5-Fluorouracil In Vitro. *Coatings* **2020**, *10*, 446. [CrossRef]
103. Synthesis and characterization of magnetic nanoparticles of cobalt ferrite coated with silica. *Biointerface Res. Appl. Chem.* **2019**, *10*, 4908–4913. [CrossRef]
104. Takahashi, Y.; Yamaoka, K.; Nishikawa, M.; Takakura, Y. Quantitative and Temporal Analysis of Gene Silencing in Tumor Cells Induced by Small Interfering RNA or Short Hairpin RNA Expressed from Plasmid Vectors. *J. Pharm. Sci.* **2009**, *98*, 74–80. [CrossRef] [PubMed]

105. Xu, C.-F.; Wang, J. Delivery systems for siRNA drug development in cancer therapy. *Asian J. Pharm. Sci.* **2015**, *10*, 1–12. [CrossRef]
106. Boettcher, M.; McManus, M.T. Choosing the Right Tool for the Job: RNAi, TALEN, or CRISPR. *Mol. Cell* **2015**, *58*, 575–585. [CrossRef]
107. Unniyampurath, U.; Pilankatta, R.; Krishnan, M.N. RNA Interference in the Age of CRISPR: Will CRISPR Interfere with RNAi? *Int. J. Mol. Sci.* **2016**, *17*, 291. [CrossRef] [PubMed]
108. Gabay, M.; Li, Y.; Felsher, D.W. MYC Activation Is a Hallmark of Cancer Initiation and Maintenance. *Cold Spring Harb. Perspect. Med.* **2014**, *4*, a014241. [CrossRef]
109. Kanamaru, T.; Takagi, T.; Takakura, Y.; Hashida, M. Biological Effects and Cellular Uptake of c-mycAntisense Oligonucleotides and Their Cationic Liposome Complexes. *J. Drug Target.* **1998**, *5*, 235–245. [CrossRef]
110. Putney, S.D.; Brown, J.; Cucco, C.; Lee, R.; Skorski, T.; Leonetti, C.; Geiser, T.; Calabretta, B.; Zupi, G.; Zon, G. Enhanced Anti-Tumor Effects with Microencapsulated c-mycAntisense Oligonucleotide. *Antisense Nucleic Acid Drug Dev.* **1999**, *9*, 451–458. [CrossRef]
111. Carroll, J.S.; Swarbrick, A.; Musgrove, E.A.; Sutherland, R.L. Mechanisms of growth arrest by c-myc antisense oligonucleotides in MCF-7 breast cancer cells: Implications for the antiproliferative effects of antiestrogens. *Cancer Res.* **2002**, *62*, 3126–3131.
112. Pan, D.; Kim, B.; Hu, G.; Gupta, D.S.; Senpan, A.; Yang, X.; Schmieder, A.; Swain, C.; A Wickline, S.; Tomasson, M.H.; et al. A strategy for combating melanoma with oncogenic c-Myc inhibitors and targeted nanotherapy. *Nanomedicine* **2015**, *10*, 241–251. [CrossRef] [PubMed]
113. Soodgupta, D.; Pan, D.; Cui, G.; Senpan, A.; Yang, X.; Lu, L.; Weilbaecher, K.N.; Prochownik, E.V.; Lanza, G.M.; Tomasson, M.H. Small Molecule MYC Inhibitor Conjugated to Integrin-Targeted Nanoparticles Extends Survival in a Mouse Model of Disseminated Multiple Myeloma. *Mol. Cancer Ther.* **2015**, *14*, 1286–1294. [CrossRef] [PubMed]
114. Esser, A.K.; Ross, M.H.; Fontana, F.; Su, X.; Gabay, A.; Fox, G.C.; Xu, Y.; Xiang, J.; Schmieder, A.H.; Yang, X.; et al. Nanotherapy delivery of c-myc inhibitor targets Protumor Macrophages and preserves Antitumor Macrophages in Breast Cancer. *Theranostics* **2020**, *10*, 7510–7526. [CrossRef] [PubMed]
115. Sajid, M.I.; Moazzam, M.; Kato, S.; Cho, K.Y.; Tiwari, R.K. Overcoming Barriers for siRNA Therapeutics: From Bench to Bedside. *Pharmaceutics* **2020**, *13*, 294. [CrossRef] [PubMed]

Publisher's Note: MDPI stays neutral with regard to jurisdictional claims in published maps and institutional affiliations.

© 2020 by the authors. Licensee MDPI, Basel, Switzerland. This article is an open access article distributed under the terms and conditions of the Creative Commons Attribution (CC BY) license (http://creativecommons.org/licenses/by/4.0/).

biomedicines

MDPI

Article

Optically Coupled PtOEP and DPA Molecules Encapsulated into PLGA-Nanoparticles for Cancer Bioimaging

Olena Vepris [1], Christina Eich [1], Yansong Feng [2], Gastón Fuentes [1,3], Hong Zhang [2], Eric L. Kaijzel [1] and Luis J. Cruz [1,*]

1 Translational Nanobiomaterials and Imaging Group, Department of Radiology, C2-S-Room 187, Leiden University Medical Center, Albinusdreef 2, 2333 ZA Leiden, The Netherlands; o.vepris@lumc.nl (O.V.); c.eich@lumc.nl (C.E.); gastonfe@biomat.uh.cu (G.F.); e.l.kaijzel@lumc.nl (E.L.K.)
2 Van 't Hoff Institute for Molecular Sciences, University of Amsterdam, Science Park 904, 1098 XH Amsterdam, The Netherlands; fengyansong@bit.edu.cn (Y.F.); h.zhang@uva.nl (H.Z.)
3 Department of Ceramic and Metallic Biomaterials, Biomaterials Center, University of Havana, Ave Universidad e/G y Ronda, Vedado, Plaza, La Habana 10400, Cuba
* Correspondence: l.j.cruz_ricondo@lumc.nl

Abstract: Triplet-triplet annihilation upconversion (TTA-UC) nanoparticles (NPs) have emerged as imaging probes and therapeutic probes in recent years due to their excellent optical properties. In contrast to lanthanide ion-doped inorganic materials, highly efficient TTA-UC can be generated by low excitation power density, which makes it suitable for clinical applications. In the present study, we used biodegradable poly(lactic-co-glycolic acid) (PLGA)-NPs as a delivery vehicle for TTA-UC based on the heavy metal porphyrin Platinum(II) octaethylporphyrin (PtOEP) and the polycyclic aromatic hydrocarbon 9,10-diphenylanthracene (DPA) as a photosensitizer/emitter pair. TTA-UC-PLGA-NPs were successfully synthesized according to an oil-in-water emulsion and solvent evaporation method. After physicochemical characterization, UC-efficacy of TTA-UC-PLGA-NPs was assessed in vitro and ex vivo. TTA-UC could be detected in the tumour area 96 h after in vivo administration of TTA-UC-PLGA-NPs, confirming the integrity and suitability of PLGA-NPs as a TTA-UC in vivo delivery system. Thus, this study provides proof-of-concept that the advantageous properties of PLGA can be combined with the unique optical properties of TTA-UC for the development of advanced nanocarriers for simultaneous in vivo molecular imaging and drug delivery.

Keywords: photon upconversion; triplet-triplet annihilation; in vivo imaging; PLGA; nanoparticles

Citation: Vepris, O.; Eich, C.; Feng, Y.; Fuentes, G.; Zhang, H.; Kaijzel, E.L.; Cruz, L.J. Optically Coupled PtOEP and DPA Molecules Encapsulated into PLGA-Nanoparticles for Cancer Bioimaging. *Biomedicines* **2022**, *10*, 1070. https://doi.org/10.3390/biomedicines10051070

Academic Editor: Simon J Allison

Received: 15 March 2022
Accepted: 24 April 2022
Published: 5 May 2022

Publisher's Note: MDPI stays neutral with regard to jurisdictional claims in published maps and institutional affiliations.

Copyright: © 2022 by the authors. Licensee MDPI, Basel, Switzerland. This article is an open access article distributed under the terms and conditions of the Creative Commons Attribution (CC BY) license (https://creativecommons.org/licenses/by/4.0/).

1. Introduction

Since the introduction of imaging techniques for the diagnosis and treatment of different pathologies, great progress has been achieved in the field of medical imaging, and the variety of imaging agents has become more sophisticated in terms of efficacy, safety and target specificity [1].

Fluorescence imaging represents a non-invasive imaging technique that can be used to visualize molecular processes and structures in a living organism. Due to advances in biology, organic chemistry and materials science, several types of imaging probes for fluorescent imaging can be distinguished, including small organic dyes, fluorescent proteins and nanoparticles (NPs). NPs are widely used for fluorescent imaging purposes, because of the feasibility of incorporating both hydrophilic and hydrophobic substances, stability, rich contrast and versatility due to surface modification chemistry. For fluorescent materials, there are two kinds of photoluminescence mechanisms, i.e., photon down- and up-conversion (UC) [2]. The conventional down-conversion process normally absorbs one high energy photon and emits a low energy photon, e.g., a Stokes-shift emission. However, down-conversion fluorescence imaging with short-wavelength excitation suffers from autofluorescence and low signal-to-noise ratio and can cause photo damage to living organisms [3–5]. In the last decade, photon UC has become state of the art in the field of in vitro and in vivo biomedical imaging. Since photon UC is a process that converts low-energy NIR light into high-energy light in the near-infrared (NIR) or visible range [6,7], this process overcomes some of the disadvantages faced by conventional down-conversion probes, with the advantages of (i) large hypsochromic shift, (ii) sharp emission peak, (iii) long luminescence shelf-life, (iv) high photostability, (v) absence of photo damage to living organisms and (vi) high penetration power in biological tissues [8–10]. Photon UC is a combination of photophysical processes, and it can be achieved via different energy transfer mechanisms using different materials, such as rare-earth metals, metalloporphyrins and organic polyaromatic hydrocarbons [6,7].

Triplet-triplet annihilation (TTA) is a special form of UC, resulting in superior optical properties, such as intense absorption of excitation light and high UC quantum yield, while using low excitation power density [7,11,12]; the latter makes TTA-UC particularly interesting for biomedical applications. In TTA-UC, the energy transfer between a sensitizer and an emitter molecule takes place through a series of subsequent nonradiative mechanisms, such as intersystem crossing (ISC) and triplet-triplet energy transfer (TTET, Figure 1) [7,11,12].

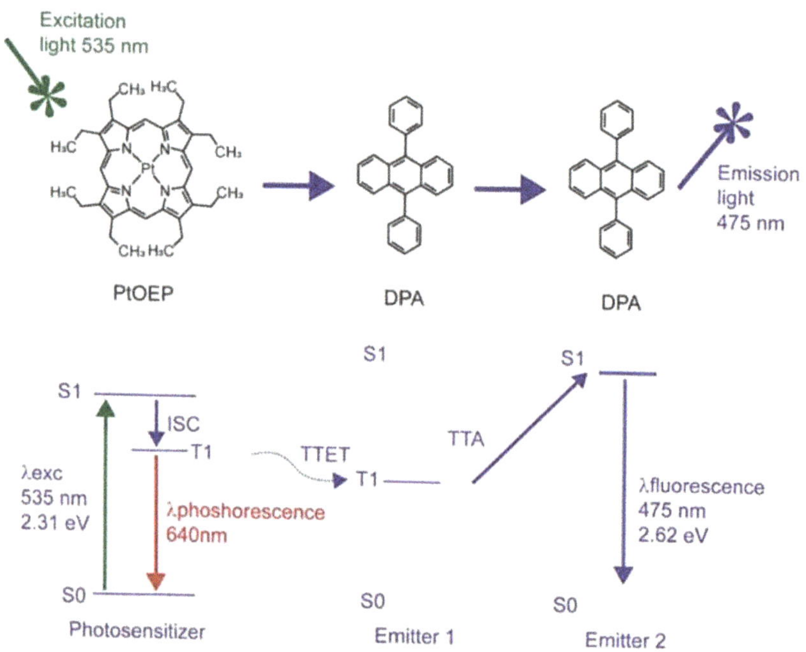

Figure 1. Jablonski diagram describing the energy levels of the electrons involved in energy transfer between PtOEP and DPA molecules. The energy values are calculated based on theoretical data.

When electromagnetic radiation "hits" a sensitizer molecule, its electrons will be promoted from the ground to a singlet excited state and then transferred to a triplet excited state nonradiatively. The released energy that accompanies the transition of the triplet excited electron to the ground singlet state will be transferred to an emitter/acceptor molecule to pump the latter to an excited triplet state. If two of these triplet excited state molecules are close enough, the interaction between the two will cause the transition of one of them to the ground singlet state and the other to a singlet excited state, i.e., the collision results in an annihilation of the triplet excited state of one acceptor molecule and promotion of another one to the singlet excited state (Figure 1).

The UC process is a highly improbable event, according to the quantum mechanical principle, since it consists of a series of spin-forbidden transitions [13]. This means that the UC process should be boosted under favourable conditions that promote those transitions. For instance, to promote ISC, which is the transition of a molecule from a singlet electronic excited state to a triplet electron state, it is advantageous to choose molecules with strong spin-orbit coupling and an expanded π conjugated system. Metalloporphyrins, such as the green absorbing platinum octaethylporphyrin (PtOEP), possess both requirements due to the presence of a transition metal that promotes a spin-forbidden transition and a cyclic π conjugated system that enhances ISC yield [14–16]. More recently, lanthanide complexes of porphyrinoids inside of nanomicelles and mesoporous silica NPs were employed for in vivo imaging in HeLA cells, demonstrating the potential of porphyrinoids as sensitizers in TTA-UC [17].

In the present study, we describe an approach whereby TTA-UC is combined with biocompatible poly(lactic-co-glycolic acid) (PLGA) to create a UC-PLGA-NP system as a potential tool for cancer visualization. NPs, such as those made of biodegradable and FDA-approved PLGA, protect their payload from premature degradation [18–24], are well described and chemically adaptable and can guide their payload inside target cells in vitro and in vivo [23–26].

To generate TTA-UC, we selected PtOEP as the photosensitizer and 9,10-diphenylanthracene (DPA) as the annihilator; a combination that has been reported to result in highly efficient UC yield [27,28]. PtOEP and DPA were successfully encapsulated into the hydrophobic core of PLGA-NPs. Phosphorescence and UC fluorescence signals resulting from TTA-UC-PLGA-NPs could be detected in vitro, in vivo and ex vivo up to 96 h after injection. We conclude that PLGA provides a suitable environment for TTA-UC, which could be utilized for simultaneous optical imaging and drug delivery in vivo.

2. Materials and Methods

2.1. Materials

PtOEP, DPA, DCM and DMF were purchased from Sigma-Aldrich®®, Zwijndrecht, The Netherlands. PLGA 50:50 (PLA/PGA) and water-soluble surfactant, polyvinyl alcohol (PVA), were purchased from Evonik industries AG, Essen, Germany. MTS (3-(4,5-dimethylthiazol-2-yl)-5-(3-carboxymethoxyphenyl)-2-(4-sulfophenyl)-2H-tetrazolium, inner salt) reagent was purchased from Promega, Leiden, The Netherlands. Histomount™ was purchased from Agar Scientific, Stansted, UK. All aqueous solutions were prepared with Milli-Q water.

2.2. Synthesis of TTA-UC PLGA-NPs Loaded with Metalloporphyrins and Polycyclic Aromatic Hydrocarbons

The TTA-UC-PLGA-NPs were synthesized according to an oil-in-water emulsion and a solvent evaporation method. Briefly, 100 mg of PLGA and 1 mg of each chromophore were dissolved in 3 mL of DCM. The mixture of chromophores and PLGA polymer was emulsified under sonication (ultrasound tip Branson, Sonifier 250), with 25 mL of water containing 500 mg of PVA for 120 s. The newly formed emulsion was left overnight in stirring at 4 °C in order to remove the organic solvent. The PLGA-NPs were collected by centrifugation at 14,800 rpm for 20 min, washed three times and freeze-dried. The NPs were stored at 4 °C and rehydrated prior to use.

2.3. Evaluation of Physicochemical Properties of the TTA-UC-PLGA-NPs

Z-average size, PDI and zeta potential of TTA-UC-PLGA-NPs were measured using a Malvern ZetaSizer 2000, software version number: ZetaSizer 7.03 (Malvern Panalytical Ltd., Malvern, UK). Fixed scattering angle of 90° at 633 nm was set up for the analysis. The measurements were performed after reconstitution of freeze-dried NPs in Milli-Q water.

2.4. Quantification PtOEP and DPA Encapsulation Efficiency

The encapsulation efficiency of chromophores was determined by UV/VIS spectrophotometry using an Ultrospec 2100 pro by Amersham Bioscience. Biodegradable PLGA-NPs were hydrolysed overnight with 0,8 M NaOH at 37 °C. Concentration of dyes was calculated using a calibration curve built with known amounts of PtOEP and DPA.

2.5. Cell Culture

The human ovarian carcinoma cell line OVCAR-3 and the breast cancer cell line MCF-7 (both cell lines, from ATCC, Manassas, VA, USA) were cultured in RPMI (Roswell Park Memorial Institute 1640 Medium) medium supplemented with 10% foetal calf serum. The cells were left to attach to the bottom of the plate for 4 h before treatment with NPs. After 4 h, the NPs were added to the cells at various concentrations of 25, 50, 100, and 200 μg/mL. DMSO at 25% was used as a positive control. The cells treated with NPs were incubated at 37 °C under 5% CO_2 for 72 h. After 72 h, the medium was changed, and 20 μL of MTS reagent was added to each well. The ELISA reader (Molecular Devices VERSAmax Tunable Microplate Reader, Software: SoftMax Pro v5.4.1) was used to measure the absorbance value of MTS product at 490 nm when the colour of the medium changed from yellow to light brown. The following formula was applied to calculate the viability of cell growth: cell viability (%) = (mean of absorbance value of treated sample/mean of absorbance of negative (live) control) × 100.

2.6. In Vitro Uptake of TTA-UC-PLGA-NPs by OVCAR-3 Cells

Fluorescence imaging of fixed OVCAR-3 cells incubated with TTA-UC-PLGA-NPs was performed with a Leica DMRA microscope using a PL APO 63x/1.32-0.6 oil objective. DAPI and Cy5 filters were used for nucleus and membrane detection, respectively, using the following protocol: Adherent cells were detached with 0.2% trypsin in PBS (Gibco by Life Technologies, Bleiswijk, The Netherlands) and seeded in 8-well chamber slides (BD Biosciences, Franklin Lakes, NJ, USA) at a concentration of 2×10^4 cells/well. After 4 h, the TTA-UC-PLGA-NPs were added at a concentration of 200 μg/mL for 2 h at 37 °C under 5% of CO_2. At the end of the incubation time, the cells were washed with PBS, fixed in 4% of paraformaldehyde (PFA) and stained with DiD labelling solution according to manufacturer's protocol (Invitrogen, The Netherlands). The staining was completed adding Vectashield mounting medium containing DAPI.

2.7. Measurement of UC Process

Confocal microscopy imaging of TTA-UC-PLGA-NPs was performed by a Leica SP8 X WLL (White Light Laser) laser scanning microscope. To this end, TTA-UC-PLGA-NPs were dissolved at 5 μg/μL in distilled water and dropped onto a glass slide and imaged. Emissions were collected between 650–670 nm and 430–475 nm; the excitation was provided at 535 nm.

2.8. Stability Measurement of TTA-UC-PLGA-NPs in Solution

In order to evaluate the stability of the TTA-UC system inside the PLGA core, the TTA-UC-PLGA-NPs were dissolved in PBS and left for 0, 24 and 72 h shaking (300 rpm) at 37 °C. At each time point, phosphorescence and UC-luminescence were measured using a spectrofluorometer (Fluorolog). The suspension of TTA-UC-PLGA-NPs was excited at 535 nm, and the emission signals were captured at 433 and 641 nm.

2.9. Animals

All the animal studies were in conformity with the animal management protocols and were approved by the Leiden University Animal Experimental Committee; studies were performed in accordance with the national legislation of the Netherlands and in compliance with the 'Code of Practice Use of Laboratory Animals in Cancer Research' (Inspectie W&V, July 1999). Female FVB mice were purchased from Charles River Laboratories

(L'Arbresle Cedex, France). Animals were housed at 22 °C and 50% humidity, with free access to food and water, and maintained under standard 12 h light/12 h dark cycles. All animal experiments were assessed according to the ethics of animal research and approved by the Animal Welfare Committee of Leiden University Medical Center, the Netherlands. All mice received humane care and were kept in compliance with the Code of Practice Use of Laboratory Animals in Cancer Research (Inspectie W&V, July 1999). All analytical procedures were performed under isoflurane gas anaesthesia (3% induction, 1.5–2% maintenance) in 70% pressurized air and 30% O_2, unless stated differently.

2.10. *In Vivo Monitoring of TTA-UC-PLGA-NPs*

In vivo imaging was performed using the IVIS spectrum Preclinical Imaging System (Caliper LS, Hopkinton, MA, USA). The images were analysed with Living Image 4.3.0 software. First, the tumour was induced by injecting 1×10^6 MCF cells in 100 µL PBS subcutaneously into the back of the mice. When the tumours reached the volume of approximately 125 mm^3, 0.5 mg TTA-UC-PLGA-NPs in 100 µL PBS were injected into the tail vein of each mouse. At 3, 24, 72 and 96 h post injection, optical imaging of treated mice was performed. Biodistribution kinetics at each time point were measured by quantifying the fluorescence intensity in pre-set regions of interest (ROIs) at the tumour site, expressed as the average radiant efficiency in $(p/sec/cm^2/sr/)/(\mu W/cm^2)$.

2.11. *Measurement of TTA-UC Ex Vivo*

After 96 h, the mice were sacrificed and subcutaneous tumours were surgically removed. Freshly isolated tumours were placed in tissue moulds and covered with Tissue-Tek®® O.C.T.™ (Sakura). For snap freezing, the tumours were left on dry ice for some minutes and then stored at −80 °C. Cryosectioning of tumours was performed using CryoStar™ NX70 at a working temperature between −25 and −30 °C; the thickness of the sections was fixed between 5 and 14 µm. Before microscopic analysis of the tumours, the sections were washed in Milli-Q water, dried under the hood and mounted in medium containing Mowiol (Sigma) and 2.5% DABCO (Sigma).

2.12. *Statistical Data Analysis*

Graph Pad Prism software version 7 was used to perform statistical analysis. The Mann–Whitney test was applied in all experiments.

3. Results and Discussion

3.1. *Preparation, Physicochemical Characterization and Cytotoxicity of TTA-UC-PLGA-NPs*

In the present study, TTA-UC-PLGA-NPs were successfully synthesized by applying an oil-in-water emulsion and solvent evaporation method, as previously described [23,24,26,29]. Briefly, 100 mg of PLGA and 1 mg of each chromophore were dissolved in 3 mL of dichloromethane (DCM). The mixture of chromophores and PLGA polymer was emulsified under sonication in the presence of PVA. After removal of the organic solvent, the TTA-UC-PLGA-NPs were collected by centrifugation, washed and freeze-dried. The oil-in-water emulsion and solvent evaporation method led to an encapsulation efficiency (EE) of PtOEP and DPA molecules of 24% and 40%, respectively, as determined by nanodrop measurement of dissolved NPs (Table 1).

Table 1. Physicochemical properties of NPs. Data are presented as average ± SD (N = 3).

NPs	Size (nm)	ζ (mV)	PDI	Loading (µg/mg)	Encapsulation Efficiency (%)
Blank PLGA	180 ± 80	−24 ± 6	0.2	-	-
TTA-UC PLGA	200 ± 50	−31 ± 5	0.4	PtOEP: 244.0 DPA: 396.0	PtOEP: 24.4 DPA: 39.6

Due to the high sensitivity of the porphyrin triplet states to oxygen quenching, the actual lifetime of the triplet state depends on the solvent, its purity and its oxygen content [30]. To protect the optical properties of chromophores, PtOEP and DPA were dissolved in DCM. To ensure that encapsulation into PLGA-NPs did not affect the optical properties of PtOEP and DPA, TTA-UC-PLGA-NPs were dissolved in dimethylformamide (DMF), analysed by absorption curve analysis and compared to pure compounds (Figure 2).

The absorption spectrum of PtOEP depicts the characteristic shape of porphyrin absorption: a strong absorption peak at 380 nm (Soret band) was accompanied by a weaker absorption peak located at 501 and 533 nm (Q-band) (Figure 2A) [31]. DPA, the emitter for the TTA-UC system, showed several absorption peaks between 350 nm and 400 nm (Figure 2B).

When the TTA-UC-PLGA-NPs were dissolved in DMF, the absorption spectrum showed the combined characteristic optical profiles of PtOEP and DPA (Figure 2C). In conclusion, absorption spectrum analysis of TTA-UC-PLGA-NPs confirmed the successful encapsulation of both chromophores into the PLGA polymer, while the optical properties of PtOEP and DPA were maintained.

The size of the NPs was determined by dynamic light scattering (DLS) analysis as ~200 nm in diameter (Table 1, Figure 3A), and the polydispersity index (PDI) value of 0.4 obtained from the DLS measurement indicated a homogeneous size distribution (Table 1). Compared to blank control PLGA-NPs, the diameter of the NPs slightly increased upon encapsulation of PtOEP and DPA.

NPs smaller than 10 nm are known to be eliminated by renal excretion [32], while NPs with sizes ranging from 50 to 300 nm show extended circulation times in the blood stream compared to NPs of larger size [33]. Hence, TTA-UC-PLGA-NPs of 200 nm have the ideal size for prolonged systemic circulation times; an important prerequisite to reach the tumour site. In addition, ZetaSizer measurement showed that the surface charge of the NPs was on average −30 mV (Table 1, Figure 3B).

The surface charge is a crucial factor to determine the stability of the NPs in solution, but also affects cellular uptake, biodistribution and cytotoxicity. A high surface charge guarantees an electrostatic stabilization of the NPs due to a strong surface repulsion between NPs of the same charge. According to the measured value, it is possible to assume that the NPs are stable in suspension for a prolonged period of time [34].

Next, we investigated the optical properties of the TTA-UC system after encapsulation into PLGA-NPs. To this purpose, TTA-UC-PLGA-NPs were dissolved in water and excited at 535 nm (Figure 3C). The detected TTA-UC signal at 433 nm was similar to that previously reported for PtOEP/DPA systems [35], thus confirming the integrity of the TTA-UC system after encapsulation into PLGA.

The charge of UC-NPs has been shown to affect the intracellular localization and cellular cytotoxicity [31]. Positively charged UC-NPs localize to mitochondria, while negatively charged UC-NPs preferentially localize to lysosomes and the cytoplasm, which is associated with lower cellular cytotoxicity [36]. To assess the potential toxicity of our NPs, TTA-UC-PLGA-NPs were incubated at different concentrations (25–200 μg/mL) for 72 h with OVCAR-3 ovarian cancer cells, and the cellular toxicity was determined by MTS assay (Figure 4). TTA-UC-PLGA-NPs were compared to empty control PLGA-NPs. No toxicity of TTA-UC-PLGA-NPs or empty PLGA-NPs on OVCAR-3 cells was observed after 72 h (Figure 4).

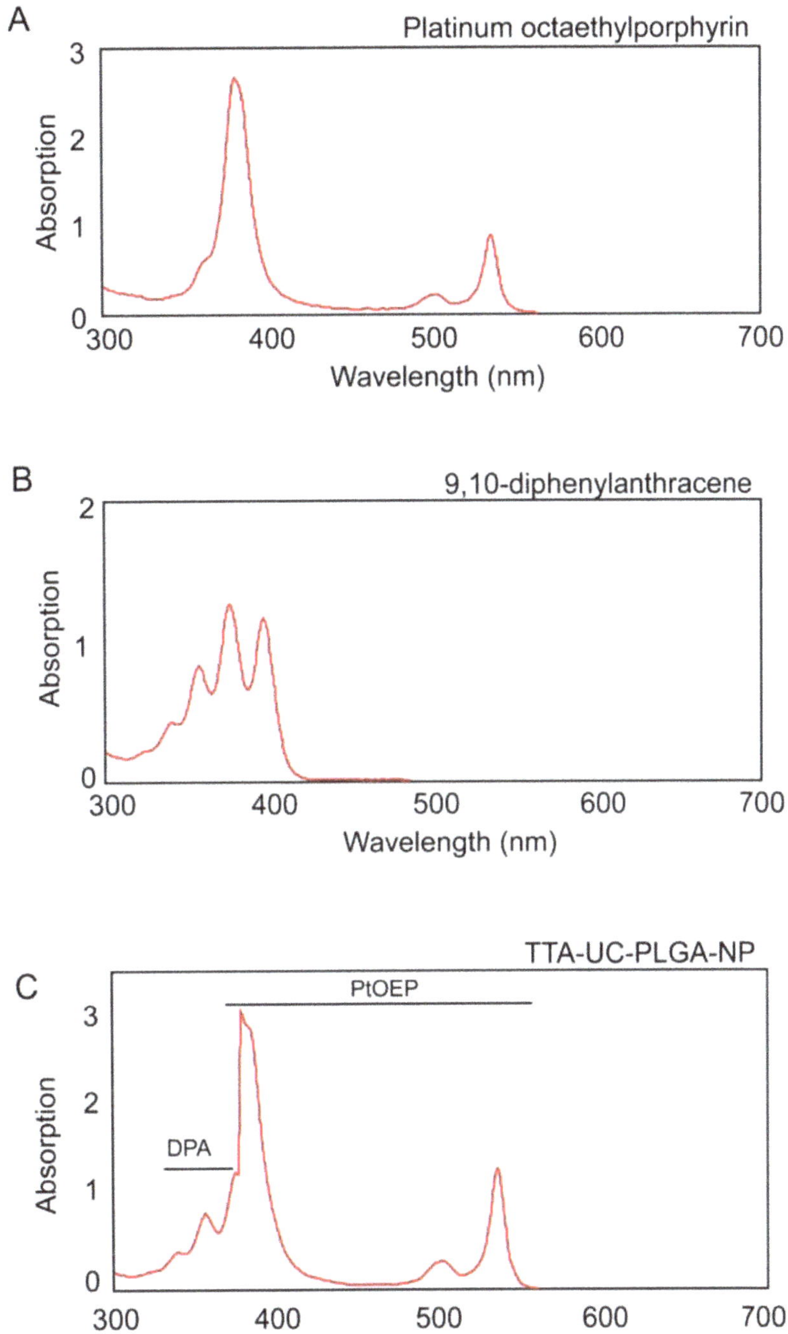

Figure 2. Individual absorption spectra of (**A**) platinum octaethylporphyrin (PtOEP) and (**B**) 9,8-diphenylanthracene (DPA), each dissolved in dimethylformamide (DMF). (**C**) TTA-UC-PLGA-NPs were hydrolysed overnight in 0.8 M NaOH at 37 °C, and the absorption spectrum of co-encapsulated PtOEP and DPA was measured.

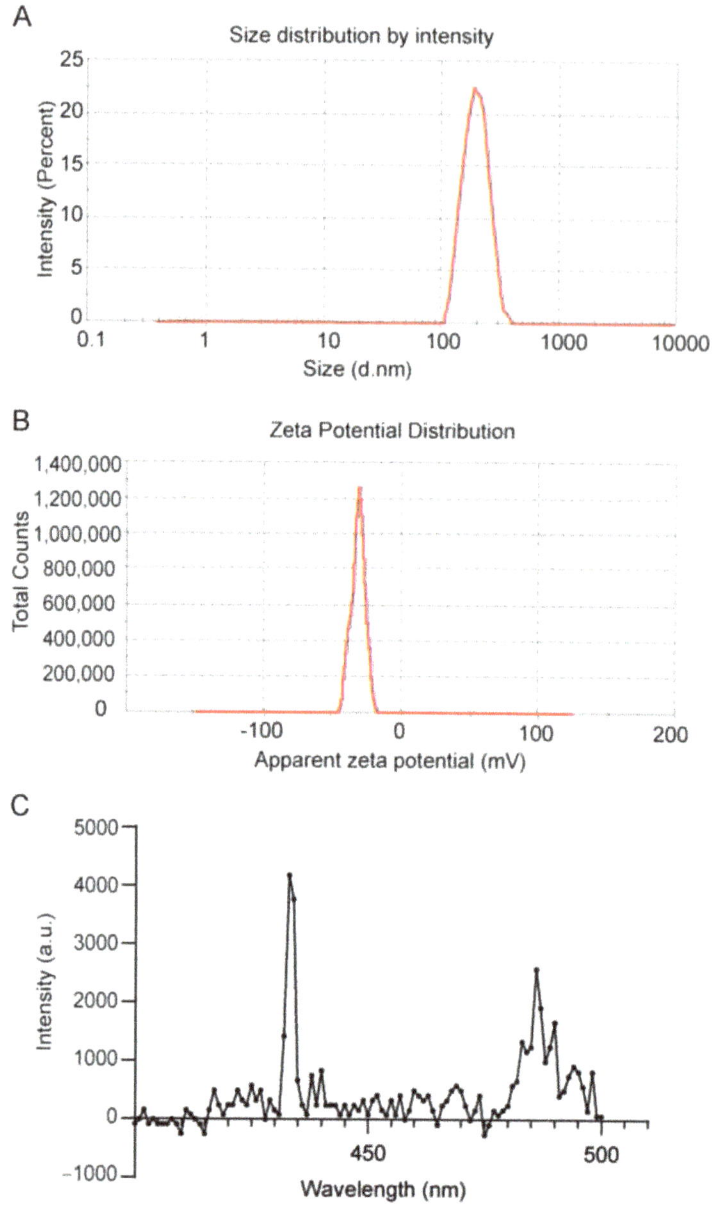

Figure 3. Physicochemical characterization of TTA-UC-PLGA-NPs. Representative (**A**) dynamic light scattering and (**B**) zeta-potential measurement of TTA-UC-PLGA-NPs reconstituted in water after freeze-drying. The average size of the nanoparticles was 198 nm in diameter, and the average zeta potential value was −31 mV. (**C**) TTA-UC-PLGA-NPs were dissolved in water and directly analysed. The NPs in solution were excited at 535 nm, and the emission was collected using a photofluorometer.

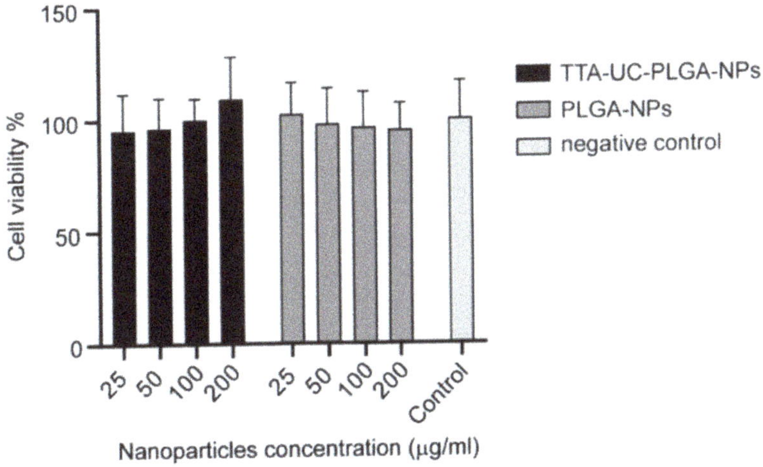

Figure 4. Assessment of cellular cytotoxicity of TTA-UC-PLGA-NPs. OVCAR-3 cells were incubated with 25, 50, 100 and 200 µg/mL TTA-UC-PLGA-NPs or control PLGA-NPs for 72 h. Untreated cells were used as negative control. The cell viability was assessed after 72 h by MTS assay.

3.2. Cellular Uptake of TTA-UC-PLGA-NPs Characterized by Fluorescence Microscopy

To assess the cellular uptake of TTA-UC-PLGA-NPs, OVCAR-3 cells were incubated for 2 h with 200 µg/mL TTA-UC-PLGA-NPs and analysed by fluorescent microscopy. In order to confirm the intracellular localization of the NPs, the cells were counterstained with DAPI ($\lambda_{excitation}$ = 340–380 nm, $\lambda_{emission}$ = 425 nm) and DiD ($\lambda_{excitation}$ = 676–688 nm, $\lambda_{emission}$ = 700–742 nm) for nucleus and membrane detection, respectively.

To detect the TTA-UC-PLGA-NPs, the samples were excited at 542–582 nm, and the phosphorescence signal was collected at 604–644 nm (Figure 5). TTA-UC-PLGA-NPs were successfully taken up by OVCAR-3 cells. The overlay of the membrane staining with the phosphorescence signal of the TTA-UC-PLGA-NPs confirmed the intracellular localization of the NPs (Figure 5).

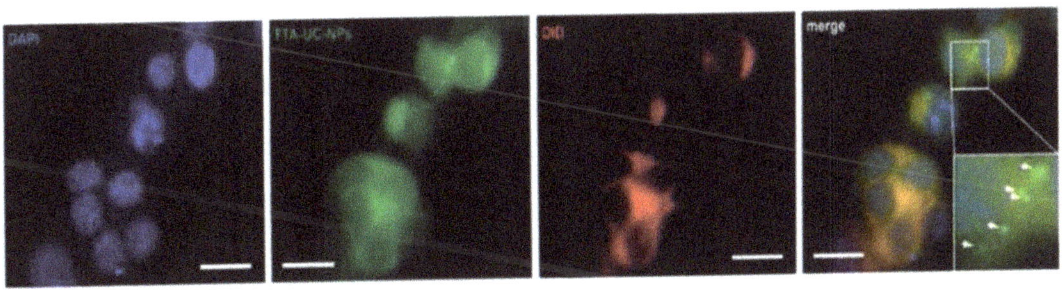

Figure 5. Cellular uptake of TTA-UC-PLGA-NPs. OVCAR-3 cells were incubated with 200 µg/mL TTA-UC-PLGA-NPs for 2 h at 37 °C and analysed by fluorescence microscopy. The cells were fixed, and the nucleus was stained with DAPI (blue) and the cell membrane was stained with DiD (red). The phosphorescence signal (green) of the TTA-UC-PLGA-NPs was collected. The following filter cubes were used to discriminate DAPI, DiD and TTA-UC-PLGA-NP-signals: DAPI, λexc = 340–380 nm and λemi = 425 nm; DiD, λexc = 676–688 nm, λemi = 700–742 nm; TTA-UC-PLGA-NPs, λexc = 542–582 nm, λemi 604–644 nm. Scale bar = 20 µm.

3.3. *Characterization of TTA-UC-PLGA-NPs UC Properties*

The ability of TTA-UC-PLGA-NPs to generate the UC process was assessed by confocal microscopy (Figure 6). The NPs were excited at a wavelength of 535 nm, and the emission signals were collected between 430–475 nm and 650–670 nm. In our study, the emission shift towards shorter wavelengths (hypsochromic shift) went from green (535 nm) to blue (430–475 nm) (Figure 6). This means that the energy associated with the emission wavelength was higher than the energy associated with the excitation wavelength, as a consequence of the anti-Stokes shift in the TTA-UC process [37]. On the contrary, the change in emission spectrum towards longer wavelengths (bathochromic shift), in our case 650–670 nm, is a characteristic of all conventional luminescence processes (Figure 6). Using confocal microscopy, we could image the single TTA-UC-PLGA-NPs that created dual emission wavelengths: the red emission from the phosphorescence of PtOEP and the blue emission as a result of the TTA-UC process. Thus, we confirmed the preparation of functional TTA-UC-PLGA-NPs.

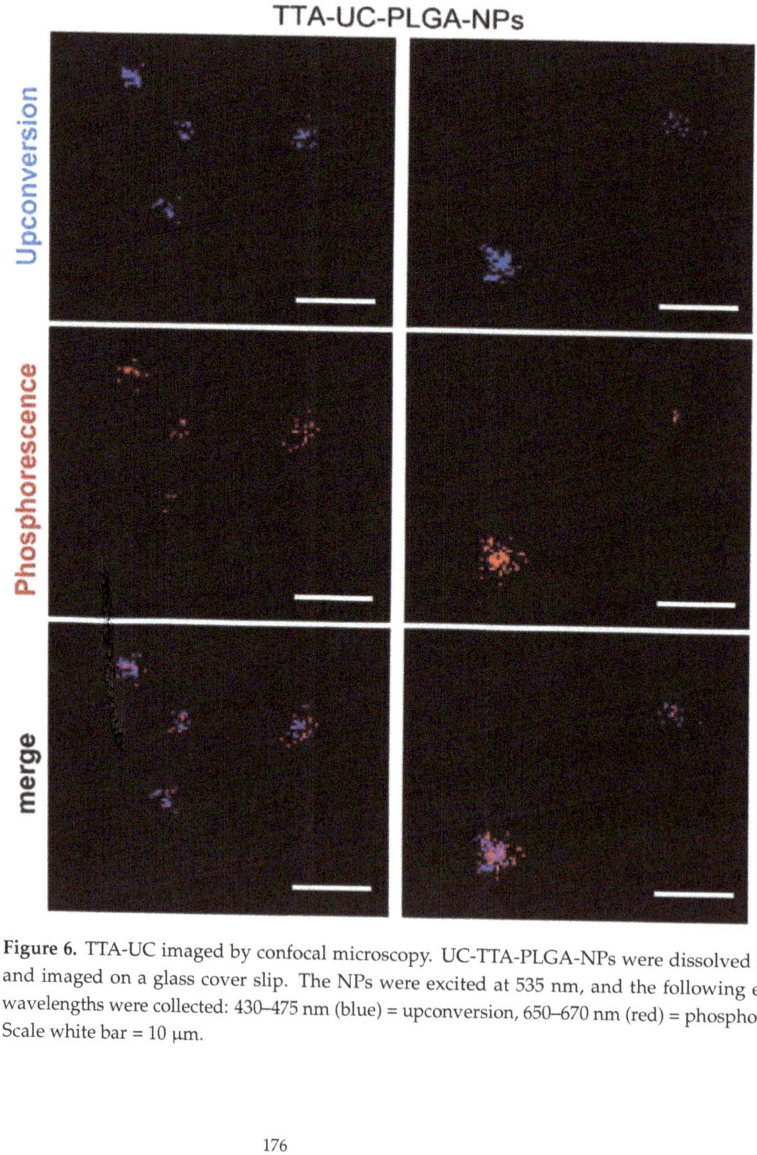

Figure 6. TTA-UC imaged by confocal microscopy. UC-TTA-PLGA-NPs were dissolved in water and imaged on a glass cover slip. The NPs were excited at 535 nm, and the following emission wavelengths were collected: 430–475 nm (blue) = upconversion, 650–670 nm (red) = phosphorescence. Scale white bar = 10 μm.

The dual colour feature of TTA-UC-PLGA-NPs was useful to improve signal-to-noise ratio and to estimate the ratio of the UC/phosphorescence signal. The overall efficiency of the TTA-UC process depends on each elementary step (ISC, TTET, TTA), on the concentration of emitter molecules and on the presence of other competing molecules, such as elementary oxygen [11,30]. Additional challenges are encountered with when TTA-UC is integrated in an NP system. Aggregation of photosensitizers and emitters can have detrimental effects on the optical properties of TTA-UC-NPs, and as a consequence, the loss of mobility of photosensitizers/emitters can limit the UC process [11,38]. Thus, creating a highly efficient UC-NP system without any secondary decay remains an enormous challenge for practical application [39].

3.4. Assessment of TTA-UC-PLGA-NP Optical Properties in Solution

Next, we investigated the optical properties of TTA-UC-PLGA-NPs over time. To this purpose, TTA-UC-PLGA-NPs were dissolved in PBS and incubated at 37 °C in shaking mode. At time point 0, 24 and 72 h, the dispersion of TTA-UC-PLGA-NPs was excited at 535 nm, and the TTA-UC and phosphorescence processes were measured at 433 and at 642 nm, respectively (Figure 7).

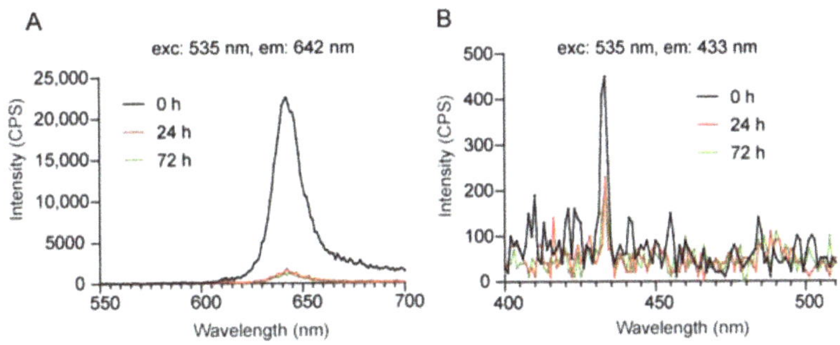

Figure 7. Assessment of optical properties of TTA-UC-PLGA-NPs in solution. TTA-UC-PLGA-NPs were dissolved in water and incubated at 37 °C for 72 h. At time point 0, 24 and 72 h, samples were analysed. The NPs in solution were excited at 535 nm, and the emission was collected using a photofluorometer. (**A**) Phosphorescence and (**B**) TTA-UC emission graphs of TTA-UC-PLGA-NPs at various time points.

The strongest phosphorescence and TTA-UC signals were detected at time point 0 (0 h post dissolution) (Figure 7A). After 24 h, the phosphorescence and TTA-UC signals decreased but could still be detected after 72 h incubation at 37 °C. Thus, the data demonstrates that TTA-UC-PLGA-NPs are suitable for imaging processes lasting for at least 72 h. As PLGA-NPs have been reported to be stable under physiological conditions for extended periods of time [40], the initial decrease in TTA-UC/phosphorescence intensity within the first 24 h observed here is likely the result of diffusion of the chromophores out of the PLGA-core.

3.5. In Vivo Monitoring of TTA-UC-PLGA-NPs

Real-time imaging of FVB mice was carried out with the goal to trace TTA-UC-PLGA-NPs in cancer cells in vivo. To this end, the IVIS imaging system was employed, which allowed measurement of the bathochromic emission captured at 640 nm (but not the UC process).

FVB mice were inoculated in the back with 1×10^6 human breast adenocarcinoma cells (MCF). When the tumours reached a volume of approximately 125 mm^3, TTA-UC-PLGA-NPs were injected intravenously in the tail vein. At different time points post injection (3, 24, 48, 72 and 96 h), the mice were imaged using a whole-body fluorescent imaging system. In vivo imaging data showed that already at 3 h post injection, TTA-UC-PLGA-NPs accumulated in the tumour area (Figure 8A).

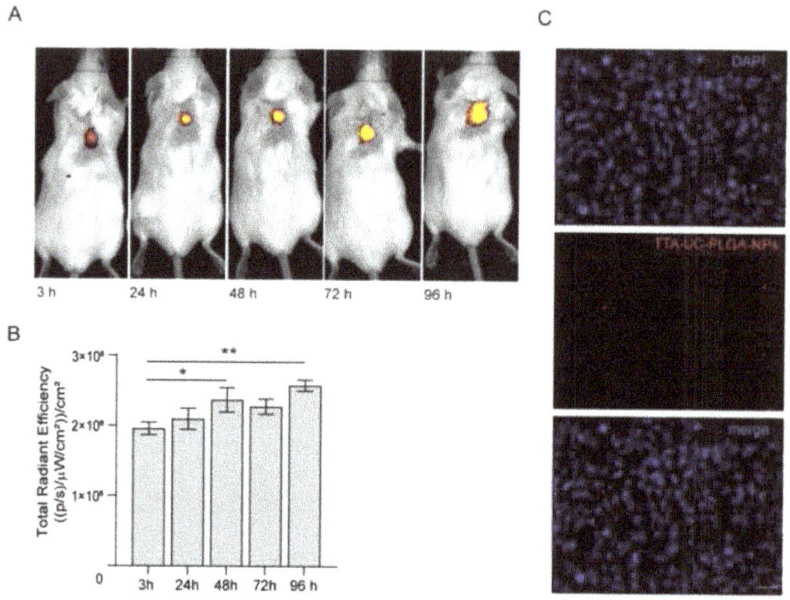

Figure 8. In vivo imaging of TTA-UC-PLGA-NPs. (**A**) FVB mice were inoculated in the back with 1×10^6 MCF-7 cells. When the tumours reached a size of 125 mm^3, mice were injected intravenously in the tail vein with 0.5 mg of TTA-UC-PLGA-NPs and imaged 3, 24, 48, 72 and 98 h post injection using the IVIS imaging system. (**A**) TTA-UC-PLGA-NPs were excited at a wavelength of 535 nm, and the emission (phosphorescence) was collected at 640 nm. (**B**) The total radiance efficiencies were calculated in the tumour areas, and the values were compared. Statistical analysis was performed using the Mann–Whitney test, * $p = 0.026$, ** $p = 0.0022$ (6 mice per group were used). (**C**) MCF-tumours were excised 96 h post injection of TTA-UC-PLGA-NPs and tumour cryosections were imaged by fluorescent microscopy. The following filter cubes were used to discriminate DAPI (blue) and TTA-UC-PLGA-NPs (red) signals: DAPI, λexc = 340–380 nm and λemi = 425 nm; TTA-UC-PLGA-NPs, λexc = 542–582 nm, λemi 604–644 nm. Scale bar = 20 μm.

The mice were monitored up to 96 h post injection. As the IVIS imaging system is not equipped to detect the emission between 430–450 nm, we instead measured the phosphorescence signal, which could be detected during all imaged time points (Figure 8A,B). Quantification of the signal showed a significant increase in phosphorescence signal in the tumour area in the first 48 h ($p = 0.026$) and after 96 h ($p = 0.0022$), indicating that more TTA-UC-PLGA-NPs accumulated at the tumour site over time.

The presence of TTA-UC-PLGA-NPs in the tumour at endpoint was confirmed by fluorescent imaging of tumour cryosections (Figure 8C). Moreover, the data showed that once accumulated in the tumour, the NPs remained at the tumour site. Since the NPs did not present any specific targeting moiety for breast adenocarcinoma cancer cells, TTA-UC-PLGA-NPs most likely accumulated at the tumour site via the EPR effect. Tumour blood vessels are naturally leaky, with large openings up to 1.5 mm, thereby promoting the passive accumulation of NPs in tumours [41,42].

3.6. Measurement of TTA-UC Process Ex Vivo

At 96 h post TTA-UC-PLGA-NP injection, breast adenocarcinoma tumours were excised, and the presence of TTA-UC phenomenon was analysed ex vivo. To this end, isolated tumours were further processed for cryosectioning and imaged using a confocal microscope. The sections were excited at 514 nm, and emissions were collected between 430–475 nm (to observe TTA-UC) and 650–670 nm (to observe phosphorescence) (Figure 9).

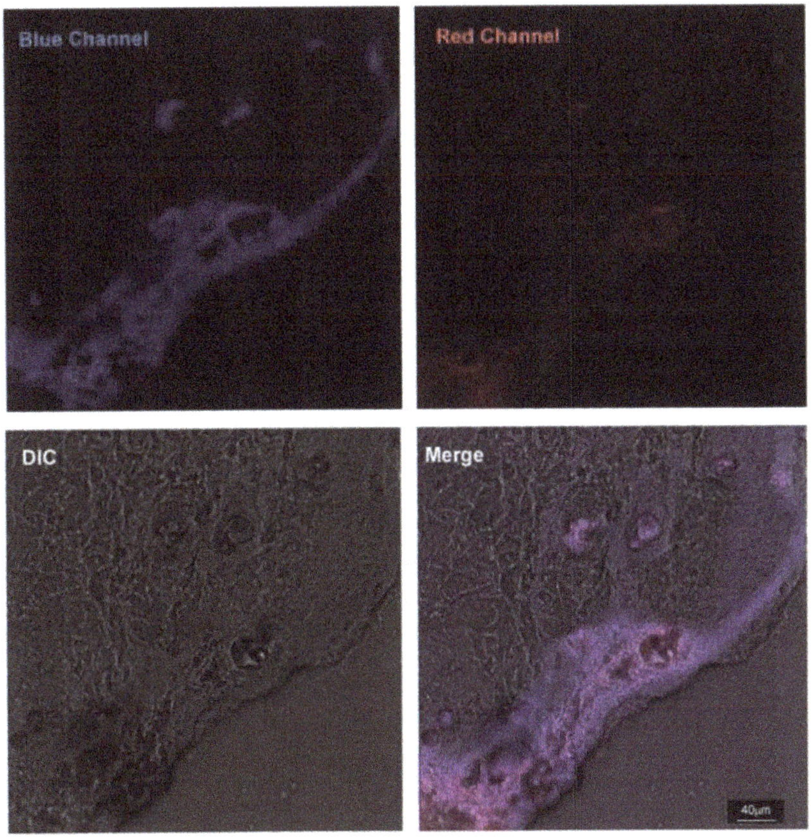

Figure 9. Ex vivo imaging of TTA-UC-PLGA-NPs. MCF-tumours were excised 96 h post injection of TTA-UC-PLGA-NPs and tumour cryosections were imaged by confocal microscopy. All cryosections were excited at 514 nm, and the upconversion signal (blue channel) was collected between 430–475 nm and the phosphorescence signal (red channel) between 650–670 nm. Scale bar = 40 μm.

Strikingly, TTA-UC and phosphorescence signals could be detected in the tumour margins and vessel-like structures 96 h post injection of TTA-UC-PLGA-NPs. The staining pattern suggests that TTA-UC-PLGA-NPs diffused into the tumour via the margins and entered into deeper tumour areas via leakages in the tumour vasculature. The data further confirmed that, despite initial diffusion of chromophores out of the PLGA copolymers (Figure 7), the integrity of TTA-UC-PLGA-NPs was maintained for at least 96 h in vivo.

4. Conclusions

In the present study, we developed a TTA-UC-PLGA-NP system by incorporating the polycyclic aromatic hydrocarbon DPA and the metalporphyrin PtOEP into a core of hydrophobic PLGA-copolymers. In vitro and ex vivo studies confirmed the presence of TTA-UC phenomenon resulting from TTA-UC-PLGA-NPs. TTA-UC could be detected 96 h post injection of TTA-UC-PLGA-NPs in the tumour area in vivo, confirming the integrity and suitability of PLGA-NPs as in vivo delivery system for TTA-UC, while maintaining the optical integrity of the TTA-UC process. This study provides proof-of-concept evidence on the feasibility of combining the advantageous properties of PLGA for drug delivery with state-of-the-art TTA-UC for simultaneous optical in vivo imaging. The presented methodology could form the basis for novel TTA-UC systems encapsulated into PLGA, which can be excited in the NIR-I or NIR-II range for clinical translation.

Author Contributions: O.V., C.E. and L.J.C. conceived the study and were in charge of the study methodology; O.V., C.E., H.Z. and L.J.C. were involved in formal analysis and investigation; O.V, C.E., Y.F. and G.F. were involved in data curation; O.V., C.E., Y.F., G.F. and E.L.K. were in charge of drafting the manuscript; O.V., C.E., G.F., H.Z., E.L.K. and L.J.C. were involved in writing, reviewing and editing of the manuscript; C.E. and L.J.C. provided the study resources and were in charge of supervision, project administration and funding acquisition. All authors have read and agreed to the published version of the manuscript.

Funding: O.V. and L.J.C. were supported by project grants from the European Commission: Marie Sklodowska Curie grant agreement CANCER (777682), PRISAR2 (872860), ACORN (807281), SIMICA (852985), BIOSAFETY (952520), PAVE (861190), CAST (857894), NOVA-MRI (859908), PIANO (956477). C.E. was supported by the H2020-WIDESPREAD-2018-03 SIMICA (852985—) project grant from the European Commission, and the Dutch PPS allowance made available by Health~Holland, Top Sector Life Sciences & Health for the project NANOCAST 2.

Institutional Review Board Statement: All animal experiments were assessed according to the ethics of animal research and approved by the Animal Welfare Committee of Leiden University Medical Center, the Netherlands. All mice received humane care and were kept in compliance with the Code of Practice Use of Laboratory Animals in Cancer Research (Inspectie W&V, July 1999).

Informed Consent Statement: Not applicable.

Data Availability Statement: Not applicable.

Conflicts of Interest: The authors declare no conflict of interest.

References

1. Kinch, M.S.; Woodard, P.K. Analysis of FDA-approved imaging agents. *Drug Discov. Today* **2017**, *22*, 1077–1083. [CrossRef] [PubMed]
2. Pratiwi, F.W.; Kuo, C.W.; Chen, B.-C.; Chen, P. Recent advances in the use of fluorescent nanoparticles for bioimaging. *Nanomedicine* **2019**, *14*, 1759–1769. [CrossRef]
3. Li, Z.; Sun, Q.; Zhu, Y.; Tan, B.; Xu, Z.P.; Dou, S.X. Ultra-small fluorescent inorganic nanoparticles for bioimaging. *J. Mater. Chem. B* **2014**, *2*, 2793–2818. [CrossRef] [PubMed]
4. Rubio-Camacho, M.; Alacid, Y.; Mallavia, R.; Martínez-Tomé, M.J.; Mateo, C.R. Polyfluorene-Based Multicolor Fluorescent Nanoparticles Activated by Temperature for Bioimaging and Drug Delivery. *Nanomaterials* **2019**, *9*, 1485. [CrossRef] [PubMed]
5. Thangudu, S.; Kalluru, P.; Vankayala, R. Preparation, Cytotoxicity, and In Vitro Bioimaging of Water Soluble and Highly Fluorescent Palladium Nanoclusters. *Bioengineering* **2020**, *7*, 20. [CrossRef] [PubMed]
6. Chen, G.; Qiu, H.; Prasad, P.N.; Chen, X. Upconversion Nanoparticles: Design, Nanochemistry, and Applications in Theranostics. *Chem. Rev.* **2014**, *114*, 5161–5214. [CrossRef]
7. Zuo, J.; Tu, L.; Li, Q.; Feng, Y.; Que, I.; Zhang, Y.; Liu, X.; Xue, B.; Cruz, L.J.; Chang, Y.; et al. Near Infrared Light Sensitive Ultraviolet–Blue Nanophotoswitch for Imaging-Guided "Off–On" Therapy. *ACS Nano* **2018**, *12*, 3217–3225. [CrossRef]
8. Nguyen, P.-D.; Son, S.J.; Min, J. Upconversion Nanoparticles in Bioassays, Optical Imaging and Therapy. *J. Nanosci. Nanotechnol.* **2014**, *14*, 157–174. [CrossRef]
9. Singh, R.; Dumlupinar, G.; Andersson-Engels, S.; Melgar, S. Emerging applications of upconverting nanoparticles in intestinal infection and colorectal cancer. *Int. J. Nanomed.* **2019**, *14*, 1027–1038. [CrossRef]
10. Rauch, M.P.; Knowles, R.R. Applications and Prospects for Triplet-Triplet Annihilation Photon Upconversion. *CHIMIA Int. J. Chem.* **2018**, *72*, 501–507. [CrossRef]
11. Huang, L.; Kakadiaris, E.; Vaneckova, T.; Huang, K.; Vaculovicova, M.; Han, G. Designing next generation of photon upconversion: Recent advances in organic triplet-triplet annihilation upconversion nanoparticles. *Biomaterials* **2019**, *201*, 77–86. [CrossRef] [PubMed]
12. Zhou, J.; Liu, Q.; Feng, W.; Sun, Y.; Li, F. Upconversion Luminescent Materials: Advances and Applications. *Chem. Rev.* **2015**, *115*, 395–465. [CrossRef] [PubMed]
13. Zhu, X.; Su, Q.; Feng, W.; Li, F. Anti-Stokes shift luminescent materials for bio-applications. *Chem. Soc. Rev.* **2017**, *46*, 1025–1039. [CrossRef] [PubMed]

14. Borisov, S.M.; Saf, R.; Fischer, R.; Klimant, I. Synthesis and Properties of New Phosphorescent Red Light-Excitable Platinum(II) and Palladium(II) Complexes with Schiff Bases for Oxygen Sensing and Triplet–Triplet Annihilation-Based Upconversion. *Inorg. Chem.* **2013**, *52*, 1206–1216. [CrossRef] [PubMed]
15. Han, J.; Duan, P.; Li, X.; Liu, M. Amplification of Circularly Polarized Luminescence through Triplet–Triplet Annihilation-Based Photon Upconversion. *J. Am. Chem. Soc.* **2017**, *139*, 9783–9786. [CrossRef]
16. Sguerra, F.; Bulach, V.; Hosseini, M.W. Molecular tectonics: Zinc coordination networks based on centric and acentric porphyrins bearing pyridyl units. *Dalton Trans.* **2012**, *41*, 14683–14689. [CrossRef]
17. Yang, Z.-S.; Ning, Y.; Yin, H.-Y.; Zhang, J.-L. Lutetium(iii) porphyrinoids as effective triplet photosensitizers for photon upconversion based on triplet–triplet annihilation (TTA). *Inorg. Chem. Front.* **2018**, *5*, 2291–2299. [CrossRef]
18. Kim, J.; Hong, J.; Lee, J.; Fakhraei Lahiji, S.; Kim, Y.-H. Recent advances in tumor microenvironment-targeted nanomedicine delivery approaches to overcome limitations of immune checkpoint blockade-based immunotherapy. *J. Control. Release* **2021**, *332*, 109–126. [CrossRef]
19. Phung, C.D.; Tran, T.H.; Kim, J.O. Engineered nanoparticles to enhance natural killer cell activity towards onco-immunotherapy: A review. *Arch. Pharm. Res.* **2020**, *43*, 32–45. [CrossRef]
20. Li, W.; Liu, Z.; Fontana, F.; Ding, Y.; Liu, D.; Hirvonen, J.T.; Santos, H.A. Tailoring Porous Silicon for Biomedical Applications: From Drug Delivery to Cancer Immunotherapy. *Adv. Mater.* **2018**, *30*, 1703740. [CrossRef]
21. Liu, L.; Li, H.; Wang, J.; Zhang, J.; Liang, X.-J.; Guo, W.; Gu, Z. Leveraging macrophages for cancer theranostics. *Adv. Drug Deliv. Rev.* **2022**, *183*, 114136. [CrossRef] [PubMed]
22. Rueda, F.; Eich, C.; Cordobilla, B.; Domingo, P.; Acosta, G.; Albericio, F.; Cruz, L.J.; Domingo, J.C. Effect of TLR ligands co-encapsulated with multiepitopic antigen in nanoliposomes targeted to human DCs via Fc receptor for cancer vaccines. *Immunobiology* **2017**, *222*, 989–997. [CrossRef] [PubMed]
23. Da Silva, C.G.; Camps, M.G.M.; Li, T.M.W.Y.; Chan, A.B.; Ossendorp, F.; Cruz, L.J. Co-delivery of immunomodulators in biodegradable nanoparticles improves therapeutic efficacy of cancer vaccines. *Biomaterials* **2019**, *220*, 119417. [CrossRef] [PubMed]
24. Da Silva, C.G.; Camps, M.G.M.; Li, T.M.W.Y.; Zerrillo, L.; Löwik, C.W.; Ossendorp, F.; Cruz, L.J. Effective chemoimmunotherapy by co-delivery of doxorubicin and immune adjuvants in biodegradable nanoparticles. *Theranostics* **2019**, *9*, 6485–6500. [CrossRef]
25. Jarak, I.; Pereira-Silva, M.; Santos, A.C.; Veiga, F.; Cabral, H.; Figueiras, A. Multifunctional polymeric micelle-based nucleic acid delivery: Current advances and future perspectives. *Appl. Mater. Today* **2021**, *25*, 101217. [CrossRef]
26. Cruz, L.J.; Tacken, P.J.; Eich, C.; Rueda, F.; Torensma, R.; Figdor, C.G. Controlled release of antigen and Toll-like receptor ligands from PLGA nanoparticles enhances immunogenicity. *Nanomedicine* **2017**, *12*, 491–510. [CrossRef]
27. Duan, P.; Yanai, N.; Kimizuka, N. Photon Upconverting Liquids: Matrix-Free Molecular Upconversion Systems Functioning in Air. *J. Am. Chem. Soc.* **2013**, *135*, 19056–19059. [CrossRef]
28. Duan, P.; Yanai, N.; Nagatomi, H.; Kimizuka, N. Photon Upconversion in Supramolecular Gel Matrixes: Spontaneous Accumulation of Light-Harvesting Donor–Acceptor Arrays in Nanofibers and Acquired Air Stability. *J. Am. Chem. Soc.* **2015**, *137*, 1887–1894. [CrossRef]
29. Pilkington, E.H.; Suys, E.J.A.; Trevaskis, N.L.; Wheatley, A.K.; Zukancic, D.; Algarni, A.; Al-Wassiti, H.; Davis, T.P.; Pouton, C.W.; Kent, S.J.; et al. From influenza to COVID-19: Lipid nanoparticle mRNA vaccines at the frontiers of infectious diseases. *Acta Biomater.* **2021**, *131*, 16–40. [CrossRef]
30. Dzebo, D.; Moth-Poulsen, K.; Albinsson, B. Robust triplet–triplet annihilation photon upconversion by efficient oxygen scavenging. *Photochem. Photobiol. Sci.* **2017**, *16*, 1327–1334. [CrossRef]
31. Bansal, A.K.; Holzer, W.; Penzkofer, A.; Tsuboi, T. Absorption and emission spectroscopic characterization of platinum-octaethyl-porphyrin (PtOEP). *Chem. Phys.* **2006**, *330*, 118–129. [CrossRef]
32. Zuckerman, J.E.; Choi, C.H.J.; Han, H.; Davis, M.E. Polycation-siRNA nanoparticles can disassemble at the kidney glomerular basement membrane. *Proc. Natl. Acad. Sci. USA* **2012**, *109*, 3137–3142. [CrossRef] [PubMed]
33. Bae, Y.H.; Park, K. Targeted drug delivery to tumors: Myths, reality and possibility. *J. Control. Release* **2011**, *153*, 198–205. [CrossRef]
34. Jeevanandam, J.; Barhoum, A.; Chan, Y.S.; Dufresne, A.; Danquah, M.K. Review on nanoparticles and nanostructured materials: History, sources, toxicity and regulations. *Beilstein J. Nanotechnol.* **2018**, *9*, 1050–1074. [CrossRef] [PubMed]
35. Aulin, Y.V.; van Sebille, M.; Moes, M.; Grozema, F.C. Photochemical upconversion in metal-based octaethyl porphyrin–diphenylanthracene systems. *RSC Adv.* **2015**, *5*, 107896–107903. [CrossRef]
36. Li, X.; Tang, Y.; Xu, L.; Kong, X.; Zhang, L.; Chang, Y.; Zhao, H.; Zhang, H.; Liu, X. Dependence between cytotoxicity and dynamic subcellular localization of up-conversion nanoparticles with different surface charges. *RSC Adv.* **2017**, *7*, 33502–33509. [CrossRef]
37. Haase, M.; Schäfer, H. Upconverting Nanoparticles. *Angew. Chem. Int. Ed.* **2011**, *50*, 5808–5829. [CrossRef]
38. Schmidt, T.W.; Castellano, F.N. Photochemical Upconversion: The Primacy of Kinetics. *J. Phys. Chem. Lett.* **2014**, *5*, 4062–4072. [CrossRef]
39. Ogawa, T.; Yanai, N.; Monguzzi, A.; Kimizuka, N. Highly Efficient Photon Upconversion in Self-Assembled Light-Harvesting Molecular Systems. *Sci. Rep.* **2015**, *5*, 10882. [CrossRef]
40. Cruz, L.J.; van Dijk, T.; Vepris, O.; Li, T.; Schomann, T.; Baldazzi, F.; Kurita, R.; Nakamura, Y.; Grosveld, F.; Philipsen, S.; et al. PLGA-Nanoparticles for Intracellular Delivery of the CRISPR-Complex to Elevate Fetal Globin Expression in Erythroid Cells. *Biomaterials* **2021**, *268*, 120580. [CrossRef]

41. Kalyane, D.; Raval, N.; Maheshwari, R.; Tambe, V.; Kalia, K.; Tekade, R.K. Employment of enhanced permeability and retention effect (EPR): Nanoparticle-based precision tools for targeting of therapeutic and diagnostic agent in cancer. *Mater. Sci. Eng. C* **2019**, *98*, 1252–1276. [CrossRef] [PubMed]
42. Hashizume, H.; Baluk, P.; Morikawa, S.; McLean, J.W.; Thurston, G.; Roberge, S.; Jain, R.K.; McDonald, D.M. Openings between Defective Endothelial Cells Explain Tumor Vessel Leakiness. *Am. J. Pathol.* **2000**, *156*, 1363–1380. [CrossRef]

biomedicines

MDPI

Article

Predicting Agents That Can Overcome 5-FU Resistance in Colorectal Cancers via Pharmacogenomic Analysis

Tsui-Chin Huang [1,2,3,4,5], Kuan-Chieh Peng [1], Tzu-Ting Kuo [2], Li-Chun Lin [2], Bai-Chia Liu [6], Shu-Ping Ye [1], Chien-Chou Chu [1], Shih-Min Hsia [6,7,8,9] and Hsin-Yi Chang [1,3,6,10,*]

1 Graduate Institute of Cancer Biology and Drug Discovery, College of Medical Science and Technology, Taipei Medical University, Taipei 11031, Taiwan; tsuichin@tmu.edu.tw (T.-C.H.); pkj@tmu.edu.tw (K.-C.P.); sabrina.yeh1603@gmail.com (S.-P.Y.); chienchou02@hotmail.com (C.-C.C.)
2 PhD Program for Cancer Molecular Biology and Drug Discovery, College of Medical Science and Technology, Taipei Medical University and Academia Sinica, Taipei 11031, Taiwan; d621107003@tmu.edu.tw (T.-T.K.); d621105001@tmu.edu.tw (L.-C.L.)
3 Master Program in Clinical Pharmacogenomics and Pharmacoproteomics, College of Pharmacy, Taipei Medical University, Taipei 11031, Taiwan
4 TMU Research Center of Cancer Translational Medicine, Taipei Medical University, Taipei 11031, Taiwan
5 Cancer Center, Wan Fang Hospital, Taipei Medical University, Taipei 11031, Taiwan
6 Graduate Institute of Metabolism and Obesity Sciences, College of Nutrition, Taipei Medical University, Taipei 11031, Taiwan; deamily@tmu.edu.tw (B.-C.L.); bryanhsia@tmu.edu.tw (S.-M.H.)
7 School of Nutrition and Health Sciences, College of Nutrition, Taipei Medical University, Taipei 11031, Taiwan
8 School of Food and Safety, Taipei Medical University, Taipei 11031, Taiwan
9 Nutrition Research Center, Taipei Medical University Hospital, Taipei 11031, Taiwan
10 Taipei Medical University and Affiliated Hospitals Pancreatic Cancer Groups, Taipei Cancer Center, Taipei Medical University, Taipei 11031, Taiwan
* Correspondence: hsinyi.chang@tmu.edu.tw; Tel.: +886-2-26972035 (ext. 133)

Citation: Huang, T.-C.; Peng, K.-C.; Kuo, T.-T.; Lin, L.-C.; Liu, B.-C.; Ye, S.-P.; Chu, C.-C.; Hsia, S.-M.; Chang, H.-Y. Predicting Agents That Can Overcome 5-FU Resistance in Colorectal Cancers via Pharmacogenomic Analysis. *Biomedicines* **2021**, *9*, 882. https://doi.org/10.3390/biomedicines9080882

Academic Editor: Simon J. Allison

Received: 29 June 2021
Accepted: 22 July 2021
Published: 24 July 2021

Publisher's Note: MDPI stays neutral with regard to jurisdictional claims in published maps and institutional affiliations.

Copyright: © 2021 by the authors. Licensee MDPI, Basel, Switzerland. This article is an open access article distributed under the terms and conditions of the Creative Commons Attribution (CC BY) license (https://creativecommons.org/licenses/by/4.0/).

Abstract: 5-Fluorouracil (5-FU) is one of several chemotherapeutic agents in clinical use as a standard of care to treat colorectal cancers (CRCs). As an antimetabolite, 5-FU inhibits thymidylate synthase to disrupt the synthesis and repair of DNA and RNA. However, only a small proportion of patients benefit from 5-FU treatment due to the development of drug resistance. This study applied pharmacogenomic analysis using two public resources, the Genomics of Drug Sensitivity in Cancer (GDSC) and the Connectivity Map, to predict agents overcoming 5-FU resistance in CRC cells based on their genetic background or gene expression profile. Based on the genetic status of adenomatous polyposis coli (APC), the most frequent mutated gene found in CRC, we found that combining a MEK inhibitor with 5-FU exhibited synergism effects on CRC cells with APC truncations. While considering the gene expression in 5-FU resistant cells, we demonstrated that targeting ROCK is a potential avenue to restore 5-FU response to resistant cells with wild-type APC background. Our results reveal MEK signaling plays a pivotal role in loss-of-function, APC-mediated 5-FU resistance, and ROCK activation serves as a signature in APC-independent 5-FU resistance. Through the use of these available database resources, we highlight possible approaches to predict potential drugs for combinatorial therapy for patients developing resistance to 5-FU treatment.

Keywords: 5-FU resistance; colorectal cancer; drug repurposing; Genomics of Drug Sensitivity in Cancer; Connectivity Map

1. Introduction

Colorectal cancer (CRC) is the second leading cause of cancer death, with an estimated 147,950 new CRC cases and 53,200 deaths in the United States in 2020 [1]. Administration of 5-fluorouracil (5-FU) has been one of the standard chemotherapy regimens both alone and in combination with other agents for colon cancers for almost five decades. 5-FU is an antimetabolite that inhibits thymidylate synthase to disrupt synthesis and repair of DNA and RNA, resulting in cell death. However, 5-FU has been found to be only 20% effective, and the response rate of advanced CRC patients treated with 5-FU/leucovorin is about 10–15% [2,3]. A notable proportion of the patients show tumor recurrence after 5-FU treatment that is mainly caused by drug-resistant cancer cells.

The most common mutations in colon cancer are found in adenomatous polyposis coli (APC) and resulted in inactivated truncation forms. APC controls the activity of β-catenin, particularly in the Wingless/WNT signaling pathway. APC protein typically builds a complex with glycogen synthase kinase 3β (GSK-3β) and AXIN through physical interactions with the 20 AA and SAMP repeats. This complex is then able to bind β-catenin in the cytoplasm, where β-catenin dissociates from adherent contacts between cells. β-catenin is subsequently phosphorylated by GSK-3β followed by casein kinase (CK1)-triggered initial phosphorylation. Phosphorylated β-catenin is then targeted for ubiquitination and degraded by the cellular proteasome. This prevents it from translocating into the nucleus, where it acts as a transcription factor and forms a complex with TCF4, promoting transcription of the proto-oncogene and cell cycle regulator c-MYC, the G1/S-regulating Cyclin D, the gene encoding the matrix-degrading metalloproteinase, matrilysin, the AP-1 transcription factors c-JUN and FRA1, and the urokinase-type plasminogen activator receptor genes [4–7]. In addition to regulating β-catenin, APC is responsible for various cellular processes, including cytoskeletal integrity, cellular adhesion, and DNA repair [8–11]. The versatile role of APC suggests its mutations would contribute to a wide range of controlling or modulating cellular physiology, including 5-FU resistance in CRC. Recent studies revealed that APC mutations play a critical role in colorectal cancer patients acquiring resistance to clinical 5-fluorouracil treatment, indicating a limited CRC treatment outcome remains to be improved [12].

Several studies have applied transcriptome profiling techniques such as RNA-seq and microarray on mRNA and miRNA expression levels in CRC cells established with 5-FU resistance [13–15]. Furthermore, valuable resources of systematic assessments on drug responses and multiplexed genomic manipulation data across model cancer cell lines are now distributed to the cancer research community, facilitating our understanding of drug resistance in CRC. To predict potential drugs to overcome 5-FU resistance, we performed the analysis in a genetic background-dependent manner.

We utilized the APC status-related drug responses from the Genomics of Drug Sensitivity in Cancer (GDSC) [16]. We validated the proposed drug, MEK inhibitors, to overcome the APC dysfunction-dependent 5-FU resistance in six CRC cell lines. To correlate our finding to the clinical evidence, we compared the phosphoprotein levels obtained from The Cancer Genome Atlas (TCGA) database [17] with the APC truncation status in cBioPortal [18,19]. We confirmed that CRC patients carrying truncated APC exhibited higher phosphorylations in MAPK1 and MAPK2 than patients bearing full-length APC.

For the APC-independent 5-FU resistance, we applied a cellular context-based prediction on HCT116 cells, which express wild-type APC. We analyzed the gene expression profiles acquired from 5-FU resistant HCT116 cells and their parental line to identify the differential gene expression. By reversely matching the 5-FU gene signatures to drug-induced patterns in the Connectivity Map database [20], we found that the RHO kinase inhibitor can sensitize resistant HCT116 cells to 5-FU treatment.

2. Materials and Methods

2.1. Retrieval of the Effect of APC Mutation on Drug IC50s from Genomics of Drug Sensitivity in Cancer (GDSC)

The effect of APC mutation on IC50 values in response to drugs in Pan-Cancer analysis from GDSC1 and GDSC2 and the drug annotation were downloaded from two screenings, GDSC web portal (https://www.cancerrxgene.org/ (accessed on 21 July 2019). The positive values of the effect of APC mutation on drug sensitivity (represented as IC50 effect) indicate drug resistance, and the negative values show drug sensitivity. We selected the drug responses that were commonly tested in both screenings and ranked the drug by the average calculated effect size of each category. The distribution of effects was categorized by the mechanism of action of drugs based on the target pathway.

2.2. Cell Culture and Cell Viability Measurement

Human colorectal carcinoma cells HCT116, HT29, DLD1, and SW620 were obtained from the American Type Culture Collection (ATCC, Manassas, VA, USA). All cells were maintained in RPMI 1640 (Gibco Laboratories, Grand Island, NY, USA) supplemented with 10% fetal bovine serum (Gibco Laboratories) at 37 °C in a humidified incubator with 5% CO_2. To establish a 5-FU-resistant HCT116 subline, we repeatedly exposed expose HCT116 cells to stepwise increasing concentrations of 5-FU over a period of ~12 months to obtain 5-FU-resistant HCT116 (HCT116-5FUR) with IC_{50} over 50 μM. All cells were verified as mycoplasma-free by a PCR-based detection.

2.3. Constructs and Transfection

pCMV-Neo-Bam APC (Addgene plasmid #16507), pCMV-Neo-Bam APC 1-1309 (Addgene plasmid #16509), pCMV-Neo-Bam APC 1-1941 (Addgene plasmid #16510), pCMV-Neo-Bam APC 1-2644 (Addgene plasmid #16511) were gifts from Bert Vogelstein. HCT116 cells were seeded into 6-well plates and transfected with 2 μg of vector pCMV harboring various truncated APCs, from 1-1309, 1-1941, 1-2644, full-length APC, or an empty one using lipofectamine 3000 (Invitrogen, Grand Island, NY, USA) for 48 h before drug treatment.

2.4. MTS Assay

A total of 2500 cells were seeded into a 96-well plate for 16 h and treated with 5-FU (Sigma-Aldrich, St. Louis, MO, USA), PD-0325901 (Cayman Chemical, Ann Arbor, MI, USA), 3-(4-Pyridyl)indole (RHO-kinase-inhibitor-III, Rockout) (Cayman Chemical), and BMS-754807 (Cayman Chemical) at indicated concentrations from 1000 × stocks to a final concentration of DMSO as 0.1%. Cell viability was measured by MTS assay using the CellTiter 96® AQueous One Solution Cell Proliferation Assay kit (Promega, Madison, WI, USA) according to the manufacturer's instruction. The absorbance was detected at 490 nm with an Epoch Microplate Spectrophotometer (BioTek Instruments, Winooski, VT, USA).

2.5. Gene Expression Dataset and Data Analysis

The expression dataset of HCT116 parental and HCT116 5-FU resistant cells (E-MEXP-390) [21] was obtained from the Array Express database (http://www.ebi.ac.uk/arrayexpress/ (accessed on 7 October 2019)). The Affymetrix CEL file data was processed by the RMA (Robust Multichip Average) method, including background correction, quantile normalization, and log2 transformation, and the differentially expressed genes were defined using limma package [22]. Overexpressed genes were inverted and searched against the perturbation-induced gene expression profiles in the Connectivity Maps 2.0.

2.6. Functional Enrichment Analysis

Differentially expressed genes were submitted to the Database for Annotation, Visualization, and Integrated Discovery (DAVID) v6.8 (https://david.ncifcrf.gov/home.jsp (accessed on 12 November, 2019)) [23] for functional enrichment analysis using Gene Ontology terms, according to their molecular function (MF), biological process (BP), and cellular component (CC). Human Genome U133 Plus 2 Array was selected as backgrounds. The p-value was corrected for multiple testing by Bonferroni correction.

2.7. Western Blot Analysis

PBS washed cells were harvested and lysed in RIPA buffer (20 mM Tris-HCl (pH 7.5), 150 mM NaCl, 1 mM EDTA, 1 mM EGTA, 1% NP-40, and 1% sodium deoxycholate) containing the Halt Protease and Phosphatase Inhibitor Cocktail (Thermo Fisher, Waltham, MA, USA). Protein concentration was determined using the T-Pro BCA Protein Assay Kit (T-Pro biotechnology, Taipei, Taiwan). Protein samples were separated by SDS-PAGE, transferred onto PVDF membranes (Millipore, Billerica, MA, USA), and blocked with Block PRO blocking buffer (Visual Protein Biotechnology Corporation, Taipei, Taiwan). Blot was incubated with a primary antibody followed by incubation of horseradish peroxidase-conjugated secondary antibodies. Protein level was determined by enhanced chemiluminescence (Advansta, Menlo Park, CA, USA) detection. Antibodies probing p-AKT (S473), AKT, p-ERK1/2 (Y202/Y204), ERK1/2, and GAPDH (Cell Signaling Technology, Inc., Danvers, MA, USA) were used.

2.8. Clinical Data Analysis

Clinical data of CRC patients investigated by The Cancer Genome Atlas (TCGA) program were obtained from cBioPortal (http://www.cbioportal.org/index.do (accessed on 20 April 2018) [18,19]. Data of protein expression and phosphorylation levels were retrieved from antibody-based reverse-phase protein array (RPPA) profiling of CRC patients. The differences between patients with truncated APC and with wild-type APC were compared using Student's *t*-test followed by the Benjamini–Hochberg procedure to control the false discovery rate.

2.9. Statistical Analysis

All experiments were performed at least three times independently. Data were expressed as mean ± SD. Unpaired two-tailed t-tests were used for the comparison of two groups, and *p*-values < 0.05 were considered significant. To calculate the drug combination effect, the Combination Index (CI) was assessed using CompuSyn software version 1.0 (Combosyn, Inc, Paramus, NJ, USA) [24]. A CI greater, lesser, or similar than 1 indicates antagonism, synergism, or additive effect, respectively.

3. Results

3.1. Cell Overexpressing 1-1309 Truncated APC Acquires 5-FU Drug Resistance

To examine whether APC truncation participates in 5-FU resistance in CRC, we transfected APC mutants (Figure 1A) lacking EB1/DLG-binding (APC 1-2644), microtubule- and AXIN-binding (APC 1-1941), and β-Catenin binding and downregulation (APC 1-1309) domains, respectively, into HCT-116 cells that carry full-length APC. We measured the cell viability by MTS assay after 48 h of 5-FU treatment and compared that of cells transfected with truncated APCs to untransfected cells, cells transfected with empty vector (pCMV), or cells overexpressing wild-type APC. We found that only the 1-1309 truncated APC lacking β-catenin binding/downregulation sites significantly exhibited tolerance to 5-FU treatment (Figure 1B), suggesting that loss of binding of β-catenin or disrupting the local structure of nearby DRI domain is crucial for 5-FU sensitivity.

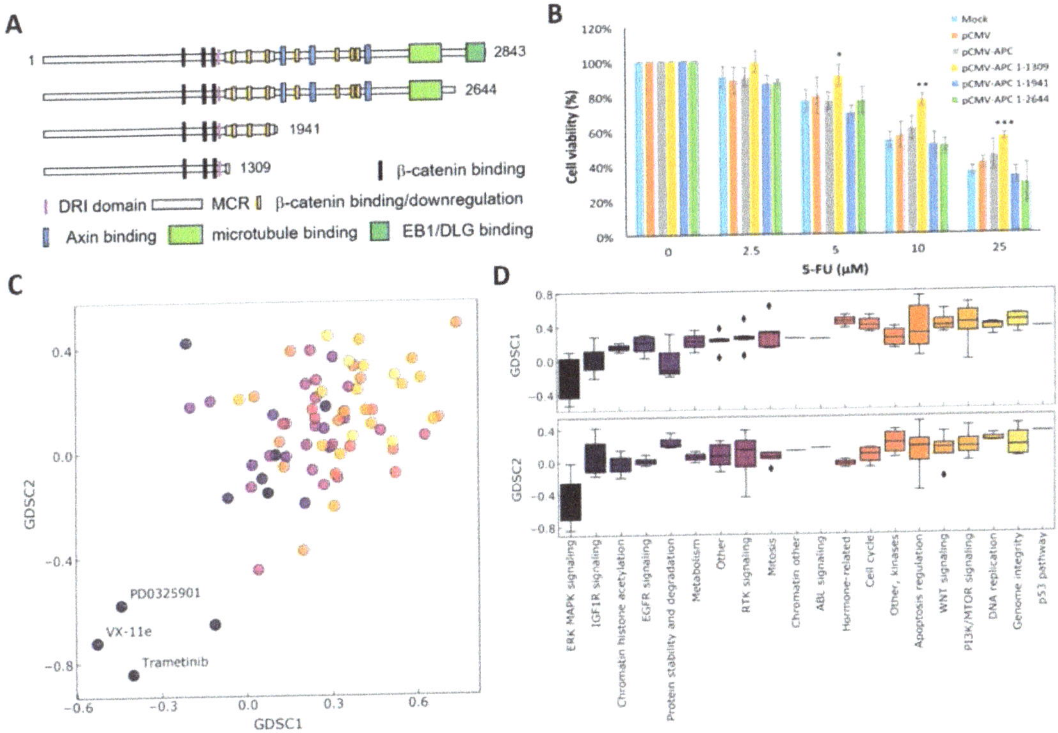

Figure 1. APC status and drug response: (**A**) C-terminal protein domains of wild-type (1-2843) and truncated APCs. Domains of APC are subsequently deleted as indicated: EB1/DLG-binding domains (1-2644), AXIN-binding (1-1941), and mutation cluster region (MRC) and β-Catenin binding/downregulation sites (1-1309). (**B**) HCT116 cells were untransfected (Mock) or transfected with pCMV harboring various truncated APCs illustrated in (**A**) with lipofectamine 3000 for 48 h. The expressing length of APC is indicated. pCMV-APC represents the full-length APC, and pCMV indicates the empty vector. After transfection, 2500 cells were seeded into 96-well plates for 16 h and treated with 5-FU at the concentration of 2.5, 5, 10, and 25 μM for 48 h. Cell viability was then accessed by MTS assay and normalized to DMSO vehicle control. *, $p < 0.05$; **, $p < 0.01$; ***, $p < 0.001$. (**C**) APC status correlated drug sensitivity prediction obtained from Genomics of Drug Sensitivity in Cancer (GDSC) database. The APC mutation effects on drug sensitivity were plotted from two independent screening, GDSC1 and GDSC2. Each drug is colored by its targeting pathway in terms of the mechanism of action. Top drugs targeting ERK MAPK signaling contributed to increased sensitivity are indicated. (**D**) The effect of APC mutation on drug response is categorized by the target pathway and ranked by the mean of effects from GDSC1 and GDSC2. The color code is the same as (**C**). A negative effect indicates improved sensitivity.

3.2. APC Mutations and Drug Sensitivity

To understand whether and to what extent the APC mutation correlates with drug sensitivity, we extracted the data from GDSC [16] to survey the correlation between APC mutation and drug sensitivity from the IC50 values of 448 anti-cancer drugs to 851 and 720 cancer cell lines from GDSC1 and GDSC2, respectively (Figure 1C). The differences between mean IC50 from cells with APC mutants and cells with wild-type APC were used to present the effect of APC mutation on drug sensitivity. The negative value indicates increased drug sensitivity of a given drug to cells with APC mutation. To gain the global response of anti-cancer drugs to APC mutation, we classify the effects according to the drug targeting pathway (Figure 1D). Compounds targeting the p53 pathway, genome integrity, DNA replication, and PI3K/mTOR signaling are the most frequent unresponsive drugs to APC-mutated cells, echoing the lower sensitivity to 5-FU treatment in cells carrying the

1-1309 APC mutant (Figure 1B) and supporting 5-FU sensitivity in p53-dependent cells with wild-type APC (Figure 2C). On the other hand, cells with APC mutations are sensitive to drugs targeting ERK/MAPK signaling, mainly MEK inhibitors, intriguing us to combine these drugs to increase the drug response to 5-FU.

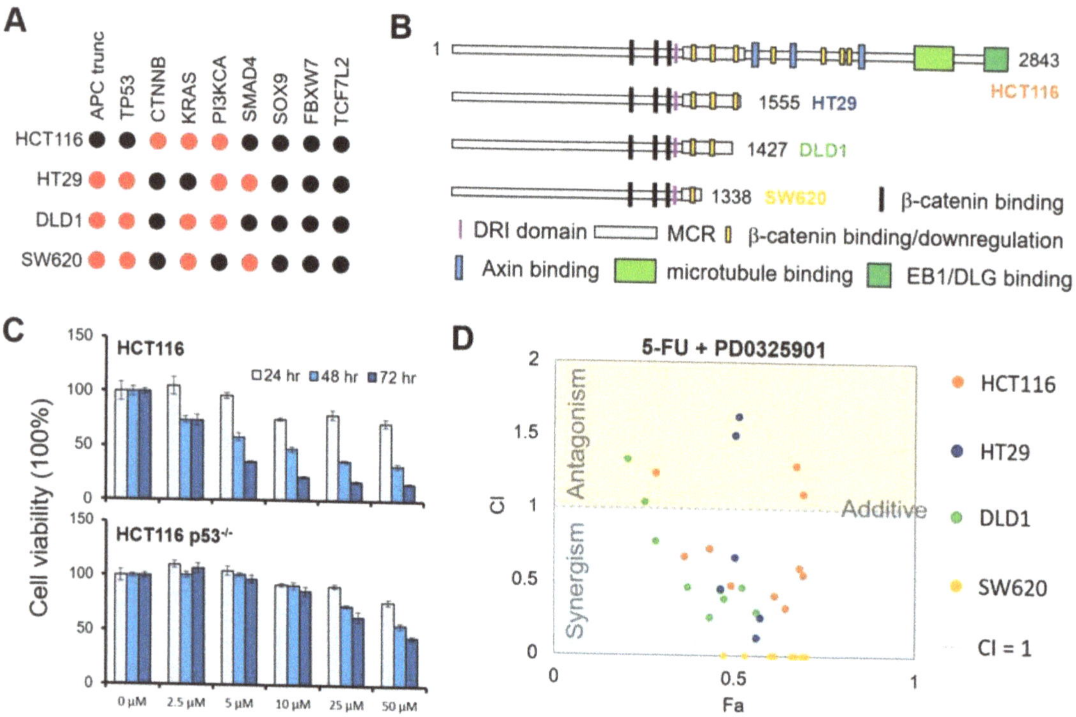

Figure 2. Combinatorial effects of 5-FU with targeting p53 pathway or ERK/MAKP signaling: (**A**) The genetic status of frequently mutated genes in CRC cell lines was used. Wild-type and genes harboring mutations are colored in black and red, respectively. (**B**) APC status in cell lines was tested. Cell lines were colored corresponding to the data in (**D**). (**C**) Inhibition of p53-mediated pathway contributed to 5-FU resistance. P53 null HCT116 ($p53^{-/-}$) cells or parental HCT116 cells were treated with 5-FU for 24, 48, and 72 h at indicated concentrations. Cell viability was measured by MTS assay and normalized to DMSO treated control. (D) Quantitation of synergism and antagonism in MEK inhibitor and 5-FU combination. Fraction affected (Fa) versus Combination Index (CI) plots were generated by CompuSyn. CI indicates a greater/synergism (CI < 1), lesser/antagonism (CI > 1), or similar (CI = 1) effect than the expected additive effect.

3.3. MEK Inhibitor PD-0325901 Sensitizes APC Mutation Containing CRC Cells to 5-FU Treatment

To validate whether targeting the p53 pathway causes drug resistance in cells with wild-type APC, we treated p53 null HCT116 cells and their parental cells with 5-FU in a dose- and time-dependent manner. We found that inhibiting p53 signaling under an APC wild-type background enhanced 5-FU resistance, confirming the synergic role of p53 in APC mutation (Figure 2C). Next, we validate whether ERK/MAPK singling inhibitors can overcome APC mutation caused by 5-FU resistance. We selected one of the MEK inhibitors, PD-0325901, and examined its cytotoxicity in four colon cancer cell lines. Each cell line has different truncated APCs (Figure 2B) except for HCT116 cells, which express full-length APC but harbor gain-of-function mutations in β-Catenin (Figure 2A). We treated the cells with PD-0325901 alone or combined with 5-FU in dose-dependent manners. As expected, SW620 cells harboring the shortest truncated APC showed the highest survival rate under 5-FU treatment, suggesting 5-FU-inhibited cell proliferation is correlated with APC-interacting downstream events (Figure 2D and Figure S1). To testify the hypothesized roles of MEK activity to compensate for 5-FU sensitivity, we treated these CRC cells with PD-0325901 in combination with 5-FU. To examine the drug effects, we applied the Combination Index (CI), a standard measure of combination effect. A CI indicates a greater (CI < 1), lesser (CI > 1), or similar (CI = 1) effect than the expected single additive effect expected from the knowledge of the effects of each drug individually. We found that the MEK inhibitor PD-0325901 displayed cell line-dependent effects with a strong synergic effect (CI < 0.1) in SW620 cells, a moderate synergic effect in DLD1 and HCT116 cells, and an antagonistic effect in HT29 cells (Figure 2D), suggesting the activation of MEK1/2 might be involved in but not required for developing 5-FU resistance.

3.4. ERK1/2 Phosphorylation Is Induced in Patients with Truncated APC

To explore what molecular clues were in line with APC truncation clinically, we compared the protein expression and phosphorylation from CRC patients baring wild-type or truncated APC from TCGA. We found that pERK1/2 (pT202/Y204 on MAPK1 and MAPK3) were the most upregulated proteins with 36-fold induction in patients carrying truncated APC compared with patients baring wild-type APC (Figure 3, q-value = 6.13×10^{-7}), demonstrating that APC truncation is associated with ERK1/2. We suggest that APC status might serve as a molecular signature for CRC patients to exhibit a positive response from the MEK inhibitor in combination with 5-FU.

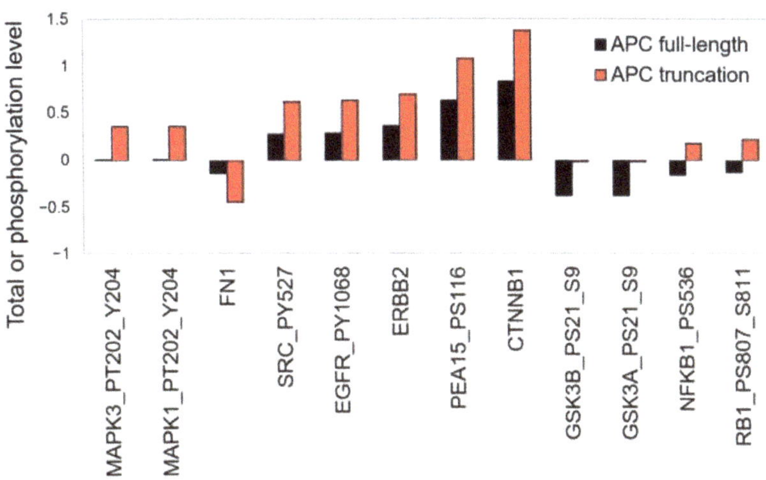

Figure 3. ERK1/2 phosphorylation in the TCGA cohort is associated with APC truncation. The protein expression and phosphorylation levels of CRC patients (COAD in TCGA) were retrieved and compared between APC full-length (black bars) and truncated APC (red bars) patients. Data with a q-value (adjusted *p*-value by Benjamini–Hochberg correction) less than 0.05 are shown and ranked based on fold changes. Phosphorylation information is annotated by official gene symbol followed by phosphorylation sites.

3.5. Function of Extracellular Ligand Binding to Regulate Cell Mobility Was Upregulated in 5-FU Resistant Cells

Next, we sought to discover the molecular signatures in APC-dependent 5-FU resistance for combinatorial therapy. Consequently, we analyzed the gene expression profiles of APC-wild-type HCT-116 cells and the derived 5-FU resistant subline (Figure 4). There were 869 genes differentially expressed, with 456 upregulated and 413 downregulated. The differentially expressed genes were then subjected to functional enrichment analysis by DAVID. The selectively enriched functions of differentially expressed genes are listed in

Tables S1 and S2, respectively. We found that the downregulated genes in 5-FU-resistant cells were involved in RNA processing and metabolism as well as folic acid metabolism (Table S2), suggesting that 5-FU resistance might be caused by abrogating the nucleoside biosynthesis pathway, which 5-FU interrupts. On the other hand, the top enriched terms in upregulated genes are receptor binding, cell migration, and extracellular space from receptor binding in MF, BP, and CC, respectively (Table S1).

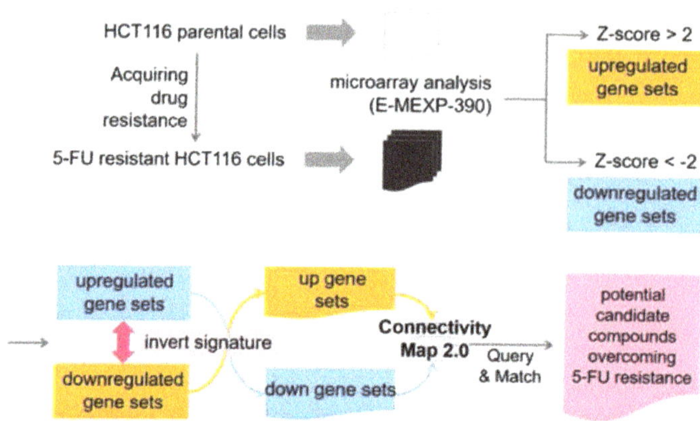

Figure 4. Pattern matching analysis to reveal drugs able to overcome APC-independent 5-FU resistance. The differentially expressed genes from 5-FU-resistant and parental HCT116 cells were defined and inverted to match the drug-responsive gene expression pattern in the Connectivity Map 2.0 database. The drugs triggering negatively correlated gene expression patterns are reported as potential candidate compounds overcoming 5-FU resistance.

3.6. Reversed Expression Predicts ROCK Activation in 5-FU Resistant Cells

To discover possible inhibitors to restore 5-FU drug sensitivity in cells with wild-type APC, we inversed the differential gene regulation of 5-FU-resistant cells to parental ones. Subsequently, we submitted the list to query the Connectivity Map database (Figure 5). The query results generated a list of potent candidate drugs whose effects matched the expression signature inverted to 5-FU. We listed the top 10 results of pharmacological perturbations in Table 1. The inhibition on ROCK demonstrated the highest correlation in inverse gene expression of 5-FU resistant cells, showing the participation of the ROCK pathway in 5-FU resistance establishment. As the major downstream effector of small GTPase RhoA, ROCK is involved in the regulation of actomyosin cytoskeleton to promote contractile force generation for cell mobility [25,26]. This feature was in agreement with our previous findings that metastatic activity is gained in 5-FU resistant cells through receptor binding from extracellular stimulation, which promotes cell migration (Table S1).

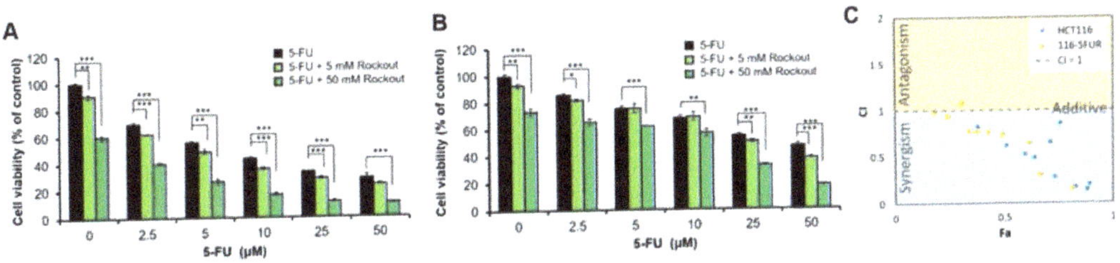

Figure 5. ROCK inhibitor synergizes 5-FU effects on resistant cells: (**A**) HCT116 and (**B**) 116-5FUR cells were treated with Rockout and 5-FU at the indicated doses. We measured the cell viability by MTS assay and normalized it to DMSO vehicle control. Grass green and green bars are cell viability of ROCK inhibitor Rockout-treated cells at different dosages compared with 5-FU alone treated cell viability. *, $p < 0.05$; **, $p < 0.01$; ***, $p < 0.001$. (**C**) Quantitation of synergism and antagonism in Rockout inhibitor and 5-FU combination. Fa versus CI plots were generated by CompuSyn. CI indicates a greater/synergism (CI < 1), lesser/antagonism (CI > 1), or similar (CI = 1) effect than the expected additive effect.

Table 1. Predicted drugs list for setback gene expression of drug-resistant traits.

Rank	Perturbation	Target
1	RHO-kinase-inhibitor-III [Rockout]	ROCK
2	dexamethasone	steroid
3	PKCbeta-inhibitor	PKC8β
4	CYT-997	microtubule polymerization
5	AR-C133057XX	NOS2
6	NF-449	P2 receptors
7	SB-216763	GSK3 inhibitor
8	emetine	40S ribosome
9	AC-55649	RARβ2
10	FK-866	NAMPT

3.7. ROCK and IGF1R Inhibitions Overcome Resistance to 5-FU Treatment

To confirm the result of predicted drugs overcoming acquired 5-FU resistance, we independently established a 5-FU-resistant HCT116 cell line (116-5FUR). The IC_{50} at 72 h of 116-5FUR and parental cell line is 2.15 μM and 22.52 μM, respectively. We treated the 116-5FUR and parental cells with 5-FU in combination with ROCK inhibitor Rockout. Rockout presented efficient cytotoxicity in both HCT116 and 116-5FUR cells (Figure 5). As predicted, 116-5FUR cells responded to Rockout alone. Moreover, they were sensitive to 5-FU in the presence of Rockout (Figure 5B). We showed that 5 μM of Rockout sensitized the 116-5FUR cells to 5-FU at an IC50 of 14.8 μM, more than a three-fold reduction compared with the IC50 of 5-FU in 116-5FUR cells (50 μM). The synergistic effect of Rockout and 5-FU was observed in 116-5FUR and parental cells (Figure 5C), suggesting the RhoA/ROCK axis was activated in 116-5FUR cells to promote their migratory ability through extracellular signaling. IGF1R activation can trigger ROCK indirectly by forming a complex with leukemia-associated RhoGEF-12 (LARG), which in turn activates RhoA [27]. We also observed that IGF1R downstream signaling, including AKT and ERK1/2 phosphorylation, were activated in 116-5FUR cells and even higher while treating with 5-FU (Figure S2A), implying IGF1R activation might be involved in 5-FU resistance. To investigate whether blockade of IGF1R signaling improved the 5-FU response in 116-5FUR cells, we treated the cells with a combination of the 5-FU and IGF1R inhibitor BMS-754807, revealing that 1 μM of BMS-754807 sensitized 116-5FUR cells to 5-FU at an IC50 of 6.53 μM, which is comparable with that of parental cells with a 5-FU IC50 (Figure S2B).

4. Discussion

Overactivation of WNT signaling through β-catenin is predominantly involved in the development of CRC. About 80% of CRC patients carrying loss-of-function mutations in

APC and about 5% carrying activating mutations in β-catenin are reported in The Cancer Genome Atlas (TCGA) project [28]. Most APC mutations result in C-termini truncated forms, suggesting that the protein conformation and protein–protein interactions of APC are critical to maintaining its normal function. For instance, the DNA repair inhibitory (DRI) domain of APC interacts with two DNA repair molecules, DNA polymerase β (POLβ) and flap endonuclease 1 (FEN1), to block the POLβ-directed base excision repair (BER) pathway and lead to apoptosis [11]. Here, we reported that overexpression of truncated APC 1-1309 (Figure 1) and cell lines with shorter APC forms (Figure 2) exhibited higher tolerance to 5-FU treatment. The result suggests that the DRI domain and downregulation of the β-catenin region in APC contribute to 5-FU sensitivity.

Deming et al. [29] indicated that mutation in PIK3CA simultaneously occurred with APC mutation in CRC, highlighting the resistance on PI3K inhibitors in APC-mutated cells. GDSC analysis also indicates that except for p53-mediated DNA repair and genome instability, PI3K/mTOR signaling ranked as the fourth resistant pathway related to APC mutation (Figure 1D). We postulated this might be due to the co-occurrence of the PI3K mutation with mutated APC. On the other hand, we showed that the MEK inhibitor exhibited a positive response in APC-mutated and β-catenin-activated cells (Figure 1C,D and Figure 2D), suggesting that the APC mutation might trigger mitogen-activated signaling to promote cell survival. This data implies that 5-FU induced several pathways contributing to drug resistance by increasing genome instability and intracellular proliferation signaling.

Previous studies demonstrate that APC abolished β-catenin regulated TCF4 transcription via RAF1/MEK/ERK signaling [30,31]. Loss of APC is required for KRAS-driven CRC transformation [31] and leads to activation of the RAS-mediated cell survival pathway [32]. Moreover, Frizzleds (FZ1) ligand WNT3A, which initiates canonical WNT/β-catenin signaling, has been shown to activate ERK and stimulate cell proliferation via the RAF1/MEK/ERK pathway [33,34]. The overactivation of MEK can also be observed clinically. The ERK1/2 phosphorylation ranked as the top phosphorylation event in patients with APC truncation compared with those harboring wild-type APC in TCGA database (Figure 3). These findings demonstrate that the mitogen-activated signaling is rewired to activate and promote cell proliferation in APC-mutated cells, supporting our results of the synergic effect of MEK inhibitor and 5-FU (Figure 2D).

In the scenario of APC-independent drug resistance, we established 5-FU-resistant HCT116 cells by culturing the parental cells in the presence of an increasing dose of 5-FU over one year. To recapture the progress of developing drug resistance, we compared the gene expression profiles of 5-FU sensitive and resistant HCT116 cells. To our surprise, a p-value of 2.26×10^{-54} in the molecular function of receptor binding was remarkedly significant than any other terms in resistant cells (Table S2). Therefore, we analyzed the distribution of genes in the top five molecular functions (data not shown) to determine which is the most dominant receptor type. We found that genes were distinguished enriched in growth factor activity and cytokine receptor binding. We also noticed that growth factors, including NOV, BMP4, MDK, GDF15, VEGFC, NTF3, and LIF, were previously reported to promote cancer development, especially in cancer metastasis in various cancers, including CRC [35–40]. We suggest that cancer metastasis might develop after acquiring 5-FU resistance through the secretion of extracellular ligands modulating the microenvironment homeostasis in vivo.

The tumor microenvironment (TME) specifies critical proto-oncogenic niches to support tumorigenesis through cytokines and growth factors [41,42]. Among the growth factor activity activated in HCT116-5FUR transcriptomics analysis, IGR1R is one of the possible upstream of ROCK signaling. Concomitantly, 116-5FUR was responsive to 5-FU in the presence of the ROCK inhibitor and IGF1R inhibitor (Figure 5 and Figure S1), emphasizing the plasticity of TME in establishing drug resistance. These findings support our results that targeting TME factors improved 5-FU sensitivity in CRC cells.

5. Conclusions

For the APC mutation background, we applied a genetic background-based prediction to generate a list of sensitive and resistant drugs. For APC wild-type, we performed a gene expression-based method to predict drugs to reverse the 5-FU resistance phenotype. We validate that both ways provide meaningful information for combinatorial therapy and illustrate a considerable direction to implement two strategies while preparing cocktail treatment in cancer therapy. We suggest that the MEK inhibitor is the option for patients with APC truncation and ROCK inhibitor for patients with wild-type APC when 5-FU resistance develops. Taken together, we provide two independent approaches, genetic background- and gene expression-based analysis, to predict drugs that may be suitable for overcoming drug resistance, especially highlighting the importance of personalized medicine.

Supplementary Materials: The following are available online at https://www.mdpi.com/article/10.3390/biomedicines9080882/s1, Table S1: Functional enrichment analysis of upregulated genes in 5-FU resistant HCT116 cells, Table S2: Functional enrichment analysis of downregulated genes in 5-FU resistant HCT116 cells, Figure S1: MEK inhibitor PD-0325901 is cytotoxic to APC-mutated CRC cells, Figure S2: IGF1R inhibitor increases the response of resistant cells to 5-FU.

Author Contributions: Conceptualization, H.-Y.C. and T.-C.H.; methodology, T.-C.H., K.-C.P. and H.-Y.C.; formal analysis, T.-C.H., K.-C.P., T.-T.K., L.-C.L., B.-C.L., S.-P.Y., C.-C.C., S.-M.H. and H.-Y.C.; writing—original draft preparation, T.-C.H., K.-C.P. and H.-Y.C.; writing—review and editing, H.-Y.C.; supervision, H.-Y.C.; project administration, H.-Y.C.; funding acquisition, T.-C.H. and H.-Y.C. All authors have read and agreed to the published version of the manuscript.

Funding: This research was funded by the health and welfare surcharge of tobacco products (grant number: MOHW110-TDU-B-212-144020), the TMU Research Center of Cancer Translational Medicine from The Featured Areas Research Center Program within the framework of the Higher Education Sprout Project by the Ministry of Education in Taiwan (grant number: DP2-110-21121-03-C-03-01 and DP2-110-21121-03-C-03-02), the Ministry of Science and Technology, Taiwan (grant number: 109-2320-B-038-048), and the Taiwan Ministry of Health and Welfare (MOHW110-TDU-B-212-124007). The APC was funded by DP2-110-21121-03-C-03-02.

Institutional Review Board Statement: Not applicable.

Informed Consent Statement: Not applicable.

Data Availability Statement: Data can be available upon request.

Conflicts of Interest: The authors declare no conflict of interest.

References

1. Siegel, R.L.; Miller, K.D.; Goding Sauer, A.; Fedewa, S.A.; Butterly, L.F.; Anderson, J.C.; Cercek, A.; Smith, R.A.; Jemal, A. Colorectal cancer statistics, 2020. *CA A Cancer J. Clin.* **2020**, *70*, 145–164. [CrossRef]
2. Longley, D.B.; Harkin, D.P.; Johnston, P.G. 5-fluorouracil: Mechanisms of action and clinical strategies. *Nat. Rev. Cancer* **2003**, *3*, 330–338. [CrossRef]
3. Chen, M.L.; Fang, C.H.; Liang, L.S.; Dai, L.H.; Wang, X.K. A meta-analysis of chemotherapy regimen fluorouracil/leucovorin/oxaliplatin compared with fluorouracil/leucovorin in treating advanced colorectal cancer. *Surg. Oncol.* **2010**, *19*, 38–45. [CrossRef]
4. He, T.-C.; Sparks, A.B.; Rago, C.; Hermeking, H.; Zawel, L.; Da Costa, L.T.; Morin, P.J.; Vogelstein, B.; Kinzler, K.W. Identification of c-MYC as a target of the APC pathway. *Science* **1998**, *281*, 1509–1512. [CrossRef] [PubMed]
5. Shtutman, M.; Zhurinsky, J.; Simcha, I.; Albanese, C.; D'Amico, M.; Pestell, R.; Ben-Ze'ev, A. The cyclin D1 gene is a target of the β-catenin/LEF-1 pathway. *Proc. Natl. Acad. Sci. USA* **1999**, *96*, 5522–5527. [CrossRef]
6. Brabletz, T.; Jung, A.; Dag, S.; Hlubek, F.; Kirchner, T. β-Catenin regulates the expression of the matrix metalloproteinase-7 in human colorectal cancer. *Am. J. Pathol.* **1999**, *155*, 1033–1038. [CrossRef]
7. Mann, B.; Gelos, M.; Siedow, A.; Hanski, M.; Gratchev, A.; Ilyas, M.; Bodmer, W.; Moyer, M.; Riecken, E.; Buhr, H. Target genes of β-catenin–T cell-factor/lymphoid-enhancer-factor signaling in human colorectal carcinomas. *Proc. Natl. Acad. Sci. USA* **1999**, *96*, 1603–1608. [CrossRef]
8. Munemitsu, S.; Souza, B.; Müller, O.; Albert, I.; Rubinfeld, B.; Polakis, P. The APC gene product associates with microtubules in vivo and promotes their assembly in vitro. *Cancer Res.* **1994**, *54*, 3676–3681. [PubMed]

9. Smith, K.J.; Levy, D.B.; Maupin, P.; Pollard, T.D.; Vogelstein, B.; Kinzler, K.W. Wild-type but not mutant APC associates with the microtubule cytoskeleton. *Cancer Res.* **1994**, *54*, 3672–3675.
10. Hülsken, J.; Birchmeier, W.; Behrens, J. E-cadherin and APC compete for the interaction with beta-catenin and the cytoskeleton. *J. Cell Biol.* **1994**, *127*, 2061–2069. [CrossRef] [PubMed]
11. Jaiswal, A.S.; Narayan, S. A novel function of adenomatous polyposis coli (APC) in regulating DNA repair. *Cancer Lett.* **2008**, *271*, 272–280. [CrossRef]
12. Das, D.; Preet, R.; Mohapatra, P.; Satapathy, S.R.; Siddharth, S.; Tamir, T.; Jain, V.; Bharatam, P.V.; Wyatt, M.D.; Kundu, C.N. 5-Fluorouracil mediated anti-cancer activity in colon cancer cells is through the induction of Adenomatous Polyposis Coli: Implication of the long-patch base excision repair pathway. *DNA Repair* **2014**, *24*, 15–25. [CrossRef]
13. Griffith, M.; Griffith, O.L.; Mwenifumbo, J.; Goya, R.; Morrissy, A.S.; Morin, R.D.; Corbett, R.; Tang, M.J.; Hou, Y.-C.; Pugh, T.J. Alternative expression analysis by RNA sequencing. *Nat. Methods* **2010**, *7*, 843–847. [CrossRef] [PubMed]
14. Kurokawa, K.; Tanahashi, T.; Iima, T.; Yamamoto, Y.; Akaike, Y.; Nishida, K.; Masuda, K.; Kuwano, Y.; Murakami, Y.; Fukushima, M. Role of miR-19b and its target mRNAs in 5-fluorouracil resistance in colon cancer cells. *J. Gastroenterol.* **2012**, *47*, 883–895. [CrossRef]
15. Allen, W.L.; Stevenson, L.; Coyle, V.M.; Jithesh, P.V.; Proutski, I.; Carson, G.; Gordon, M.A.; Lenz, H.-J.D.; Van Schaeybroeck, S.; Longley, D.B. A systems biology approach identifies SART1 as a novel determinant of both 5-fluorouracil and SN38 drug resistance in colorectal cancer. *Mol. Cancer Ther.* **2012**, *11*, 119–131. [CrossRef] [PubMed]
16. Yang, W.; Soares, J.; Greninger, P.; Edelman, E.J.; Lightfoot, H.; Forbes, S.; Bindal, N.; Beare, D.; Smith, J.A.; Thompson, I.R. Genomics of Drug Sensitivity in Cancer (GDSC): A resource for therapeutic biomarker discovery in cancer cells. *Nucleic Acids Res.* **2013**, *41*, D955–D961. [CrossRef]
17. Tomczak, K.; Czerwińska, P.; Wiznerowicz, M. The Cancer Genome Atlas (TCGA): An immeasurable source of knowledge. *Contemp. Oncol.* **2015**, *19*, A68. [CrossRef] [PubMed]
18. Cerami, E.; Gao, J.; Dogrusoz, U.; Gross, B.E.; Sumer, S.O.; Aksoy, B.A.; Jacobsen, A.; Byrne, C.J.; Heuer, M.L.; Larsson, E. The cBio cancer genomics portal: An open platform for exploring multidimensional cancer genomics data. *Cancer Discov.* **2012**, *2*, 401–404. [CrossRef]
19. Gao, J.; Aksoy, B.A.; Dogrusoz, U.; Dresdner, G.; Gross, B.; Sumer, S.O.; Sun, Y.; Jacobsen, A.; Sinha, R.; Larsson, E. Integrative analysis of complex cancer genomics and clinical profiles using the cBioPortal. *Sci. Signal.* **2013**, *6*, pl1. [CrossRef]
20. Lamb, J. The Connectivity Map: A new tool for biomedical research. *Nat. Rev. Cancer* **2007**, *7*, 54–60. [CrossRef]
21. Boyer, J.; McLean, E.G.; Aroori, S.; Wilson, P.; McCulla, A.; Carey, P.D.; Longley, D.B.; Johnston, P.G. Characterization of p53 wild-type and null isogenic colorectal cancer cell lines resistant to 5-fluorouracil, oxaliplatin, and irinotecan. *Clin. Cancer Res.* **2004**, *10*, 2158–2167. [CrossRef] [PubMed]
22. Smyth, G. Limma: Linear models for microarray data. In *Bioinformatics and Computational Biology Solutions Using R and Bioconductor*; Springer: New York, NY, USA, 2005; pp. 397–420.
23. Dennis, G.; Sherman, B.T.; Hosack, D.A.; Yang, J.; Gao, W.; Lane, H.C.; Lempicki, R.A. DAVID: Database for annotation, visualization, and integrated discovery. *Genome Biol.* **2003**, *4*, R60. [CrossRef]
24. Chou, T.-C. Drug combination studies and their synergy quantification using the Chou-Talalay method. *Cancer Res.* **2010**, *70*, 440–446. [CrossRef] [PubMed]
25. Rath, N.; Olson, M.F. Rho-associated kinases in tumorigenesis: Re-considering ROCK inhibition for cancer therapy. *EMBO Rep.* **2012**, *13*, 900–908. [CrossRef]
26. Bhadriraju, K.; Yang, M.; Ruiz, S.A.; Pirone, D.; Tan, J.; Chen, C.S. Activation of ROCK by RhoA is regulated by cell adhesion, shape, and cytoskeletal tension. *Exp. Cell Res.* **2007**, *313*, 3616–3623. [CrossRef] [PubMed]
27. Taya, S.; Inagaki, N.; Sengiku, H.; Makino, H.; Iwamatsu, A.; Urakawa, I.; Nagao, K.; Kataoka, S.; Kaibuchi, K. Direct interaction of insulin-like growth factor-1 receptor with leukemia-associated RhoGEF. *J. Cell Biol.* **2001**, *155*, 809–820. [CrossRef]
28. Network, C.G.A. Comprehensive molecular characterization of human colon and rectal cancer. *Nature* **2012**, *487*, 330–337. [CrossRef]
29. Deming, D.A.; Leystra, A.A.; Nettekoven, L.; Sievers, C.; Miller, D.; Middlebrooks, M.; Clipson, L.; Albrecht, D.; Bacher, J.; Washington, M.K. PIK3CA and APC mutations are synergistic in the development of intestinal cancers. *Oncogene* **2014**, *33*, 2245–2254. [CrossRef] [PubMed]
30. Park, K.-S.; Jeon, S.H.; Kim, S.-E.; Bahk, Y.-Y.; Holmen, S.L.; Williams, B.O.; Chung, K.-C.; Surh, Y.-J.; Choi, K.-Y. APC inhibits ERK pathway activation and cellular proliferation induced by RAS. *J. Cell Sci.* **2006**, *119*, 819–827. [CrossRef]
31. Sansom, O.J.; Meniel, V.; Wilkins, J.A.; Cole, A.M.; Oien, K.A.; Marsh, V.; Jamieson, T.J.; Guerra, C.; Ashton, G.H.; Barbacid, M. Loss of Apc allows phenotypic manifestation of the transforming properties of an endogenous K-ras oncogene in vivo. *Proc. Natl. Acad. Sci. USA* **2006**, *103*, 14122–14127. [CrossRef]
32. Malhotra, P.; Anwar, M.; Nanda, N.; Kochhar, R.; Wig, J.D.; Vaiphei, K.; Mahmood, S. Alterations in K-ras, APC and p53-multiple genetic pathway in colorectal cancer among Indians. *Tumor Biol.* **2013**, *34*, 1901–1911. [CrossRef]
33. Kim, S.-E.; Choi, K.-Y. EGF receptor is involved in WNT3a-mediated proliferation and motility of NIH3T3 cells via ERK pathway activation. *Cell. Signal.* **2007**, *19*, 1554–1564. [CrossRef]
34. Yun, M.-S.; Kim, S.-E.; Jeon, S.H.; Lee, J.-S.; Choi, K.-Y. Both ERK and Wnt/β-catenin pathways are involved in Wnt3a-induced proliferation. *J. Cell Sci.* **2005**, *118*, 313–322. [CrossRef]

35. Dobson, J.R.; Taipaleenmäki, H.; Hu, Y.-J.; Hong, D.; Van Wijnen, A.J.; Stein, J.L.; Stein, G.S.; Lian, J.B.; Pratap, J. hsa-mir-30c promotes the invasive phenotype of metastatic breast cancer cells by targeting NOV/CCN3. *Cancer Cell Int.* **2014**, *14*, 73. [CrossRef] [PubMed]
36. Cao, Y.; Slaney, C.Y.; Bidwell, B.N.; Parker, B.S.; Johnstone, C.N.; Rautela, J.; Eckhardt, B.L.; Anderson, R.L. BMP4 inhibits breast cancer metastasis by blocking myeloid-derived suppressor cell activity. *Cancer Res.* **2014**, *74*, 5091–5102. [CrossRef] [PubMed]
37. Mehta, R.S.; Song, M.; Bezawada, N.; Wu, K.; Garcia-Albeniz, X.; Morikawa, T.; Fuchs, C.S.; Ogino, S.; Giovannucci, E.L.; Chan, A.T. A prospective study of macrophage inhibitory cytokine-1 (MIC-1/GDF15) and risk of colorectal cancer. *J. Natl. Cancer Inst.* **2014**, *106*, dju016. [CrossRef] [PubMed]
38. George, M.L.; Tutton, M.G.; Janssen, F.; Arnaout, A.; Abulafi, A.M.; Eccles, S.A.; Swift, R.I. VEGF-A, VEGF-C, and VEGF-D in colorectal cancer progression. *Neoplasia* **2001**, *3*, 420–427. [CrossRef]
39. Louie, E.; Chen, X.; Coomes, A.; Ji, K.; Tsirka, S.; Chen, E. Neurotrophin-3 modulates breast cancer cells and the microenvironment to promote the growth of breast cancer brain metastasis. *Oncogene* **2013**, *32*, 4064–4077. [CrossRef] [PubMed]
40. Yu, H.; Yue, X.; Zhao, Y.; Li, X.; Wu, L.; Zhang, C.; Liu, Z.; Lin, K.; Xu-Monette, Z.Y.; Young, K.H. LIF negatively regulates tumour-suppressor p53 through Stat3/ID1/MDM2 in colorectal cancers. *Nat. Commun.* **2014**, *5*, 5218. [CrossRef] [PubMed]
41. Joyce, J.A.; Fearon, D.T. T cell exclusion, immune privilege, and the tumor microenvironment. *Science* **2015**, *348*, 74–80. [CrossRef]
42. Peddareddigari, V.G.; Wang, D.; DuBois, R.N. The tumor microenvironment in colorectal carcinogenesis. *Cancer Microenviron.* **2010**, *3*, 149–166. [CrossRef] [PubMed]

MDPI
St. Alban-Anlage 66
4052 Basel
Switzerland
Tel. +41 61 683 77 34
Fax +41 61 302 89 18
www.mdpi.com

Biomedicines Editorial Office
E-mail: biomedicines@mdpi.com
www.mdpi.com/journal/biomedicines